# Agricultural Biotechnology

## Genetic Engineering for a Food Cause

# Agricultural Biotechnology

## Genetic Engineering for a Food Cause

**Srividhya Venkataraman**
Department of Cell and Systems Biology, University of Toronto, Toronto, ON, Canada

**Kathleen Hefferon**
Department of Microbiology, Cornell University, Ithaca, NY, United States

ELSEVIER
ACADEMIC PRESS
An imprint of Elsevier

Academic Press is an imprint of Elsevier
125 London Wall, London EC2Y 5AS, United Kingdom
525 B Street, Suite 1650, San Diego, CA 92101, United States
50 Hampshire Street, 5th Floor, Cambridge, MA 02139, United States
The Boulevard, Langford Lane, Kidlington, Oxford OX5 1GB, United Kingdom



ISBN: 978-0-323-88476-1

For information on all Academic Press publications visit our website at https://www.elsevier.com/books-and-journals

*Publisher:* Stacy Masucci
*Acquisitions Editor:* Kattie Washington
*Editorial Project Manager:* Sara Pianavilla
*Production Project Manager:* Fahmida Sultana
*Cover Designer:* Matthew Limbert

Typeset by TNQ Technologies

# Dedication

This book is dedicated to my beloved husband, Murali, whose constant, irreplaceable encouragement and support has enabled me to reach great goals in life both personally and professionally. I also dedicate this book to my respected and beloved mentors, Prof. Appaji Rao and Prof. Savithri at the Indian Institute of Science, Bangalore, with whom I had the honor of being a PhD student. The rigorous scientific training, invaluable research, and personal support that they have provided me has stood with me throughout all these years I have functioned as a Scientist in Canada. They trained me not just to be a good scientist but to be a better person as well. Also, this work would not have been accomplished without the valuable help of my seniors, Prof. Bhaskar Anna and Prof. Usha Arunachalam Akka. I further wish to thank Prof. Mounir Abouhaidar and Prof. Kathleen Hefferon at the University of Toronto for their esteemed mentorship and support. Above all, I am grateful to Kathy for launching my career as a writer and for her invaluable support.

# Contents

# Author biography

Dr. Srividhya Venkataraman obtained her PhD from the Indian Institute of Science, Bangalore, where she won the Young Scientist Award for her PhD work. After a brief experience at the Pennsylvania State University, USA, and at US Vitamins Pharmaceuticals, she moved to Toronto where she worked as a Canadian Institute of Health Research (CIHR) Fellow in Regenerative Medicine at the Toronto General Hospital and the Toronto Medical Laboratories. She then held an assignment as a Postdoctoral Associate at the University of Guelph. She is currently a Visiting Scientist at the Department of Cell and Systems Biology, University of Toronto. She has over 40 publications and has contributed a book. She also functions as Expert Selector for the Molecular and Cellular Medicine Board, Medical Research Council, UK. Her primary research interests are in the areas of viruses, virus-like particles, food security, genetic engineering, and biotechnology. She lives with her family in Toronto, Canada.

Dr. Kathleen Hefferon received her PhD from the Department of Medical Biophysics, University of Toronto, and completed her postdoctoral fellowship at Cornell University. Kathleen has published multiple research papers, chapters, and reviews and has written three books. Kathleen is the Fulbright Canada Research Chair of Global Food Security and has been a Visiting Professor at the University of Toronto over the past year. Her research interests include virus expression vectors, food security agricultural biotechnology, and global health. She is part of the faculty for the Department of Microbiology at Cornell University.

Chapter 1

# Introduction

The world is fast approaching a breaking point unless significant changes are quickly made. According to the Food and Agricultural Organization (FAO), over 800 million people, most of them residing in sub-Saharan Africa or Southeast Asia, suffer from undernutrition (a lack of sufficient calories). An additional two billion of the world's people experience "hidden hunger" or malnutrition. Global population is expected to increase to close to 10 billion over the next 30 years, making the prospect of food insecurity even more bleak. At the same time, the Earth is facing global warming, with an unprecedented amount of greenhouse gas emissions such as methane and $CO_2$ reaching the atmosphere to levels never seen before in the history of this planet. Agricultural methods used in the past will not enable us to feed everyone without converting more wilderness into arable farmland. Solutions that embrace energy sources, which are not linked to fossil fuel usage, as well as improved crops that can increase yield and yet be tolerant of climate shocks, will be essential to our success.

As we move toward a hotter and hungrier planet, the United Nations have set out the Sustainable Development Goals, with a deadline to be reached by the year 2030. These goals are outlined in Fig. 1.1, and success will rely heavily on our stewardship of the planet over the next decade. The most realistic way of meeting these goals holistically will be through the responsible use of agricultural biotechnology.

Although genetically modified (GM) organisms (GMOs) have been available commercially for over 20 years, they remain as controversial as ever. According to the International Service for the Acquisition of Agri-biotech Applications (ISAAA), the global area of crops bred using biotechnology has reached approximately 200 million hectares; more than half of this land is used by smallholder farmers in developing countries. First-generation GM crops increased yields and saved costs by providing resistance traits to biotic stresses, such as insects, weeds, and plant pathogens. Second-generation biotech crops have targeted abiotic stresses such as heat, drought, and flooding. The world is entering a stage of third-generation GM crops, with nutritional traits that are highly valued by consumers and the food and manufacturing industries, such as biofortified, nutritionally enhanced food crops or other crops to reduce food waste. Agricultural biotechnology can address hunger, nutrition, and biodiversity, increase yields, and reduce the

**Agricultural Biotechnology. https://doi.org/10.1016/B978-0-323-88476-1.00010-2**

**FIGURE 1.1 Sustainable development goals outlined by the United Nations for betterment of mankind.** These goals such as alleviation of poverty, undernutrition, and malnutrition and achievement of gender equality, industrialization, better climate, reduction of greenhouse gas emissions, etc. have been set by the United Nations so as to be achieved by the year 2030.

amount of farmland required. GM and genome-edited crops can improve economic outcomes for the rural poor and environmental outcomes through decreased agricultural inputs. Genetically engineered crops, produced by genetically modified and genome editing technologies, are part of a continuum of new breeding technologies, which are discussed below and illustrated in Fig. 1.2.

It is important to acknowledge that many of the varieties of crops that are available today have had their genomes altered by a technology that existed long before the advent of recombinant DNA. For example, mutagenesis breeding originates from mutation research that originated in the 1930s and includes the introduction of random mutations to plant cuttings using chemical or irradiation mutagenesis. Explants can then be propagated from these mutation events into new crop varieties. There are more than 3000 varieties of crops that have been generated through mutagenesis breeding, according to the Mutant Variety Database, including the popular ruby red grapefruit (https://mvd.iaea.org). Genetic engineering of crops for improved nutritional and agronomic traits is essential and critical for modern plant science. This can include the introduction of a novel trait into a crop by either adding new genes or making small changes to pre-existing genes that are already part of the crop genome. Novel forms of genetic material are incorporated into the plant genome via Agrobacterium-mediated transformation or particle bombardment (gene gun). A well-known example of a transgenic crop is Golden Rice, which expresses beta-carotene and contains genes derived from different species, which together reconstitute the beta-carotene pathway. Cisgenic crops express

**FIGURE 1.2 Crop modification technologies.** Several crop modification technologies such as genome modification and the recently developed genome editing schemes have been employed to produce better crop quality and yield in order to address food insecurity in developing and underdeveloped nations. *Image from Biology Fortified Inc.*

genes from closely related plants rather than other species. These can allow farmers to regain useful traits that have been lost over years of crop breeding. For example, the Wheat Stem Rust Initiative has created cisgenic versions of wheat containing multiple resistance genes to the fungal pathogen Ugg99 using wheat relatives. RNAi, or RNA interference technology, represents another form of genetic engineering that is gaining in popularity. Plants can be engineered to express the antisense RNA version of a specific plant or pathogen gene, and then expression of the gene targeted is blocked via gene silencing. For example, genetically modified papaya grown in Hawaii confers resistance to papaya ringspot virus by expressing an antisense RNA to the viral genome. The most recent technology used in agricultural biotechnology toolkit is genome editing. Genome editing does not require the introduction of new gene sequences; rather, it may direct only one or a few nucleotide changes within a plant genome. Genome editing technologies such as CRISPR-Cas9 make use of various defense strategies derived by bacteria to target specific sequences of DNA and cleave them at these sites with Cas nucleases. The technology is then able to take advantage of DNA repair mechanisms found already in the cells of all organisms. By repairing the sites of cleavage, specialized changes that will be carried through the genome of the "edited" organism to subsequent generations can be produced. Genome editing is an extremely promising new tool for plant breeding, as many complex traits such as crop yields, photosynthesis and drought resistance involve multiple genes and therefore would be difficult to achieve through conventional breeding alone. Taken together, these technologies can advance agriculture in ways that were never believed possible and at speeds unheard of a few decades ago. As microbial communities within and surrounding plants can play a profound role on plant physiological outcomes, genetic modification, and genome editing are being implemented in microbes as well.

This Volume 1 of our book on 'Agricultural Biotechnology' addresses important issues including phytoremediation, gene drive management, biofortified crops, genome-edited crops, plant-based biopharmaceuticals, engineering crops resistant to biotic stresses, strategies to address climate change, novel ways to generate food proteins using biotechnology to reduce greenhouse gas emissions, as well as engineered microbes for improved soil and plant health. The forthcoming Volume 2 of this book will address other significant topics such as engineering crops resistant to abiotic stresses, repurposing food and agricultural waste using biotechnology, synthetic microbiology and microbial symbiosis (e.g., probiotics, Wolbachia, etc.), metabolic engineering of plants for secondary metabolite production, sustainability through agricultural biotechnology, economic/ecological benefit of GM crops, biofuel production and agricultural biotechnology, development of molecular biomarkers in plants and plant molecular breeding, public perception, intellectual property issues and the future of biotechnology in agricultural practices. In all, both volumes cover all the major aspects of agricultural biotechnology and this work could be considered as a positive contribution to this field.

Chapter 2

# Engineered microbes for improved soil and plant health

## 1. Introduction

Environmental sustainability refers to the responsible maintenance of natural resources so as not to jeopardize the ability of future generations to meet their needs. This can include selective logging/reforestation, the use of alternative forms of energy such as solar and wind to avoid the burning of fossil fuels, and the recycling of materials such as metals and minerals. This also includes the degradation or repurposing of plastic and other forms of waste and the reduction of practices leading to greenhouse gas emissions, including those used in agriculture. To this end, several significant hurdles must be addressed within the next few decades for damage to be adequately mitigated before it is too late.

For example, preventing the production of plastic waste has become an urgent global task. According to "*Our World in Data,*" the accumulation of plastic waste is having a negative effect on our wildlife and oceans. While more plastic per capita is being generated in high-income countries, most of marine plastic comes from low- to middle-income countries. These statistics are due to differences in waste management practices. Currently, there are approximately 380 million metric tons of plastic created each year and 8.3 billion tons of plastic waste in the world today. As another example, global greenhouse gas emissions reached 35,000 million metric tons in 2018 and continue to climb. Along these lines, global energy—related carbon dioxide emissions rose by 6% in 2021 to 36.3 billion tonnes (Fig. 2.1). While much of energy-related carbon emissions include the use of coal, oil, and other fossil fuels, greenhouse gas emissions are also created when industrial products such as cement and chemicals such as acetone or isopropanol are generated. Even agricultural practices, such as the use of tractors (which require fossil fuel), soil tillage, and livestock maintenance release GHG emissions. Fertilizer production and overuse further exacerbate the problem.

The use of biotechnology, such as altering the genomes of plants or the metabolic pathways of microbes, is one promising method for mitigating these hurdles. Plants have been genetically engineered to produce new bioplastic

**Agricultural Biotechnology. https://doi.org/10.1016/B978-0-323-88476-1.00004-7**

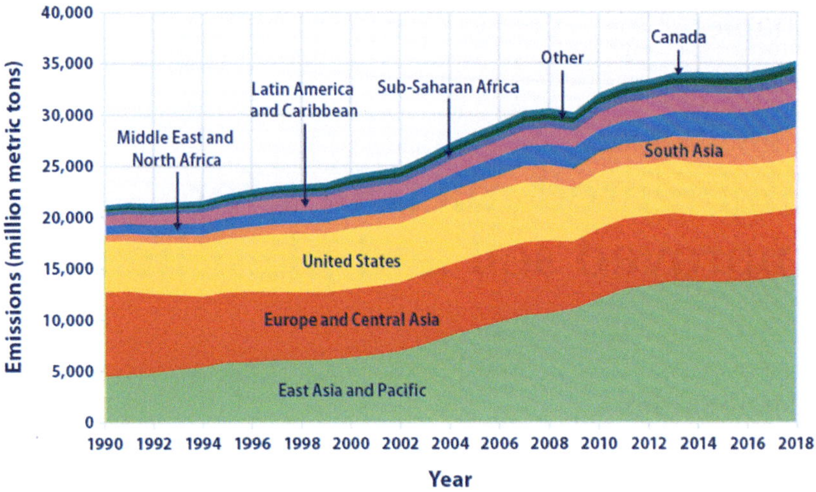


**FIGURE 2.1 Graph showing the amount of greenhouse gas emissions generated globally on a yearly basis.** Emission produced in different continents of the world is increasing every year.

materials, store carbon, and even fix their own nitrogen. Microbes such as bacteria and yeast have been engineered to carry out novel reactions that will be essential for maintaining or restoring the health of our planet. A plethora of technologies are involved in these processes, ranging from artificial intelligence (AI) to genome editing to bioinformatics. In many cases, it takes a combination of several to achieve the desired result.

In this chapter, we discuss the use of microbes and plants for a variety of purposes that culminate in a more environmentally sustainable planet. The chapter opens with a description of how plants and microbes can be used to break down industrial waste, including methane and carbon dioxide. The chapter next discusses the use of engineered microbes and transgenic plants to fix nitrogen into the soil, to avoid the overuse of fertilizers and the eutrophication of our waterways. The chapter then describes how microbes can be used to breakdown plastics, and both plants and microbes can be harnessed to produce environmentally sustainable bioplastics. Finally, the chapter touches upon how plants and microbes can be used to fix carbon from the air into biomass. The chapter concludes with a description of how these technologies should be a key to our future.

## 2. Engineered plants and microbes that break down industrial waste

When most people hear about oil, as in petroleum, they think of what gets refined to produce gasoline. From clothes to dentures to lipstick and even toothpaste, the number of household products made from chemicals based on fossil fuels numbers in the thousands.[1] Without the availability of oil to make petrochemicals, our everyday lives would be much more austere, and many products, such as the plastics used in cars and other consumer goods, would be sorely missed or far more expensive.[2]

A possible solution is the manufacturing of products using "green chemistry," in which production pathways are altered in such a way that carbon emissions are reduced or even eliminated. Synthetic biology, the convergence of advances in chemistry, biology, computer science, and engineering that enables us to go from idea to product faster, cheaper, and with greater precision, is one such approach. If the thousands of chemicals derived from petroleum, natural gas, and coal—including fuels, plastics, and industrial chemicals—could be synthesized instead using microorganisms, the annual savings in global greenhouse gas emissions could be substantial, and it could lead the way to a more sustainable manufacturing sector.

Employing bacteria and other microorganisms to synthesize new chemicals to be used in manufacturing has been achievable for some time, but often at an undesirable carbon cost. Microorganisms such as *E. coli* and yeast produce a few commodity chemicals such as isobutanol and lactic acid, but they use sugar as their starting material and thus emit substantial amounts of $CO_2$ waste.[3] Generating chemicals for household products with this approach involves increased use of both land and energy to produce the sugar feedstocks, which further increases the carbon footprint.

However, this trend may be ripe for change. Ref. 4 recently published in Nature Biotechnology describes how a specific class of bacteria, known as acetogens, can be used for the large-scale production of important industrial chemicals using a process that is actually carbon-*negative*. Unlike traditional production processes, which result in release of greenhouse gases, their fermentation process actually *fixes* carbon. Acetogens can live on one-carbon molecules such as $CO_2$, the prototypic greenhouse gas, and can incorporate them into more complex and desirable molecules required for the manufacturing of myriad products.

In the past, acetogens have proven to be extremely difficult to genetically engineer. The authors of this new study have overcome this hurdle, as well as others such as the need for sugar feedstocks, and at the manufacturing scale. They produce the commodity chemicals acetone and isopropanol (IPA) in a carbon-negative fashion and as such, provide a roadmap to produce other industrially relevant chemicals used using sustainable feedstocks. Acetone and IPA alone have a global market of over $10 billion.

The authors concluded that "engineered acetogens enable sustainable, high-efficiency, high-selectivity chemicals production," and best of all, they "expect that [their] approach can be readily adapted to a wide range of commodity chemicals." This approach is an important portent for an increasingly warming planet, especially as other potential feedstock sources beyond fossilized carbon are employed. These could include gasification of forest and agricultural residues, for example, as well as non-recyclable municipal solid waste.[5] Using such carbon sources, an array of other products, such as jet fuel or diesel, could be amenable to similar methods of production.

What if we were to produce all petrochemicals through acetogens using synthetic biology? The statistics are striking. Globally, chemical manufacturing emits a half a billion tons of $CO_2$ each year and is the third largest $CO_2$ emitting industrial sector (behind cement and steel and iron mills).[6,7] These sectors have been notoriously difficult to decarbonize. Microorganisms that render chemical manufacturing greener by creating a circular economy that uses waste instead of sugar as feedstocks could be a critical advance.

Recently, scientists at Rothamsted Research genetically engineered camelina plants to produce chemicals known as vinyl phenols, which are needed to make a variety of household products such as plastic for the LCD screens of TVs and cellphones and which have traditionally been made from fossil fuels. Camelina plants are grown for their oilseed; in this instance, they were metabolically engineered to produce the desired chemicals and then grown and harvested from the field. The materials produced in these plants could both lower the carbon footprint of chemical production and reduce our dependence on fossil fuels.[8]

Breaking the link between fossil fuels and manufacturing in a way that does not negatively affect the global economy is a pivotal step toward achieving carbon neutrality. We predict that genetically engineered bacteria and plants that produce the substrates for our household products in a way that is carbon neutral (or negative) are the future.

## 3. Engineered plants and microbes that break down plastics

Plastic is a product of petroleum, and about one quarter of every barrel of oil is used for plastic making. This makes modern life continually dependent on fossil fuels. A little under 400 million metric tons of plastic is produced globally every year, and much of it ends up in landfills or the natural environment, where it take hundreds of years to decompose into "microplastics," which can accumulate to all places on earth and can even be found within humans. Single-use plastics, such as water bottles, grocery bags, and food packaging to drinking straws, make up approximately 50% of the world's yearly estimated 380 million tons of global plastic waste. Plastic debris found in the Great Pacific Garbage Patch is currently twice the size of Texas.

Efforts are being made to address the enormous problem of plastic waste. One way would be to engineer novel enzymes in microbes that could break plastics down.[9] Such research has been explored by Ref. 10 who used artificial intelligence to generate a novel enzyme that can break down polyethylene terephthalate, or PET, the plastic commonly used to make fabrics and food and beverage packaging, into components that can be recycled into more plastic. These efforts could result in a plastic circular economy. The work initiated with the discovery of a number of microbial hydrolases known as PETases, produced by specific bacteria, which could break down chains of molecules that compose plastic. These naturally found enzymes required high temperatures, and so the researchers used AI to design variants of the enzyme that could function efficiently at close to room temperature. This new enzyme could break down plastic bottles and containers within 1–2 days.

Another way to address plastic pollution is to find a new way to produce it that is green and sustainable, thus removing our dependence of fossil fuel production.[11] To this end, scientists have been exploring opportunities to develop environmentally friendly plastics. Bioplastics produced from microbes or plants is one way to address this. For example, photosynthetic cyanobacteria, or blue-green algae, which utilize sunlight and $CO_2$ captured from power plants, can produce carbon storage molecules, which then provide the raw material for making bioplastics. Researchers can then use artificial intelligence (AI) to screen biopolymers made in these microbes and determine which algal genes are necessary to generate them. With this knowledge, gene expression can be increased or decreased to increase yield or chemical properties of the biopolymers, as well as virtually examine what will be required for large-scale production. The outcome of this work will provide a novel way to remove and store carbon from the atmosphere, as well as reduce the burden of plastic in the environment.

In a similar fashion, the company Biohm is using a newly identified fungus to break down plastic waste. Other scientists from the University of Edinburgh have recently used a lab-engineered version of the bacteria *E. coli* to transform terephthalic acid, a molecule derived from PET, into the culinary flavoring vanillin.[12] Another research team at the Helmholtz Center for Environmental Research-UFZ in Leipzig is using a Pseudonomas sp TDA1, a type of bacteria originally found in a local rubbish dump to break down polyurethane. Approximately half of the plastic is used to increase the microbes biomass, while the other half is released as carbon dioxide. Similarly, the French company Carios has reengineered an enzyme found in compost heap isolated bacteria to break down plastic.

Plants depend upon beneficial interactions between roots and root-associated microorganisms for growth promotion, disease suppression, and nutrient availability. Crops require nitrogen to grow, and although there is an abundance of it in the atmosphere, it is not available in a form that plants can readily use. Instead, the roots of most crops take up nitrogen in the form of ammonia, either as synthetic fertilizer or in biological waste such as manure or compost. The availability of nitrogen from any of those sources increases crop yield, which is in itself an important step toward achieving food security.[13]

Before synthetic fertilizers became available, feeding the world's population relied solely upon an endless demand for manure to such an extent that wars have been fought over it. For example, in 1864, a naval conflict broke out between Spain and Peru over the Chincha Islands, which were covered in guano deposits said to be over 100 feet high.

Some plants take up nitrogen with the help of soil bacteria that actually "fix" nitrogen gas ($N_2$) from the atmosphere, converting it into ammonia, which the plant's roots can use. This process of biological nitrogen fixation (BNF) is tightly regulated by the bacteria that perform this action, so that the process is switched off if sufficient levels of nitrogen are available in a form that can be taken up by the plant.[14] These microorganisms are found associated among the roots of certain leguminous crops, such as soybean, alfalfa, or pea, which contain nodules within their root systems where these processes are carried out. Crops such as these can be rotated in the field with others, which lack this ability, such as corn or wheat, for example, to obtain higher yields and for maintenance of soil health.

Synthetic fertilizers are now essential to feeding the world's population, but they are costly and their production requires fossil fuels. While saving and enhancing the lives of countless people, synthetic fertilizers have had other, undesirable effects, especially on the environment. Excess nitrogen in the form of ammonia can make its way out of farmland and into our waterways where it collects and forms anoxic dead zones, such as the one found where the Mississippi River Delta meets the Gulf of Mexico. Another aspect of the problem is that excess ammonia left on farmers' fields can also be converted into nitrous oxide ($N_2O$), a stable, long-lived greenhouse gas with 300 times the potency of carbon dioxide. Nitrous oxide represents 6% of greenhouse gas emissions, about three quarters of which is derived from agriculture. In addition, $N_2O$ also diminishes the ozone layer, further exacerbating climate change.

A long-sought solution would be to replace synthetic fertilizers with bacteria that perform nitrogen fixation in the roots of crops other than legumes, which could reduce nitrogen runoff and the production of nitrous oxide. This has now been achieved, as described in Ref. 15. Their report described "the identification, development, and deployment of the first microbial product optimized using synthetic biology tools to enable BNF for corn (*Zea mays*) in fertilized fields, demonstrating the successful, safe commercialization of root-associated diazotrophs and realizing the potential of BNF to replace and reduce synthetic nitrogen fertilizer use in production agriculture."

The bacterium chosen for this work already contained the genes that enabled it to fix atmospheric nitrogen via an enzyme called nitrogenase, so it comes by this ability naturally. However, because nitrogenase expression is switched off in the presence of ammonia, applying synthetic fertilizer would suppress bacterial BNF. By using a genome editing approach that employs CRISPR, the bacteria were tweaked to express the enzyme in the presence of fertilizer, which boosted the levels of nitrogen in a form available to the plant. The study demonstrated that the need for applications of synthetic fertilizer was significantly less when this engineered bacterium was provided as an inoculum to the roots of nonleguminous crops such as corn. Most important, the authors showed that the system worked not just in the greenhouse but under a variety of real-world field conditions.

This elegant study could be a game-changer on many fronts. The process of generating synthetic fertilizer requires fossil fuel, and the simple act of applying it to farmers' fields via tractor use burns even more fuel and thus exacerbates greenhouse gas emissions. Fewer applications of fertilizer (or manure) means fewer opportunities for nitrogen runoff and greenhouse gas emissions to be created, while maintaining the crop yields necessary to feed a burgeoning global population.

For years, farmers and scientists have been aware of a phenomenon known as "commensalism," a long-term biological interaction in which members of one species gain benefits while those of the other species neither benefit nor are harmed; specifically, certain crops benefit from their intimate associations with their local soil microbial community. Thus, microorganisms are being developed for various agricultural applications, including as biopesticides; soil remediators to remove pollutants and biosensors to detect crop stress or assess the microclimate of fields, orchards, and vineyards. The genomes of individual microorganisms as well as entire microbial communities are being examined using computational tools, including AI, to determine which can improve crop yield, reduce pest pressure, control weeds, and deliver nutrients.

For millennia, microorganisms have been used for fermentation of food and beverages, and, more recently, as a source of pharmaceuticals and industrial enzymes. They are truly part of the story of humanity. Using them to produce higher crop yields, which would feed more people at less expense, improve farmers' bottom line, and reduce agriculture's greenhouse gas footprint, is a logical and welcome extension.

The ability of new species of plants to fix nitrogen could be a game changer for the developing world, who cannot afford modern farming inputs such as synthetic fertilizer.[16] The lack of fertilizer results in low crop yields, an endless cycle of poverty, and malnutrition. The global expenditure of fertilizers is over $100 billion per year. If low- to middle-income countries were provided with transgenic crops or microbes that could fix their own nitrogen, their yields would improve dramatically, as well as their nutritional health and well-being. It is estimated that such crops will be available within the next few years.

## 4. Engineered plants and microbes for carbon fixation

Photosynthesis is an important determinant of crop yield. A number of approaches have been made to increase the rate of photosynthesis in plants. These include using the enzymes from cyanobacteria or green algae to increase carbon sequestration and introducing the C4 pathway into C3 plants. Ref. 17 found that photosynthesis may be improved by over 40% by using genetic engineering approaches. 52 Over-expression of enzymes involved in photosynthesis, such as sedoheptulose-1,7-bisphosphatase (SBPase), can increase both photosynthetic carbon assimilation and biomass yield. Agricultural Biotechnology 421 On the other hand, tolerance to cold stress can also be a major challenge. Ref. 18 used the gene OsRBGD3 derived from the drought tolerant Indica rice cultivar N22.53 Constitutive over-expression of the OsRBGD3 in transgenic plants conferred tolerance to cold stress, in addition to significant root development and early flowering in transgenic plants.

Another way that improved planet health is being addressed is through the use of biotechnology to engineer microbes and plants to reduce carbon dioxide emissions.[19] This can take place in several ways. For example, autotrophic microbes can be used to generate biomass out of thin air, using carbon dioxide as their food source. By moving this metabolic pathway to other organisms, such as heterotrophic bacteria or yeast, it should be feasible for them to also fix carbon from the air, instead of using a glucose feedstock as they normally do.

This form of metabolic engineering was achieved using *E. coli* and *Pichia pastoris*, microbes that are commonly used in biotechnology. This was accomplished by introducing enzymes involved in carbon fixation to the microbes genome, while erasing others that are chiefly used in glucose metabolism. The engineered microbes required high concentrations of carbon dioxide to metabolize, and grew slowly; however, future research should be able to design microbes, which are more energy efficient. It is hoped that in the future, microbes such as these can also be used to clean up waste produced by other industrial processes that generate greenhouse gases as end products, so that GHG are not released into the environment, as described in an earlier section of this chapter.

In a similar fashion, plants can be engineered to improve carbon fixation of industrially made greenhouse gases. For example, a research group from the Max-Planck-Institute (MPI) has reverse-engineered a novel metabolic pathway based on a new $CO_2$-fixing enzyme that is almost 20 times faster than RuBisCo, the plant enzyme responsible for capturing $CO_2$ in plants using photosynthesis.[20] In this case, the researchers used variants of enoyl-CoA carboxylase/reductase (ECR) enzyme with the highest $CO_2$-fixation activity to successfully build a more efficient artificial $CO_2$ fixation pathway. Introducing this pathway into an organism could improve photosynthesis efficiency.

## 5. Conclusion

The technologies described in this chapter illustrate the methods underway to address issues surrounding environmental sustainability and planetary health. In the future, these metabolic processes, contained within microbes and plants, will be more efficient, rapid, and economically feasible. The discovery and implementation of novel pathways using AI along with bioinformatics will enhance research and development even more. Identification of previously unknown genes, and sourcing from naturally acquired as well as synthetic genes will be critical for this implementation. One would hope that the collective efforts of all these technologies should be sufficient to alter the disastrous trajectory toward climate change that we find ourselves currently on, into one that can pave the way for a sustainable planet that future generations can thrive upon and cherish.

## References


1. Goodluck K. *10 everyday products derived from petroleum*. 2019. Published, https://www.kcet.org/shows/earth-focus/10-everyday-products-derived-from-petroleum.
2. MacLeod R. *How the petrochemicals industry can reduce its carbon footprint*. World Economic Forum; 2019.
3. Huang J, Liu Z, Bloomer BJ, et al. Unnatural biosynthesis by an engineered microorganism with heterologously expressed natural enzymes and an artificial metalloenzyme. *Nature Chemistry* 2021;**13**(12):1186–91. https://doi.org/10.1038/s41557-021-00801-3.
4. Liew FE, Nogle R, Abdalla T, et al. Carbon-negative production of acetone and isopropanol by gas fermentation at industrial pilot scale. *Nature Biotechnology* 2022;**40**(3):335–44. https://doi.org/10.1038/s41587-021-01195-w.
5. EPA. *Energy recovery from the combustion of municipal solid waste (MSW)*. 2022. Published, https://www.epa.gov/smm/energy-recovery-combustion-municipal-solid-waste-msw.
6. IEA. *Iron and steel technology roadmap*. 2020. Published, https://www.iea.org/reports/iron-and-steel-technology-roadmap.
7. Scown CD, Keasling JD. Sustainable manufacturing with synthetic biology. *Nature Biotechnology* 2022;**40**(3):304–7. https://doi.org/10.1038/s41587-022-01248-8.
8. Menard GN, Langdon M, Bhunia RK, et al. Diverting phenylpropanoid pathway flux from sinapine to produce industrially useful 4-vinyl derivatives of hydroxycinnamic acids in Brassicaceous oilseeds. *Metabolic Engineering* 2022;**70**:196–205. https://doi.org/10.1016/j.ymben.2022.01.016.
9. Patel P. *AI-designed enzyme devours plastic trash in days. The new enzyme works at lower temperatures than previous ones, which would make it greener, faster, and cheaper*. 2022. Published, https://www.anthropocenemagazine.org/2022/05/ai-designed-enzyme-devours-plastic-waste-in-hours/.
10. Lu H, Diaz DJ, Czarnecki NJ. Machine learning-aided engineering of hydrolases for PET depolymerization. *Nature* 2022;**604**:662–7.
11. Marrone B. *Bioplastics point the way to an environmentally sustainable*. Green Future. The New Mexican; 2022.
12. Woollacott E. *The fungus and bacteria tackling plastic waste technology of business reporter*. 2021.
13. Miller H, Hefferon K. *Viewpoint: essential to feeding the world, synthetic fertilizers have negative environmental impacts. Here's how genetic engineering can change that*. European Scientist; 2022.
14. Vadakattu G, Paterson J. Free-living bacteria lift soil nitrogen supply. *Farming Ahead* 2006;**169**:40.

15. Wen A, Havens KL, Bloch SE, et al. Enabling biological nitrogen fixation for cereal crops in fertilized fields. *ACS Synthetic Biology* 2021;**10**(12):3264–77. https://doi.org/10.1021/acssynbio.1c00049.
16. Good A. Nitrogen-fixing GMO crops could reduce synthetic fertilizer use, benefit environment. *Science* 2018;**359**:6378.
17. Nowicka B, Ciura J, Szymańska R, Kruk J. Improving photosynthesis, plant productivity and abiotic stress tolerance – current trends and future perspectives. *Journal of Plant Physiology* 2018;**231**:415–33. https://doi.org/10.1016/j.jplph.2018.10.022.
18. Lenka SK, Singh AK, Muthusamy SK, Smita S, Chinnusamy V, Bansal KC. Heterologous expression of rice RNA-binding glycine-rich (RBG) gene OsRBGD3 in transgenic *Arabidopsis thaliana* confers cold stress tolerance. *Function of Plant Biology* 2019;**46**(5):482–91. https://doi.org/10.1071/FP18241.
19. Vavitsas K. *Fighting climate change by reprogramming yeast, bacteria to feast on carbon dioxide*. 2020. Published, https://geneticliteracyproject.org/2020/01/07/fighting-climate-change-by-reprogramming-yeast-bacteria-to-feast-on-carbon-dioxide/.
20. Schwander T, Von Borzyskowski LS, Burgener S, Cortina NS, Erb TJ. A synthetic pathway for the fixation of carbon dioxide in vitro. *Science* 2016;**354**(6314):900–4. https://doi.org/10.1126/science.aah5237.

Chapter 3

# Biofortified crops

## 1. Malnutrition: the hidden hunger

The ever-increasing global population faces the emerging threat of malnutrition, which is exacerbated further in developing nations where people face hunger or consume nutritionally poor food. The world population is projected to increase to 8.3 billion in 2030 as against the current level of 7.8 billion.[1] As reported by the Food and Agriculture Organization of the United Nations (UNFAO), 792.5 million of the global populace face malnutrition and the death rate caused by hunger adds up to 24,000 people each day.[2] About a third of the world's inhabitants suffer from "hidden hunger" resulting from the lack of one or all of the essential macro- or micro-nutrients such as Fe, Zn, selenium, folic acid, iodine, vitamins (A, B12, C, D, and E), lysine, and others in their diets.[3,4]

### Impacts of micronutrient deficiency

These nutrients and vitamins are essential for the growth, development, sustenance, and normal function of the human body. Micronutrient deficiency is not very harmful if it exists for short periods but in the long run may lead to diseases such as blindness (vitamin A deficiency), beriberi (vitamin B deficiency), scurvy (vitamin C deficiency), rickets (vitamin D deficiency), pellagra (niacin deficiency), anemia (Fe deficiency), or may be even fatal.[5] This situation highlights the significance of crop biofortification to provide healthy diets to the population to circumvent these issues. Most people consume cereals and pulses to meet their dietary requirements. However, cereal lacks balanced nutritional content including zinc, iron, and other vital nutrients that is the primary cause of malnutrition.

## 2. Addressing malnutrition: strategies and technologies

There are several direct and indirect strategies to augment concentrations of essential nutrients in the grains as well as the other parts of the plants to surmount malnutrition.[6] These include dietary diversification, administration of artificial supplements, and crop biofortification. Among these,

**Agricultural Biotechnology. https://doi.org/10.1016/B978-0-323-88476-1.00006-0**

biofortification happens to be the most feasible, low-cost, and long-term solution to alleviate nutrient deficiency. Biofortified crops contain all the essential minerals, fatty acids, and vitamins required by the human body to fulfill nutritional demands.[7–9] Also, biofortified crops are inherently eco-friendly because they augment uptake of nutrients from the soil and promote soil health and are therefore sustainable.[10,11] Fig. 3.1 presents a schematic of different technologies practiced for biofortification.

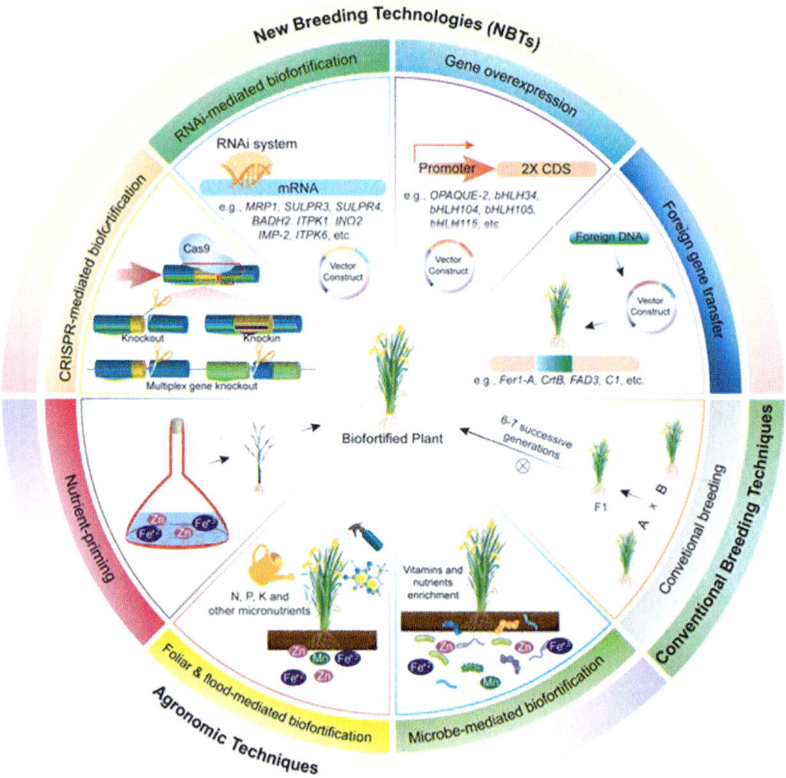


**FIGURE 3.1 Various methods for crop biofortification.** These methods are classified into: agronomic techniques using foliar application of micronutrients, mineral fertilization by flooding alongside irrigation, mineral fertilization via soil application, microbe-enabled enhanced nutrient uptake, conventional breeding techniques involving crosses between plants of contrasting phenotypes, new breeding techniques such as knockout of genes coding for antinutrient compounds, RNAi-mediated repression of antinutrient compound biosynthesis, overexpression of genes coding for micronutrient production and accumulation and gene transfer from foreign species that result in augmentation of the crop nutritional quality. *Reproduced from Shahzad R, Jamil S, Ahmad S, Nisar A, Khan S, Amina Z, Kanwal S, Aslam HMU, Gill RA and Zhou W. Biofortification of Cereals and Pulses Using New Breeding Techniques: Current and Future Perspectives.* Front. Nutr *2021;8:721728. https://doi.org/10.3389/fnut.2021.721728.*

## 3. Conventional strategies

### Microbe-enabled and foliar application of nutrients

Use of agronomic strategies toward biofortification is a cost-effective and facile method but nevertheless requires care with respect to their method of administration, type as well as environmental factors. Agronomic approach improves nutrient availability in the plants, their effectual utilization and easy mobility in addition to increasing microbial activity for successful utilization. Microbes such as Bacillus, Rhizobium, Actinomycete, Azotobacter, and some fungal strains like *P. indica* are used to augment nutrient availability and absorption.[12] Additionally, mineral nutrients can be supplied in the soil and can also be administered by foliar application. Notable among these are fertilizers based on phosphorus, potassium, and nitrogen, which are necessary for both human and plant health. Some of the other micro-minerals such as copper, iron, zinc, iodine, molybdenum, manganese, nickel, etc are also essential for crop plants. A majority of the above micronutrients are easily available in the soil and are absorbed by the plants enabling these nutrients to enter the food chain. However, at times plants are not able to absorb them and therefore are provided as mineral nutrition.[13]

### Conventional breeding practices

Another technique used for biofortification is conventional breeding, which is dependent on the genetic variability of the respective trait of interest (TOI) within the gene pool.[6] The required genes are pyramided through conventional crossings of plants followed by elaborate screening of the segregated plant populations.

### Improved crop varieties generated by conventional breeding

For instance, the National Agriculture Research Council (NARC), Pakistan, and the Wheat Research Institute (WRI), Faisalabad, have generated wheat varieties such as "Zincol" in 2015 and Akbar in 2019 that are biofortified with increased iron and zinc contents, respectively [[14] Lowe et al., 2018]. Likewise, the Bangladesh Rice Research Institute (BRRI) produced rice varieties, Dhan 62, 64, 72, and Jalmagna with high iron and zinc composition, respectively.[15] Several common varieties of beans such as RWR2154, RWV2887, 1129, 3317, 3006, and 33,166, MAC42, MAC44, CAB2, RWR2245, PVA1438, CODMLB 3,2 and 001, HM21-7 and Cuarentino with elevated iron and zinc contents were released to mitigate malnutrition. Lentil varieties including Barimasur 8, 4, 7, and 5, Khajurah2, Alemaya, L4704, ILL 7723, Idlib2, Idlib3, Sisir, Simal, and Shital with augmented zinc content were generated and released.[6,16]

### Crop wild relatives as sources for nutritional traits

Crop wild relatives (CWRs) contain important genes that code for essential micronutrients. To achieve standard protein requirements and nutritional concentrations, induction of genetic variations from CWRs to modern crops is essential.[17] The most effectual example of hybridization using CWRs is the introgression of the Gpc-B1 locus in *Triticum aestivum* from the wild emmer (*Triticum turgidum* ssp. dicoccoides) through chromosomal substitution. This introgression was greatly successful as it produced wheat plants with enhanced nutrient profile (Zn, Fe, Mn) without impacting crop yield.[18]

## 4. Use of plant growth promoting organisms to enhance crop biofortification

Rhizobia, actinomycetes, diazotrophic bacteria, and mycorrhizal fungi are beneficial soil microbes that are symbiotically associated with plant roots, and these organisms protect plants via several methods such as promoting nutrient mineralization and bioavailability as well as elicitation of plant growth hormones.[19] Despite their natural presence in the soil, their populations can be augmented by agricultural management strategies and inoculation. Several plant growth-promoting (PGP) soil microbes including Bacillus, Pseudomonas, and Enterobacter can be used to enhance micronutrient phytoavailability. These microbes are applied predominantly as seed inoculants and promote plant growth by eliciting the synthesis of antibiotics, growth hormones, siderophores, and chitinases in addition to inducing systemic resistance as well as mineralization.[20] PGP microbes are able to chelate iron through the synthesis of siderophores in addition to being capable of solubilizing phosphorus and inhibiting the growth of pathogens.[21,22] Therefore, they play a major role in iron fortification and fertility.

These PGP microbial organisms are normally present in the soil and in the compost as well as decomposing organic matter and afford a harmless, low-cost strategy for augmenting crop production and improving soil and environmental health.[23] Several studies have demonstrated increased concentrations of Se, Zn, and Fe in plants by the use of microbial inoculants through mycorrhizal associations.[24–26] Moreover, increased nitrogen fixation, grain yield, and plant growth have been observed in legumes including soybeans, peas, and chickpeas by colonization of microbes such as Brevibacterium sp., Pseudomonas sp., Enterobacter sp., Acinetobacter sp., and Bacillus sp. in their roots and root nodules.[27–31]

PGP actinobacteria inoculation in chickpeas enhanced the concentration of seed minerals such as Zn (13%–30%) and Fe (10%–38%) when compared to uninoculated control plants.[32] Likewise, inoculation with arbuscular mycorrhizal fungi augmented the nutritional content of chickpea grains by improving Zn and Fe concentrations in addition to augmenting protein content and yield.[33] Inoculation of PGP rhizobacteria along with Fe compound (FeSO4) in the soil enhanced the concentration of Fe up to 81% in chickpeas in comparison with control plants.[34] This suggested a putative role of microbes in the additional Fe uptake from the soil due to Fe supplementation. Inoculation of seeds prior to sowing and inoculation of field plots of pigeonpeas and chickpeas every 15 days until the stage of flowering with seven bacterial strains, namely, *B. antiquum* (SRI-158), *P. plecoglossicida* (SRI-156), *B. altitudinis* (SRI-178), *E. ludwigii* (SRI-229), *E. ludwigii* (SRI-211), *P. monteilii* (SRI-360), and *A. tandoii* (SRI-305) greatly enhanced root nodules, grain yield, and pod number when compared to that of uninoculated controls. Additionally, the harvested grains contained augmented concentrations of Zn and Fe. Fe content was increased by up to 12% and 18%, while the Zn content was increased up to 5% and 23% in pigeonpeas and chickpeas, respectively.

## 5. HarvestPlus program

The HarvestPlus program has been initiated and practiced to achieve an end to hidden hunger and micronutrient deficiencies on a global basis. This program targets the most vulnerable populations to ensure the provision of a food-based solution to curb micronutrient deficiencies. In collaboration with its partners, HarvestPlus develops crop varieties rich in vitamin A, zinc, and iron, three of the nutrients most lacking in foods as recognized by the World Health Organization (WHO).[35]

### HarvestPlus objectives

The objective of this program is to develop nutritionally enriched crops that are cost-efficient, sustainable, and evidence-based by means of traditional breeding practices and is projected to reach 1 billion people who will be fed with biofortified crops by the year 2030. Currently, this program addresses malnutrition issues in various countries such as Bangladesh, Pakistan, India, Nigeria, Uganda, Rwanda, Zimbabwe, Zambia, Congo, Caribbean, and Latin America.

### Three phases of the HarvestPlus program

The program has been divided into three distinct phases, namely, the discovery phase (2003–08), the development phase (2009–13), and the delivery phase (2014 onward). In the discovery phase, high-risk populations were identified and their food consumption practices, their plant breeding capabilities, and feasibility studies for biofortification were investigated. A pilot project was launched to provide sweet potato enriched with pro-vitamin A to populations in Africa. In the development phase, biofortified crops rich in pro-vitamin A, zinc, and iron were cultivated in the target countries followed by stability analysis testing at various locations, the nutritional profile of the generated varieties assessed, and country-wise teams were deployed for their delivery. In the delivery phase, augmented consumer demands for these biofortified crop varieties were created so as to reach the maximum malnourished populations. The HarvestPlus program also studied the long-term sustainability of the provision of these biofortified crops in the respective populations.[36]

## Success of the HarvestPlus program

Totally, 242 crop varieties have been distributed in 30 countries since the initiation of HarvestPlus and about 8.5 million smallholder agricultural breeders are currently growing these biofortified varieties and supplying them to 42.4 million of the global population. Through this program, beans enriched with iron, maize, and cassava enriched with vitamin A have been supplied to the Congo, wheat, and pearl millet enriched with iron and zinc to India, biofortified wheat rich in zinc to Pakistan, and zinc-biofortified rice to Bangladesh (https://www.harvestplus.org/where-we-work). shows some examples of successful implementation of the HarvestPlus program Table 3.1.

**TABLE 3.1** Examples of ongoing activities and accomplishments of the HarvestPlus program.

| Country | Crop | Nutrient | Varieties developed |
|---|---|---|---|
| Bangladesh | Rice | Zinc | BRRI dhan62, BRRI dhan64, BRRI dhan72, BU aromatic zinc rice |
| Democratic republic of Congo | Beans | Iron | HM 21–7 (bush), COD MLB 032 (bush), RWR 2154, RWR 2245 (bush), COD MLV 059 (bush), PVA 1438 (bush), nain de kyondo (climber), namulenga (climber), cuarentino (climber) |
| | Maize | Vitamin A | PVA-SYN12F2 (composite), SAM 4 VITA (composite), GV605 (hybrid), GV604 (hybrid) |
| | Cassava | Vitamin A | Kindisa (TMS 2001/1661), lumonu, vimpi |
| India | Wheat | Zinc | BHU-6 (chitra) and BHU-3 |
| | Pearl millet | Iron | ICMH 1201 (Shakti-1201), ICTP 8203-Fe-10–2 (dhanashakti) |
| Nigeria | Maize | Vitamin A | Sammaz 37, 38, and 39, ife hybrid 4, oba super 6, SC 510, SDM4, orange maize, PVA 2 (kapam 6) |

**TABLE 3.1 Examples of ongoing activities and accomplishments of the HarvestPlus program.—cont'd**

| Country | Crop | Nutrient | Varieties developed |
|---|---|---|---|
| | Cassava | Vitamin A | TMS 01/1368—umucass 36, TMS 07/0593—umucass 45, nr07/0220—umucass 44, TMS 07/539—umucass 46, TMS 01/1371—umucass 38, TMS 01/1412—umucass 37 |
| Pakistan | Wheat | Zinc | Zincol-2015, Akbar-2019 |
| Rwanda | Beans | Iron | RWR 2245 (bush), MAC 44 (climber), RWV 1129 (climber), RWR 2154 (bush), CAB 2 (climber), RWV 3317 (climber), MAC 42 (climber), RWV 3006 (climber), RWV 3316 (climber), RWV 2887 (climber) |
| Zambia | Sweet potato | Vitamin A | Olympia, chiwoko, twatasha, zambezi, kokota, mansa red, chingovwa |
| | Maize | Vitamin A | GV662A (kamano seed), GV665A (SeedCo), GV664A, GV671A (ZamSeed), GV573 (advanta), GV672A (afriseed) |
| Uganda | Sweet potato | Vitamin A | Ejumula, vita, kakamega, naspot 12 O, naspot 13 O, kabode |
| | Bean | Iron | Roba 1 (bush) |

Adapted from Shahzad R, Jamil S, Ahmad S, Nisar A, Khan S, Amina Z, Kanwal S, Aslam HMU, Gill RA and Zhou W. Biofortification of Cereals and Pulses Using New Breeding Techniques: Current and Future Perspectives. *Front. Nutr* 2021;8:721728. https://doi.org/10.3389/fnut.2021.721728.

## 6. New breeding technologies

New breeding technologies for crop biofortification include transgenic breeding, genome editing and RNA interference (RNAi). These genetic engineering technologies have opened more avenues for the generation of novel genetic variations that are absent in the gene pool.[5,37]

### Transgenic breeding

#### *Cereals*

#### Overexpression of genes encoding micronutrients and inhibition of antinutrient biosynthesis

An effectual, low-cost sustainable approach to address malnutrition issues is through biofortification by transgenic breeding. Transgenic crops are formulated when little genetic variability exists in the gene pool as well as wild relatives. Transgenic breeding introduces genes that either augment micronutrient accretion and bioavailability or inhibit antinutrient compounds. In cereal grains, insufficiency of micronutrients such as iron, zinc, proteins, and vitamin A can be surmounted by introducing and overexpressing genes coding for these micronutrients. For instance, in wheat grains, accumulation of high levels of iron was demonstrated through the ectopic overexpression of OsNAS2 and TaFer1-A genes although the expression of TaFer1-A was not uniform over generations.[9,38] Likewise, TaVIT gene overexpression doubled iron contents in the wheat grain.[39] When Amaranthus albumin-encoding Ama1 gene was overexpressed, it was found to overcome deficiencies of amino acids such as lysine, tyrosine, methionine, and cysteine in addition to essential proteins.[40]

Phytochrome A (PhyA) gene when overexpressed increased production of phytase and antinutrient activity while silencing of the phyA and the TaABC13 transporter gene repressed production of phytic acid and augmented zinc and iron bioavailability.[41] When SBEIIA gene was silenced, it enhanced the concentration of the less digestible amylose starch to counter overnutrition and obesity. Bacterial carotene desaturase genes PSY, Crtl, and CrTB when overexpressed in cereals augmented vitamin A concentration.[39] Further increase in carotenoid concentration was obtained when the maize phytoene synthase-encoding PSY1 gene and the bacterial Crtl gene were overexpressed.[42] Crtl and CrtB overexpression also augmented provitamin A accumulation.[43]

Biofortified rice varieties containing increased zinc, iron, vitamins A, B1, B6, and E were also generated. By overexpressing the PSY gene sourced from daffodil, vitamin A biofortification was obtained. When the *Erwinia uredovora* phytoene desaturase (Crtl) genes were overexpressed, this led to increase in vitamin A concentration.[44] Overexpression of the maize PST gene and the *Erwinia uredovora* Crtl gene also increased vitamin A concentration. In rice, varieties with augmented vitamin A concentration were yielded upon overexpression of the lycopene β-cyclase (β-lcy) and PSY genes sourced from daffodil. Similarly, the soybean Soyfer H-1 gene encoding ferritin when overexpressed led to increased Fe content, while the overexpression of the Phaseolus ferritin gene, Osfer2 and the OsNAS2 genes also improved Fe concentration.[45]

In polished rice varieties, iron biofortification was achieved when the ferritin gene was overexpressed under the control of endosperm-specific promoters' glutelin B1 (OsGluB1) and globulinb1 (OsGlb1); the NAS gene under the control of the OsActin1 promoter; the OsYSL2 gene regulated by the OsSUT1 and OsGlb1 transport promoters.[46] Likewise, zinc biofortification was accomplished when genes HvNAATA, HvNAAT-B, HvIDS3, and HvNAS1[47] were overexpressed while that of the OsNAS1, OsNAS2, and *Aspergillus flavus* phytase also improved zinc content in rice.[48] Transgenic rice varieties were developed[49] with lysine biofortification by overexpressing the endogenous RLRH1 and RLRH2 genes encoding lysine-rich histone proteins under the control of a modified rice glutelin1 promoter. Doubling of zinc content accumulation was observed upon overexpression of the rice genes OsIRTI, OsNAS1, OsNAS2, and OsNAS3. The genes Afphytase, Pvferritin, AtNAS1, and AtIRT1 were introduced into rice, which increased the zinc and iron concentrations in polished rice. A similar result was achieved when the activities of the vacuolar iron transporter OsVIT2 and OsVIT1 genes were disrupted in rice.[47]

## Golden rice and other improved crop varieties developed by transgene overexpression

In rice, improved folic acid content was reported by overexpressing the AtGTP-cyclohydrolase1 and the amino deoxy-chorismate synthase ADCS genes.[50] Golden rice was designed with high levels of provitamin A accumulation and was regarded to be the best example of biofortified transgenic crop.[51] The rice antioxidant activity is related to flavonoids, which can be enhanced by overexpressing chalcone synthase, phenylalanine ammonia lyase, maize C1, and GmFAD3 genes leading to augmented accumulation of linoleic acid and flavonoid contents.[52] Maize Psy1 gene when overexpressed in maize led to enhanced beta-carotene accretion. Also, overexpressing the dehydroascorbate reductase gene in maize augmented vitamin C accumulation by transforming the ascorbic acid oxidizing form into its reducing form.[53]

## Transgene-enabled biofortification by augmentation of micronutrients and amino acid contents

Overexpression of the aspergillus niger phy2, aspergillus phytase, and soybean ferritin gene increased iron bioavailability in maize. Potato sb401 gene when overexpressed in maize improved its lysine content. Enhancement of lysine and tryptophan accumulation was reported in maize upon expression of the antisense dsRNA targeting a-zeins. Also in maize, the gene for homogentisic acid geranyl geranyl transferase (HGGT) when overexpressed led to improved tocotrienol and tocopherol concentrations,[6] which resulted in biofortification. Further, the PSY1 gene overexpression also enhanced vitamin A concentrations in maize.[54,55]

## Development of crop varieties with improved micronutrient, vitamin, fatty acid, and fiber contents using transgene technology

Transgenic sorghum line homo 188 contains different genes such as CRT-I, PSY-1, LPA-1, and PMI for augmented pro-vitamin A content, and this line has improved nutritional value.[56] Overexpression of the zinc transporter genes led to increased zinc content in barley. Expression of the phytase and DHPS genes led to increased iron, zinc, and lysine bioavailability in barley. In barley seeds, when the 2-methyl-6-phytyl benzoquinol methyltransferase AtVTE-4 and AtVTE3 genes were co-expressed, it augmented vitamin E content. Similarly, the D6D gene when expressed in barley improved linolenic acid and stearidonic acid contents that are necessary for human health. Overexpression of the cellulose synthase-like (HvCslF) gene led to the escalation of the concentration of dietary fibers such as b-glucans.[57] High zinc content was reported in transgenic millets expressing zinc transporters. Transcriptomic studies showed that higher calcium sensor genes were associated with the uptake, accumulation, and translocation of calcium in the finger millet. Additionally, in these transgenic millets, lowered concentrations of anti-nutrients such as polyphenols, tannins, and phytic acid were reported using genome editing techniques.[58]

### Biofortification of pulses using transgenic technology

Compared to studies on cereals regarding transgenic breeding for bio-fortification, pulses have been less investigated. Modification of soybeans, common bean, chickpea, and lupines has been reported to overcome problems due to malnutrition.[59] The major objective of pulse biofortification through transgenic breeding is the enhancement of essential amino acids, zinc, and iron biofortification and decrease of antinutrient compounds. Heterologous proteins rich in sulfur-rich amino acids have been overexpressed in pulses to overcome deficiencies in these amino acids. Nutritional enrichment of cysteine was enabled in various pulse crops by the overexpression of the maize-derived 27 kDa g-zein, a cysteine-rich protein.[60] Likewise, methionine concentration was augmented in lupin and narbon bean by overexpressing S-rich proteins. Methionine concentration was increased in the Brazil nut by overexpressing aspartate kinase and the 2S albumin storage protein under the control of a seed-specific promoter, which resulted in 4 times higher methionine accumulation in the seeds.[60] In adzuki beans, transformation of the OASA1D transgene from rice increased accumulation of free tryptophan.[60]

Chickpea NAS2 and glycine max ferritin genes upon introduction and overexpression in chickpea plants, augmented iron bioavailability.[61] Cystathionine g-synthase gene overexpression in soybeans increased methionine concentration. Both methionine and cysteine contents of soybean seeds were augmented by overexpressing the maize zein protein in soybean plants. Also, O acetyl serine sulfhydrylase gene overexpression improved the seed cysteine content. Enhancement of seed lysine content was reported in transgenic soybean by upregulating aspartokinase and dihydrodipicolinic acid synthase genes.[6] Further, in soybean, enhanced pro-vitamin A accumulation was demonstrated by overexpressing the carotene desaturase gene and the bacterial PSY, crtB, crtW, and bkt1 genes.[62]

Reduction of the a-linolenic concentration was achieved in soybean by omega-3 FAD3 gene silencing using siRNA-enabled knockout. When the 16-desaturase gene was overexpressed, it augmented conversion of a-linolenic to its stable form: omega-3 fatty acids and g-linolenic acid. Synthesis of isoflavone was enhanced by the introduction of the maize C1 and the R transcription factors, which enabled gene activation.[1,6] In common beans, methionine content was increased by transformation of the Brazil nut methionine-rich storage albumin. Further, the lupin S-rich amino acid content was augmented by transformation of the respective gene from sunflower albumin.[6]

## 7. RNA interference

RNAi technology involves the repression of either the transcription or translation of a given gene through sequence-specific gene regulation enabled by a double-stranded RNA (dsRNA) molecule.[5] RNAi has opened new avenues for crop enhancement as it is efficient, stable, and highly precise and works as a better tool compared to antisense technology. The RNAi platform addresses the generation of tolerance to abiotic and biotic stresses and yield of quality food via biofortification as well as bio-elimination wherein enhanced nutritional quality and the removal of contaminants and food allergens are accomplished.[63,64] Some representative examples of the use of RNAi for biofortifying cereals and pulses are discussed below.

Phytic acid (PA) has been found to be a primary antinutrient in pulses and cereals due to its undesired ability to cause chelation of micronutrients and restriction of the bioavailability of essential nutrients. In wheat, RNAi technology was used to disrupt [inositol pentakisphosphate kinase (TaIPK1)] gene that is important for the PA biosynthesis pathway. The resulting wheat homozygous transgenic lines at the T4 stage demonstrated decreased expression of TaIPK1, diminished PA accumulation (26%−58%), and augmented grain iron, zinc, and phosphate contents.[65] Carotenoids, being pro-vitamin A precursors, are endowed with high nutritional value. Carotenoid contents in rice were increased by RNAi-driven disruption of the carotenoid cleavage dioxygenases (CCDs) genes that degrade carotenoids.

RNAi technology was used to disrupt the *OsCCD4a, OsCCD4b, and OsCCD1* genes in rice. This resulted in enhancement of carotenoid synthesis in the mutants: OsCCD4a-RNAi (1.6-fold), OsCCD4b-RNAi (1.3-fold) and OsCCD1-RNAi (1.4- fold) when compared to that of the wild-type plants.[66] In rice, lysine is the major limiting essential amino acid and the enzymes dihydrodipicolinate synthase (DHDPS) and aspartate kinase (AK) are important limiting molecules for lysine biosynthesis and are highly sensitive to a feedback mechanism. Inhibition of the DHDPS and AK enzymes by an RNAi-driven mechanism was performed in rice, which resulted in 21.7- and 6.6-fold higher lysine accumulation in the mutant lines respectively. Combining both mutations in one genotype led to 58.5-fold higher lysine levels compared to wild type plants.[67] In the maize endosperm, there is deficiency in the levels of beta-carotene and pro-vitamin A, which lead to vitamin A deficiency in the population. Nevertheless, other carotenoids such as zeaxanthin derived from beta-carotene through a 2-step hydroxylation reaction are present in enough quantities that could be precluded to augment beta-carotene concentration in the maize endosperm. Accordingly, two of the maize genes ZmBCH1 and ZmBCH2 associated with this hydroxylation reaction were disrupted by RNAi-mediated knockout to increase beta-carotene in the maize endosperm. ZmBCH2 mutants demonstrated significant enhancement in the maize grain beta-carotene contents indicating that the ZmBCH2 gene plays an important role in converting beta-carotene to other carotenoids.[68]

The digestibility of sorghum was increased by RNAi-based silencing of the expression of g-kafirin, g-kafirin 1A and g-kafirin1, 2.[6] There is potential for enhancement of pulse crops using RNAi-based biofortification. Phytate limits micronutrient bioavailability in pulses and phytic acid forms complexes with iron, copper, magnesium, and calcium, which results in the reduction of their solubility characteristics.[69] Yet other compounds such as the prebiotics, primarily fructans, and inulin absorb calcium, zinc and iron in themselves and limit phytic acid activity. Likewise, beta-carotene also augments the absorption of zinc and iron in peas, chickpeas, and lentils. The genes coding for these compounds can be overexpressed to improve the bioavailability of micronutrients.[70] Selenium has been reported to increase iodine bioavailability in peas, lentils, and chickpeas. Notwithstanding, generation of antinutrient substances should be better investigated and their gene expressions knocked out or silenced using RNAi in order to develop nutritionally enhanced pulse crops.[71]

## 8. Genome editing toward crop biofortification

Recently, genome editing of plants using sequence-specific nucleases (SSNs) has been accomplished for the targeted gene modification and stable inheritance in crops resulting in plants that are free of transgenes. Several SSN types such as ZFNs, TALENs, and the CRISPR-Cas system are employed for plant genome editing.[72] CRISPR-Cas-based genome editing involves the use of Cas9/13, RNA-guided DNA endonucleases that are guided by a short guided RNA (sgRNA) in a sequence-specific manner wherein complexes are formed at the target site to enable targeted gene editing.[63,73,74]

### Genome editing for cereal biofortification

Using genome editing, knockout of the rice OsVIT2 gene was accomplished that led to increase in Fe availability.[75] In the rice line Kitaake, targeted insertion of the 5.2 kb carotenogenesis cassette containing the PSY and Ctrl genes of maize under the control of an endosperm-specific promoter was achieved using the CRISPR-Cas9 system. The mutants obtained accrued 7.9 μg/g dry weights (DW) of beta-carotene in the rice endosperm, which is equivalent to the Kaybonnet rice variety generated via conventional transgenic breeding popularly known as Golden Rice2.[76] Table 3.2 presents some examples of investigations regarding crop biofortification achieved by means of genome editing as well as transgenic breeding technologies.

**TABLE 3.2 Examples of studies elucidating crop biofortification using genome editing and transgenic breeding.**

| Technique | Crop | Micronutrient | Genes involved | References |
|---|---|---|---|---|
| Genome editing | Maize | Carotenoid synthesis | *ZmPSY1*↑ | Zhu et al.[77] |
| | Wheat | Low gluten | *Alpha-gliadin 33-mer* ↓ | Sánchez-león et al.[78] |
| | | Grain weight | *TaGW2* ↑ | Liang et al.[79] |
| | | Fe and Zn | *TaVIT2* ↑ | Shan et al.[80] |
| | Rice | Phytic acid | *OsITPK1-6* ↓ | Jiang et al.[81] |
| | | Protein contents | *OsNRT1.1B* ↑ | Lu and zhu[82] |
| | | Amylose contents | *Wx gene* ↓, *SBEI* ↓, *SBEII* ↓, *OsWaxy* ↓ | Fiaz et al.[83]; Shao et al.[84] |
| | | Fe | *OsVIT2* ↑ | Ludwig and slamet-loedin[75] |
| | | Beta-carotene | *OsCCD4a* ↑, *OsCCD4b* ↑, *OsCYP97A4* ↑, *OsCCD7* ↑, *OsDSM2* ↑, *Osor* ↑ | Malik and maqbool[3]; Yang et al.[85] |
| | | Thiamine | *Ostpk1* ↑, *Ostpk2* ↑, *Ostpk3* ↑, *Osncs1* ↑, *OsThiC* ↑ | As proposed by Minhas et al.[86] |
| Transgenic | *Zea mays* | Iron | *GmFER* ↑, *aspergillus phytase* ↑, *aspergillus phy2* ↑ | As proposed by Garg et al.[6] |
| | | Beta-carotene | *Zmpsy 1* ↑ | Decourcelle et al.[87] |
| | | Tocopherol, tocotrienol | *HGGT* ↑ | Decourcelle et al.[87] |

*Continued*

**TABLE 3.2 Examples of studies elucidating crop biofortification using genome editing and transgenic breeding.—cont'd**

| Technique | Crop | Micronutrient | Genes involved | References |
|---|---|---|---|---|
| | *Oryza sativa* | Vitamin E | *AtTC* ↑, *AtHP* ↑ | Sundararajan et al.[88] |
| | | Vitamin B1 | *THIC* ↑, *THI1* ↑, *TH1* ↑ | Strobbe et al.[89] |
| | | Vitamin B6 | *AtPDX1.1* ↑, *AtPDX02* ↑ | Mangel et al.[90] |
| | | Iron | *AtNAS1* ↑, *AtIRT1* ↑, *Pvferritine* ↑, *Afphytase* ↑, *OsFer2*↑, *OsIRT1* ↑, *OsNAS1* ↑, *OsNAS2* ↑, *GmFER* ↑, *OsVIT1* ↓, *OsVIT2* ↓, *HvNAS1* ↑, *OsYSL2* ↑, *OsIDEF1* ↑, *OsNAC5* ↑, *OsYSL9* ↓ | Malik and maqbool[3]; Trijatmiko et al.[91]; Kawakami and bhullar[92] |
| | *Triticum aestivum* | Iron | *TaFer1-A*↑, *OsNAS2* ↑, *TaVIT2* ↑, *GmFER* ↑, *PhyA* ↑, *NAM-B1*↑ | Malik and maqbool[3]; Connorton et al.[93] |
| | | Vitamin A | *CrtI* ↑, *CrtB* ↑, *Bacterial PSY* ↑ | Wang et al.[43] |
| | *Glycine max* | Fe and zinc | *Phytase* ↑ | Malik and maqbool[3] |
| | | Methionine | *Cystathionin* g-*synthase* ↑, *maize zein protein* ↑ | Dinkins et al.[94]; Hanafy et al.[95]; Song et al.[96] |
| | | Provitamin A | *Carotene desaturase*↑, *crtB* ↑, *crtW* ↑, *bacterial PSY* ↑, *bkt1* ↑ | Kim et al.[97] |

Note: ↓ stands for downregulation/knockout of the targeted gene, whereas ↑ refers to upregulation/overexpression of the targeted gene.

Adapted from Shahzad R, Jamil S, Ahmad S, Nisar A, Khan S, Amina Z, Kanwal S, Aslam HMU, Gill RA and Zhou W. Biofortification of Cereals and Pulses Using New Breeding Techniques: Current and Future Perspectives. *Front. Nutr* 2021;8:721728. https://doi.org/10.3389/fnut.2021.721728.

The rice OsWaxy gene was targeted using multiplex genome editing at three sites to reduce AC, and 14% reduction in accumulation was reported in the mutants compared to the wild-type plants.[98] Increased nitrogen uptake in rice was accomplished by editing the OsNRT1.1B nitrogen transporter gene that is associated with protein accumulation.[82] Augmented micronutrient availability and diminished phytic acid accumulation were obtained in rice by the CRISPR-Cas9-mediated knock out of OsITPK1-6 gene.[81] In wheat, increased zinc concentration was generated by the targeted mutagenesis and editing of the TaVIT2 gene.[99] Using genome editing technology, maize biofortification was achieved via modification of various regulatory pathways. ZmALS1 and ZmALS2 acetolactate synthase genes were edited to enable accelerated synthesis of leucine, isoleucine, and valine.[100] ZmPSY1 phytoene synthase gene associated with the carotenoid synthesis pathway was edited by the CRISPR-Cas9 type II system, which demonstrated stable transformation in the progeny.[77]

## Biofortification of pulses using genome editing

Presently, there is a dearth of literature on the role of the genome editing technology for pulse biofortification. Nevertheless, there is great potential in the use of genome editing for zinc and iron biofortification. Improved zinc and iron uptake can be accomplished in all pulse crops by editing of the genes encoding ferric-chelate reductase oxidase (FRO), iron-regulated transporter (IRT), YELLOW STRIPE 1-like (YSL), zinc-regulated transporters, natural resistance—associated macrophage protein (NRAMP), and iron-regulated transporters like protein (ZIP).[59] Another strategy for using genome editing toward pulse biofortification involves targeting of antinutrient genes that determine decreased micronutrient bioavailability. Saponins are one class of antinutrient compounds, which are beneficial at low concentrations. Nevertheless, they act as antinutrients when consumed in greater quantities.[71] In *Arabidopsis thaliana*, 112 uridine diphosphate glycosyltransferases (UGTs) genes, 246 P450 genes, and 13 OSC genes are proved to be involved in saponin biosynthesis. There is a compelling need to identify crucial regulatory genes involved in the saponin biosynthesis pathway and disrupt them in order to reduce the production of saponins.[101] Likewise, genes associated with the expression of other antinutrients such as a-amylase, and protease inhibitor have to be identified and edited to decrease the synthesis of these antinutrient compounds.[69] Fig. 3.2 exemplifies the use, advantages, and disadvantages of genome editing technology for biofortification.

## Nutritionally enriched feed crops

In order to address enhancements in feed provided for poultry and livestock several nutritionally enhanced crops have been produced.[102] Animal feed crops have been produced with elevated levels of limiting amino acids such

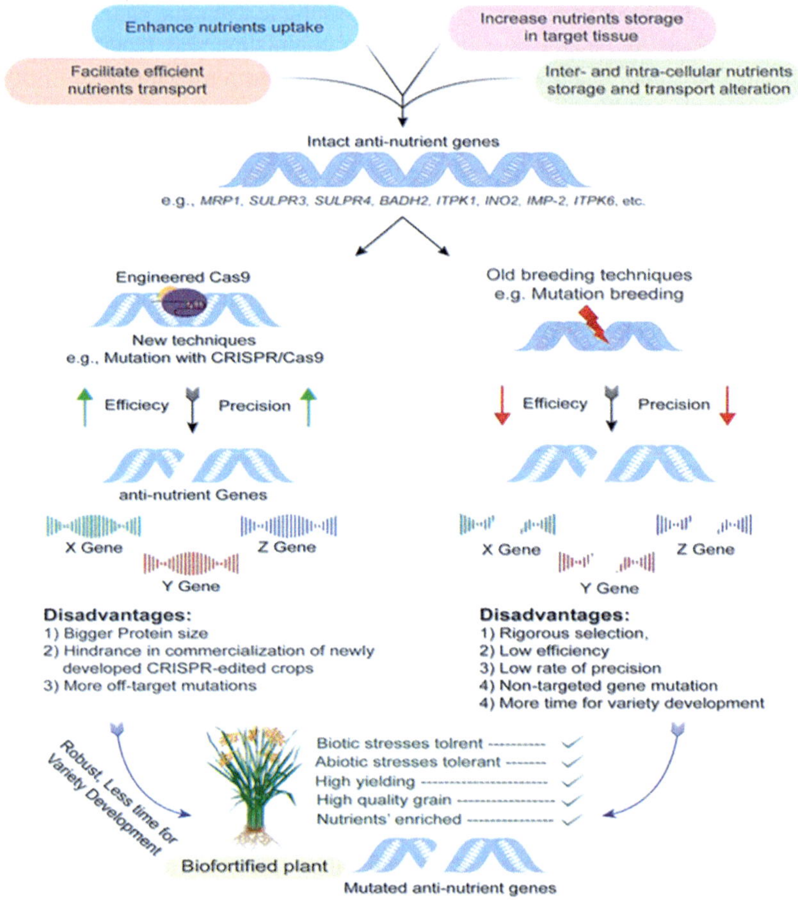


**FIGURE 3.2 CRISPR-Cas–driven targeted genome editing for crop biofortification.** First, the cultivars harboring antinutrient genes such as *BADH2, MRP1, INO2,* etc., are selected; next, the CRISPR-Cas complex modifies the respective genes without changing the genetic composition of the rest of the plant, leading to the generation of mutants in very short period of time when compared to that of conventional mutagenesis. The mutant plants thus obtained are incorporated into a breeding program for generating nutritionally enhanced varieties free of transgenes. The above illustration also compares genome editing with conventional breeding techniques: Traditional breeding has low efficiency, while the gene editing technique has more probability of off target effects. *Reproduced from Shahzad R, Jamil S, Ahmad S, Nisar A, Khan S, Amina Z, Kanwal S, Aslam HMU, Gill RA and Zhou W. Biofortification of Cereals and Pulses Using New Breeding Techniques: Current and Future Perspectives.* Front. Nutr *2021;8:721728. https://doi.org/10.3389/fnut.2021.721728.*

that fewer feed supplements would be necessary. Feed crops have also been designed toward generating manure that is more environmentally friendly. Swine and poultry provided feed with transgenic maize having higher free lysine content augmented body weight gain to levels as high as those animals that were fed with diets containing lysine provided as a supplement. Equivalent results were obtained in livestock fed with lupin and soybean.[102] Bacterial adenylylsulfate reductase and aspartate kinase genes were cloned into alfalfa plants which increased sulfur containing amino acids methionine by 60% and cysteine by 30%.[103] Also augmented abundance of some other essential amino acids such as lysine and aspartate was observed.

These strategies can improve the nutritional value of livestock feed crops. Transgenic maize capable of expressing cell wall invertase has also been generated.[104] In these transgenic plants, grain yield was significantly improved by virtue of increased grain size and number. The invertase gene when constitutively expressed augmented the total starch content within the transgenic kernels showing that the invertase gene can be used to enhance both grain quality and grain yield in these feed crops. Depletion of antinutrients in animal feed toward increasing micronutrient absorption has been investigated. Transgenic maize expressing high levels (779,800 U/kg) of the β-glucanase Bgl7AM was observed to be stable at pH 1.0–8.0, which is the normal environmental pH of the digestive tract.[105] Therefore, these transgenic plants simplify the processing of the feed, thus enabling the feed to be more amenable for consumption by livestock.

## Biofortified crops with other health benefits

In addition to biofortification of food crops through genetic engineering, crops can also be designed to possess bioreactive compounds that show enhanced health benefits or decrease the risk of cancer, heart disease, and other chronic diseases.[106,107] Seed storage oils of plants have been studied for their ability to express novel fatty acids of benefit to human health. Designer crops such as the "designer oilseed" transgenic plants have been generated via metabolic engineering to express omega-3 fatty acids that are routinely found in fish oils.[108] In this context, there is great interest in omega-3 long-chain polyunsaturated fatty acids (omega-3-FA) expression due to their dietary advantages like enhancements of brain development and function in addition to improving cardiovascular health. As a majority of omega-3 FA is obtained from marine life and in the current scenario where seas have been over-fished, plants serve as a better sustainable source of omega-3 FA. Genetically engineered algae, krill, and plants are currently being developed to express omega-3 FA to levels as high as that found in marine organisms.[109–111]

Anthocyanin, an antioxidant compound found in cranberries and blueberries has been expressed in tomato plants through metabolic engineering and the resultant tomato fruit obtained, increased the lifespans of mice susceptible to cancer by up to 30%.[112] These tomatoes were shown to contain anti-inflammatory properties and could help combat cardiovascular disease due to which they are presently used in the treatment of heart patients in Britain. In another study, tomatoes transformed with grape (*Vitis vinifera* L.) stilbene synthase gene under the regulation of a fruit-specific promoter were shown to accumulate trans-resveratrol-glucopyranoside and transresveratrol whose levels reached as high as up to 53 μg/g fresh weight.[113] Resveratrol is a beneficial bioactive compound occurring mostly in grapes and red wine, so expressing this compound in tomatoes could augment their nutritional value. Notable increases in ascorbate content and antioxidant capability were observed in these tomato lines. In a monocyte-macrophage culture under laboratory conditions these transgenic tomatoes could counteract the pro-inflammatory activity of phorbol ester.[114,115]

The isoflavone metabolic pathway found in soybean has been engineered into transgenic tobacco lines. Leaves of these tobacco plants showed increased isoflavone synthesis and extracts from these leaves when fed to estrogen-deficient mice demonstrated decreased number of osteoclasts and lowered expression of osteoclastogenic genes along with elevated uterine estrogenicity and greater total serum antioxidant levels in comparison with those of control mice.[116] Aside from the above-mentioned design and use of biofortified crops, crops expressing biopharmaceuticals such as vaccines, therapeutic proteins, and monoclonal antibodies against various human and animal diseases have been generated. This is discussed in detail in Chapter 4 of this book on "Genetically engineered plant products for health: Plant-derived recombinant biopharmaceuticals."

## 9. Regulatory issues concerned with crop varieties generated through new plant breeding techniques (NBTs)

The cultivation and subsequent commercialization of transgenic crops are subject to various regulatory processes, which are expensive and time-consuming accompanied by poor public acceptance. For instance, despite the development of Golden Rice being hailed as a stellar biotechnological achievement, several years have passed since its invention and still it is not available for cultivation on a commercial scale due to its unstable yield. Likewise, Bt brinjal (expressing *Bacillus thuringiensis* toxin) cultivation was also banned as some scientists, anti-GMO agencies, and farmers demonstrated concerns regarding its propagation. Transgenic crops have their own limitations due stringent regulatory approval procedures practiced in different geographical regions. Moreover, GM technology is time-consuming and expensive and is not popular considering the current economic and political landscape. Also, the rate of success of this technology for a given cultivar released is poor despite the painstaking efforts undertaken toward identifying a desired gene followed by its modification and expression in the target plant as well as its agronomic appraisal and biosafety assessment research.[6]

Increasing comprehension of genomics and its use in plant breeding has greatly augmented the application of modern plant biotechnology to promote the quality and quantity of crops. NBTs, by virtue of being accurate and precise, are very suitable for generating desired mutants possessing high target specificity.[5] Using NBTs, crops can be improved directly by insertion or deletion of any given segment of a gene and finally acquire desired plant traits without impacting other characteristics of the plant in a highly specific, time efficient, and cost-effective manner. These NBTs are distinct from the GM crops due to their definitive and stable mutation. Nevertheless, despite their high efficacy, controversy in regard to their implementation is still prevalent in many countries.

The major glitch in executing gene editing practices involves public perception. In this context, scientists, policy makers, and regulatory authorities are considering regulations to be imposed on genome-edited crops.[2] Their main contention is whether to classify the genome-edited crops as GMOs, which would then subject them to the GMO regulatory frameworks. This issue is complicated by the character trait being edited, the different molecular pathways employed for crop enhancement, the molecular tools used, and probable risks of the crop products obtained by genome editing.

The United States Department of Agriculture in 2012 declared that plants edited by the ZFNs and mega-nucleases must not be placed under GMO regulation and authorized the cultivation and commercialization of CRISPR-edited mushrooms and waxy corn without going through GMO regulation. Notwithstanding, some of the anti-GMO agencies are attempting to ban genome-edited crops, considering them as subject to GMO regulations and contending that the edited crop products are unnatural and could negatively impact the ecosystem.[2] Therefore, the success of genome-editing technology depends not just on the scientific- and evidence-based methodologies but also on their acceptance by regulators, nongovernmental agencies, and ultimately the consumers of these products.[63,74,117]

## 10. Conclusions and future perspectives

Member countries are being urged by the United Nations (UN) to meet the 17 set Sustainable Development Goals (SDGs) by the year 2030. Of these, SDG3 concerns assuring healthy lives and well-being for all of the world's populace. Good health practices are strongly associated with good nutrition and availability of nutritious food for consumption. Nevertheless, hidden hunger and malnutrition are major hurdles to enabling healthy living by severely impacting the global population, wherein mostly women and young children in developing countries are affected by these serious issues.[5] Global micronutrient deficiency still exists in developing nations owing to the common use of cereals that result in an imbalanced nutritional profile. When consumed in the recommended quantities, biofortified pulses, and cereals meet with the complete dietary needs and afford a balanced diet.[118]

Crop biofortification, in addition to imparting balanced nutrition to the masses, also enable governments to reach SDG3 by generating crops that are nutritionally enriched. Efficient biofortification programs must focus on the development of crop varieties containing augmented micronutrient content without impacting economic benefits to farmers.[47] The capability and effectiveness of breeding programs for biofortification are founded on the decrease in antinutrients, availability of genetic diversity, and augmented concentrations of micronutrients such as ascorbic acid (vitamin C) and amino acids (cysteine, lysine, and methionine) that can improve the absorption of essential elements.

There are several drawbacks in the use of traditional approaches to achieve crop biofortification. For instance, in conventional breeding, there is an inherent lack of genetic diversity within the gene pool, which a major shortcoming in addition to the protracted time required for this process. In a few cases, this could be overcome by crosses with distant relatives of the respective plants toward introducing the trait of interest (TOI). However, in most instances, it is hard to identify the desired trait in the distant relatives. Therefore, it is nearly impossible to augment the TOI using conventional breeding.

Genome editing affords a viable solution to most of the above stated problems. It is highly targeted, precise, and efficient in eliciting targeted modification of the gene of interest and is deemed biologically safe since the plant end product obtained using this technology is transgene-free.[72] The most recent genome editing breakthrough is (a) base editing that irreversibly converts a base at the target site without the involvement of double-stranded breaks, donor templates, dependency on HDR, and NHEJ; (b) prime editing which involves the introduction of indels and all of the base to base conversions without causing a DNA double-strand break by using prime editing guide RNA (pegRNA) to direct the Cas9 endonuclease.

Both Canada and the United States have already announced that the GE crops are free from GM crop regulations, although in the EU, this issue is still being debated with hope for positive acceptance.[63,73] GE crops by virtue of being bereft of the transgene must be regulated just like the plant products developed by conventional breeding and must reach as many consumers as possible in order to obtain maximum advantages of this technology within the briefest possible time frame.[73] The future of biofortification exists in genome editing. Targeted genetic engineering should aim toward augmented micronutrient uptake from the soil, highest level of translocation to the seed, and enhanced bioavailability. Most micronutrients are present in the aleurone layers of cereals and these layers get eliminated during milling in addition to bran and hence, they are not accessible for consumption. This issue can be remedied by (a) processing whole grains which will augment the nutritional value of products generated afterward and (b) genetic engineering of the micronutrient biosynthetic pathways which will change the deposition of micronutrients to the endosperm rather than the aleurone layer so as to not lose them during processing and milling.

## References


[1] Shahzad R, Jamil S, Ahmad S, et al. Harnessing the potential of plant transcription factors in developing climate resilient crops to improve global food security: current and future perspectives. *Saudi Journal of Biological Sciences* 2021;**28**(4):2323–41. https://doi.org/10.1016/j.sjbs.2021.01.028.

[2] Sedeek KEM, Mahas A, Mahfouz M. Plant genome engineering for targeted improvement of crop traits. *Frontiers in Plant Science* 2019;**10**. https://doi.org/10.3389/fpls.2019.00114.

[3] Malik KA, Maqbool A. Transgenic crops for biofortification. *Frontiers in Sustainable Food Systems* 2020;**4**. https://doi.org/10.3389/fsufs.2020.571402.

[4] Mayer JE, Pfeiffer WH, Beyer P. Biofortified crops to alleviate micronutrient malnutrition. *Current Opinion in Plant Biology* 2008;**11**(2):166–70. https://doi.org/10.1016/j.pbi.2008.01.007.

[5] Shahzad R, Jamil S, Ahmad S, et al. Biofortification of cereals and pulses using new breeding techniques: current and future perspectives. *Frontiers in Nutrition* 2021;**8**. https://doi.org/10.3389/fnut.2021.721728.

[6] Garg M, Sharma N, Sharma S, et al. Biofortified crops generated by breeding, Agronomy, and transgenic approaches are improving lives of millions of people around the world. *Frontiers in Nutrition* 2018;**5**. https://doi.org/10.3389/fnut.2018.00012.

[7] Tan GZH, Das Bhowmik SS, Hoang TML, et al. Investigation of baseline iron levels in australian chickpea and evaluation of a transgenic biofortification approach. *Frontiers in Plant Science* 2018;**9**. https://doi.org/10.3389/fpls.2018.00788.

[8] Singh G, Andersson MS. *Zinc-biofortified wheat: harnessing genetic diversity for improved nutritional quality*. 2017.

[9] Singh SP, Keller B, Gruissem W, Bhullar NK. Rice Nicotianamine synthase 2 expression improves dietary iron and zinc levels in wheat. *Theoretical and Applied Genetics* 2017;**130**(2):283–92. https://doi.org/10.1007/s00122-016-2808-x.

[10] Singla P, Grover K. Biofortified cereal crops: a sustainable approach for food and nutritional security. *Current Journal of Applied Science and Technology* 2017:1–13. https://doi.org/10.9734/CJAST/2017/37865.

[11] Waters BM, Sankaran RP. Moving micronutrients from the soil to the seeds: genes and physiological processes from a biofortification perspective. *Plant Science* 2011;**180**(4):562–74. https://doi.org/10.1016/j.plantsci.2010.12.003.

[12] Kaur T, Rana KL, Kour D, et al. *Microbe-mediated biofortification for micronutrients: present Status and future Challenges*. Elsevier BV; 2020. p. 1–17. https://doi.org/10.1016/b978-0-12-820528-0.00002-8.

[13] Rashid A, Ram H, Zou C, Guilherme L, Corguinha A, Guo S. Simultaneous biofortification of rice with zinc, iodine, iron and selenium through foliar treatment of a micronutrient cocktail in five countries. *Frontiers in Plant Science* 2020;**11**:589835. https://doi.org/10.3389/fpls.2020.589835.

[14] Lowe NM, Khan MJ, Broadley MR, et al. Examining the effectiveness of consuming flour made from agronomically biofortified wheat (Zincol-2016/NR-421) for improving Zn status in women in a low-resource setting in Pakistan: study protocol for a randomised, double-blind, controlled cross-over trial (BiZiFED). *BMJ Open* 2018;**8**(4):e021364. https://doi.org/10.1136/bmjopen-2017-021364.

[15] Majumder S, Datta K, Datta SK. Rice biofortification: high iron, zinc, and vitamin-A to fight against hidden hunger. *Agronomy* 2019;**9**(12):803. https://doi.org/10.3390/agronomy9120803.

[16] Kumar J, Gupta DS, Kumar S, Gupta S, Singh NP. Current knowledge on genetic biofortification in lentil. *Journal of Agricultural and Food Chemistry* 2016;**64**(33):6383–96. https://doi.org/10.1021/acs.jafc.6b02171.

[17] Engels JMM, Thormann I. Main challenges and Actions needed to improve conservation and sustainable use of our crop wild relatives. *Plants* 2020;**9**(8):968. https://doi.org/10.3390/plants9080968.

[18] Fatiukha A, Filler N, Lupo I, et al. Grain protein content and thousand kernel weight QTLs identified in a durum × wild emmer wheat mapping population tested in five environments. *Theoretical and Applied Genetics* 2020;**133**(1):119–31. https://doi.org/10.1007/s00122-019-03444-8.

[19] FAO. *The Plant Production and Protection Division (AGP)—Soil Biological Management with Beneficial Microorganisms*. Rome, Italy: FAO; 2019.

[20] Mahaffee WF, Kloepper JW. Applications of plant growth-promoting rhizobacteria in sustainable agriculture. *Soil Biota: Management in Sustainable Farming Systems* 1994:23–31.

[21] Panhwar QA, Othman R, Rahman ZA, Meon S, Ismail MR. Isolation and characterization of phosphate-solubilizing bacteria from aerobic rice. *African Journal of Biotechnology* 2012;**11**:2711–9.

[22] Sreevidya M, Gopalakrishnan S, Kudapa H, Varshney RK. Exploring plant growth-promotion actinomycetes from vermicompost and rhizosphere soil for yield enhancement in chickpea. *Brazilian Journal of Microbiology* 2016;**47**(1):85–95. https://doi.org/10.1016/j.bjm.2015.11.030.

[23] Gopalakrishnan S, Vadlamudi S, Samineni S, Sameer Kumar CV. Plant growth-promotion and biofortification of chickpea and pigeonpea through inoculation of biocontrol potential bacteria, isolated from organic soils. *SpringerPlus* 2016;**5**(1). https://doi.org/10.1186/s40064-016-3590-6.

[24] Cavagnaro TR. The role of arbuscular mycorrhizas in improving plant zinc nutrition under low soil zinc concentrations: a review. *Plant and Soil* 2008;**304**(1–2):315–25. https://doi.org/10.1007/s11104-008-9559-7.

[25] Rengel Z, Batten GD, Crowley DE. Agronomic approaches for improving the micronutrient density in edible portions of field crops. *Field Crops Research* 1999;**60**(1–2):27–40. https://doi.org/10.1016/S0378-4290(98)00131-2.

[26] Smith S, Read D. *Mycorrhizal Symbiosis*. Australia: Elsevier Ltd; 2008. https://doi.org/10.1016/B978-0-12-370526-6.X5001-6.

[27] Gopalakrishnan S, Srinivas V, Prakash B, Sathya A, Vijayabharathi R. Plant growth-promoting traits of Pseudomonas geniculata isolated from chickpea nodules. *3 Biotech* 2015;**5**(5):653–61. https://doi.org/10.1007/s13205-014-0263-4.

[28] Minorsky PV. On the inside. *Plant Physiology* 2009;**149**(4):1621–2. https://doi.org/10.1104/pp.109.900288.

[29] Soe KM, Bhromsiri A, Karladee D. E_ects of selected endophytic actinomycetes (Streptomyces sp.) and Bradyrhizobia from Myanmar on growth, nodulation, nitrogen fixation and yield of di_erent soybean varieties. *Chiang Mai University Journal of Natural Sciences* 2010;**9**:95–109.

[30] Valverde A, Burgos A, Fiscella T, et al. Differential effects of coinoculations with Pseudomonas jessenii PS06 (a phosphate-solubilizing bacterium) and Mesorhizobium ciceri C-2/2 strains on the growth and seed yield of chickpea under greenhouse and field conditions. *Plant and Soil* 2006;**287**(1–2):43–50. https://doi.org/10.1007/s11104-006-9057-8.

[31] Tokala RK, Strap JL, Jung CM, et al. Novel plant-microbe rhizosphere interaction involving Streptomyces lydicus WYEC108 and the pea plant (*Pisum sativum*). *Applied and Environmental Microbiology* 2002;**68**(5):2161–71. https://doi.org/10.1128/AEM.68.5.2161-2171.2002.

[32] Sathya A, Vijayabharati R, Srinivas V, Gopalakrishnan S. *Plant growth-promoting Actionbacteria on chickpea seed mineral Density: an Upcoming Complementary tool for sustainable biofortification strategy.* 2013.

[33] Pellegrino E, Bedini S. Enhancing ecosystem services in sustainable agriculture: biofertilization and biofortification of chickpea (*Cicer arietinum* L.) by arbuscular mycorrhizal fungi. *Soil Biology and Biochemistry* 2014;**68**:429–39. https://doi.org/10.1016/j.soilbio.2013.09.030.

[34] Khalid S, Asghar HN, Akhtar MJ, Aslam A, Zahir ZA. Biofortification of iron in chickpea by plant growth promoting rhizobacteria. *Pakistan Journal of Botany* 2015;**47**(3):1191–4. http://www.pakbs.org/pjbot/PDFs/47%283%29/48.pdf.

[35] La Frano MR, de Moura FF, Boy E, Lönnerdal B, Burri BJ. Bioavailability of iron, zinc, and provitamin A carotenoids in biofortified staple crops. *Nutrition Reviews* 2014;**72**(5):289–307. https://doi.org/10.1111/nure.12108.

[36] Bouis HE, Saltzman A. Improving nutrition through biofortification: a review of evidence from HarvestPlus, 2003 through 2016. *Global Food Security* 2017;**12**:49–58. https://doi.org/10.1016/j.gfs.2017.01.009.

[37] Van Der Straeten D, Bhullar NK, De Steur H, et al. Multiplying the efficiency and impact of biofortification through metabolic engineering. *Nature Communications* 2020;**11**(1). https://doi.org/10.1038/s41467-020-19020-4.

[38] Borg S, Brinch-Pedersen H, Tauris B, et al. Wheat ferritins: improving the iron content of the wheat grain. *Journal of Cereal Science* 2012;**56**(2):204–13. https://doi.org/10.1016/j.jcs.2012.03.005.

[39] Kumar S, Palve A, Joshi C, Srivastava RK. Crop biofortification for iron (Fe), zinc (Zn) and vitamin A with transgenic approaches. *Heliyon* 2019;**5**(6):e01914. https://doi.org/10.1016/j.heliyon.2019.e01914.

[40] Xu M, Zhao S, Zhang Y, et al. Production of marker-free transgenic rice (*Oryza sativa* L.) with improved nutritive quality expressing AmA1. *Iranian Journal of Biotechnology* 2017;**15**(2):102–10. https://doi.org/10.15171/ijb.1527.

[41] Abid N, Khatoon A, Maqbool A, et al. Transgenic expression of phytase in wheat endosperm increases bioavailability of iron and zinc in grains. *Transgenic Research* 2017;**26**(1):109–22. https://doi.org/10.1007/s11248-016-9983-z.

[42] Cong L, Wang C, Chen L, Liu H, Yang G, He G. Expression of phytoene synthase1 and carotene desaturase crtI genes result in an increase in the total carotenoids content in transgenic elite wheat (Triticum aestivum L.). *J Agric Food Chem.* 2009 Sep 23;**57**(18):8652–60.

[43] Wang C, Zeng J, Li Y, et al. Enrichment of provitamin A content in wheat (*Triticum aestivum* L.) by introduction of the bacterial carotenoid biosynthetic genes CrtB and CrtI. *Journal of Experimental Botany* 2014;**65**(9):2545–56. https://doi.org/10.1093/jxb/eru138.

[44] Oliva N, Florida Cueto-Reaño M, Trijatmiko KR, et al. Molecular characterization and safety assessment of biofortified provitamin A rice. *Scientific Reports* 2020;**10**(1). https://doi.org/10.1038/s41598-020-57669-5.

[45] Roychoudhury A, Bhowmik R. *Genetic engineering of rice to fortify micronutrients.* Springer Science and Business Media LLC; 2020. p. 563–79. https://doi.org/10.1007/978-981-15-5337-0_24.

[46] Pandit P, Sanghamitra, Pradhan SK. Molecular breeding for iron bio-fortification in rice grain: recent progress and future perspectives. *Molecular Breeding for Rice Abiotic Stress Tolerance and Nutritional Quality* 2021:315–40.

[47] Sanjeeva Rao D, Neeraja CN, Madhu Babu P, et al. Zinc biofortified rice varieties: challenges, possibilities, and progress in India. *Frontiers in Nutrition* 2020;**7**. https://doi.org/10.3389/fnut.2020.00026.

[48] ZulfiqarU M, Hussain s. *Rice research for quality Improvement: genomics and genetic engineering*. 2020. p. 605–27.

[49] Wong HW, Liu Q, Sun SSM. Biofortification of rice with lysine using endogenous histones. *Plant Molecular Biology* 2015;**87**(3):235–48. https://doi.org/10.1007/s11103-014-0272-z.

[50] Dong W, Cheng Zj, Lei Cl, et al. Overexpression of folate biosynthesis genes in rice (*Oryza sativa* L.) and evaluation of their impact on seed folate content. *Plant Foods for Human Nutrition* 2014;**69**(4):379–85. https://doi.org/10.1007/s11130-014-0450-9.

[51] Shahbaz U, Yu X-B, Akhtar W, Ndagijimana R, Rauf H. Golden rice to eradicate the vitamin A deficiency in the developing countries. *European Journal of Nutrition & Food Safety* 2020;(1):53–63. https://doi.org/10.9734/ejnfs/2020/v12i130184.

[52] Zhu Q, Tan W, Liu YG. Genetic engineering for increasing antioxidant content in rice: recent progress and future perspectives. *Molecular Breeding for Rice Abiotic Stress Tolerance and Nutritional Quality* 2021:358–81.

[53] Foyer CH, Kyndt T, Hancock RD. Vitamin C in plants: novel concepts, new perspectives, and outstanding issues. *Antioxidants and Redox Signaling* 2020;**32**(7):463–85. https://doi.org/10.1089/ars.2019.7819.

[54] Hossain F, Muthusamy V, Zunjare RU, Gupta HS. *Biofortification of maize for protein quality and Provitamin-A content*. Springer Science and Business Media LLC; 2019. p. 115–36. https://doi.org/10.1007/978-3-319-95354-0_5.

[55] Thanh ND. Provitamin a biofortification in maize through genetic engineering and marker-assisted selection. *Academia Journal of Biology* 2019;**41**(4). https://doi.org/10.15625/2615-9023/v41n4.13804.

[56] Lipkie TE, De Moura FF, Zhao ZY, et al. Bioaccessibility of carotenoids from transgenic provitamin a biofortified sorghum. *Journal of Agricultural and Food Chemistry* 2013;**61**(24):5764–71. https://doi.org/10.1021/jf305361s.

[57] Sakellariou M, Mylona PV. New uses for traditional crops: the case of barley biofortification. *Agronomy* 2020;**10**(12):1964. https://doi.org/10.3390/agronomy10121964.

[58] Vinoth A, Ravindhran R. Biofortification in millets: a sustainable approach for nutritional security. *Frontiers in Plant Science* 2017;**8**. https://doi.org/10.3389/fpls.2017.00029.

[59] Tan GZH, Das Bhowmik SS, Hoang TML, et al. Finger on the pulse: pumping iron into chickpea. *Frontiers in Plant Science* 2017;**8**. https://doi.org/10.3389/fpls.2017.01755.

[60] Kaur S, Kumari A, Singh P, Kaur L, Sharma N, Garg M. *Biofortification in pulses*. Springer Science and Business Media LLC; 2020. p. 85–103. https://doi.org/10.1007/978-981-15-2874-3_4.

[61] Tan Z. *The Characterisation of CaNAS2 and biofortification of chickpea*. 2018.

[62] Qin Y, Park SY, Oh SW, et al. Nutritional composition analysis for beta-carotene-enhanced transgenic soybeans (Glycine max L.). *Applied Biological Chemistry* 2017;**60**(3):299–309. https://doi.org/10.1007/s13765-017-0282-z.

[63] Ahmad S, Shahzad R, Jamil S, et al. *Regulatory Aspects, risk assessment, and Toxicity associated with RNAi and CRISPR methods*. Elsevier BV; 2021. p. 687–721. https://doi.org/10.1016/b978-0-12-821910-2.00013-8.

[64] Rajam MV. RNA silencing technology: a boon for crop improvement. *Journal of Biosciences* 2020;**45**(1). https://doi.org/10.1007/s12038-020-00082-x.

[65] Aggarwal S, Kumar A, Bhati KK, et al. RNAi-mediated downregulation of inositol pentakisphosphate kinase (IPK1) in wheat grains decreases phytic acid levels and increases Fe and Zn accumulation. *Frontiers in Plant Science* 2018;**9**. https://doi.org/10.3389/fpls.2018.00259.

[66] Ko MR, Song MH, Kim JK, et al. RNAi-mediated suppression of three carotenoid-cleavage dioxygenase genes, OsCCD1, 4a, and 4b, increases carotenoid content in rice. *Journal of Experimental Botany* 2018;**69**(21):5105–16. https://doi.org/10.1093/jxb/ery300.

[67] Yang QQ, Yu WH, Wu HY, Zhang CQ, Sun SSM, Liu QQ. Lysine biofortification in rice by modulating feedback inhibition of aspartate kinase and dihydrodipicolinate synthase. *Plant Biotechnology Journal* 2021;**19**(3):490–501. https://doi.org/10.1111/pbi.13478.

[68] Berman J, Zorrilla-López U, Sandmann G, Capell T, Christou P, Zhu C. The silencing of carotenoid β-hydroxylases by RNA interference in different maize genetic backgrounds increases the β-carotene content of the endosperm. *International Journal of Molecular Sciences* 2017;**18**(12):2515. https://doi.org/10.3390/ijms18122515.

[69] Robinson GHJ, Balk J, Domoney C. Improving pulse crops as a source of protein, starch and micronutrients. *Nutrition Bulletin* 2019;**44**(3):202–15. https://doi.org/10.1111/nbu.12399.

[70] Uppal C, Kaur S. *Genome engineering for nutritional Improvement in pulses*. 2021. p. 157–80. https://doi.org/10.1002/9781119672425.ch10.

[71] Kumar S, Pandey G. Biofortification of pulses and legumes to enhance nutrition. *Heliyon* 2020;**6**(3):e03682. https://doi.org/10.1016/j.heliyon.2020.e03682.

[72] Ahmad S, Wei X, Sheng Z, Hu P, Tang S. CRISPR/Cas9 for development of disease resistance in plants: recent progress, limitations and future prospects. *Briefings in Functional Genomics* 2020;**19**(1):26–39. https://doi.org/10.1093/bfgp/elz041.

[73] Ahmad S, Tang L, Shahzad R, et al. CRISPR-based crop Improvements: a way forward to achieve zero hunger. *Journal of Agricultural and Food Chemistry* 2021;**69**(30):8307–23. https://doi.org/10.1021/acs.jafc.1c02653.

[74] Jamil S, Shahzad R, Ahmad S, et al. Role of genetics, genomics, and breeding approaches to combat stripe rust of wheat. *Frontiers in Nutrition* 2020;**7**. https://doi.org/10.3389/fnut.2020.580715.

[75] Ludwig Y, Slamet-Loedin IH. Genetic biofortification to enrich rice and wheat grain iron: from genes to product. *Frontiers in Plant Science* 2019;**10**. https://doi.org/10.3389/fpls.2019.00833.

[76] Zheng X, Kuijer HNJ, Al-Babili S. Carotenoid biofortification of crops in the CRISPR era. *Trends in Biotechnology* 2021;**39**(9):857–60. https://doi.org/10.1016/j.tibtech.2020.12.003.

[77] Zhu J, Song N, Sun S, et al. Efficiency and inheritance of targeted mutagenesis in maize using CRISPR-cas9. *Journal of Genetics and Genomics* 2016;**43**(1):25–36. https://doi.org/10.1016/j.jgg.2015.10.006.

[78] Sánchez-León S, Gil-Humanes J, Ozuna CV, Giménez MJ, Sousa C, Voytas DF, et al. Low-gluten, nontransgenic wheat engineered with CRISPR/Cas9. *Plant Biotechnology Journal* 2018;**16**:902–10. https://doi.org/10.1111/pbi.12837.

[79] Liang Z, Chen K, Li T, Zhang Y, Wang Y, Zhao Q, et al. Efficient DNAfree genome editing of bread wheat using CRISPR/Cas9 ribonucleoprotein complexes. *Nature Communications* 2017;**8**:1–5. https://doi.org/10.1038/ncomms14261.

[80] Shan Q, Wang Y, Li J, Gao C. Genome editing in rice and wheat using the CRISPR/Cas system. *Nature Protocols* 2014;**9**:2395–410. https://doi.org/10.1038/nprot.2014.157.

[81] Jiang M, Liu Y, Liu Y, Tan Y, Huang J, Shu Q. Mutation of inositol 1,3,4-trisphosphate 5/6-kinase6 Impairs plant growth and phytic acid synthesis in rice. *Plants* 2019;**8**(5):114. https://doi.org/10.3390/plants8050114.

[82] Lu Y, Zhu JK. Precise editing of a target base in the rice genome using a modified CRISPR/Cas9 system. *Molecular Plant* 2017;**10**(3):523–5. https://doi.org/10.1016/j.molp.2016.11.013.

[83] Fiaz S, Ahmad S, Noor M, et al. Applications of the CRISPR/Cas9 system for rice grain quality improvement: perspectives and opportunities. *International Journal of Molecular Sciences* 2019;**20**(4):888. https://doi.org/10.3390/ijms20040888.

[84] Shao G, Xie L, Jiao G, Wei X, Sheng Z, Tang S, et al. CRISPR/CAS9-mediated editing of the fragrant gene Badh2 in rice. *Chinese Journal of Rice Science* 2017;**31**:216–22. https://doi.org/10.16819/j.1001-7216.2017.6098.

[85] Yang X, Chen L, He J, Yu W. Knocking out of carotenoid catabolic genes in rice fails to boost carotenoid accumulation, but reveals a mutation in strigolactone biosynthesis. *Plant Cell Reports* 2017;**36**:1533–45. https://doi.org/10.1007/s00299-017-2172-6.

[86] Minhas AP, Tuli R, Puri S. Pathway editing targets for thiamine biofortification in rice grains. *Frontiers in Plant Science* 2018;**9**:975. https://doi.org/10.3389/fpls.2018.00975.

[87] Decourcelle M, Perez-Fons L, Baulande S, Steiger S, Couvelard L, Hem S, et al. Combined transcript, proteome, and metabolite analysis of transgenic maize seeds engineered for enhanced carotenoid synthesis reveals pleotropic effects in core metabolism. *Journal of Experimental Botany* 2015;**66**:3141–50. https://doi.org/10.1093/jxb/erv120.

[88] Sundararajan S, Rajendran V, Sivakumar HP, Nayeem S, Chandra HM, Sharma A, et al. Enhanced vitamin E content in an Indica rice cultivar harbouring two transgenes from Arabidopsis thaliana involved in tocopherol biosynthesis pathway. *Plant Biotechnology Journal* 2021;**19**:1083. https://doi.org/10.1111/pbi.13578.

[89] Strobbe S, Verstraete J, Stove C, Van Der Straeten D. Metabolic engineering of rice endosperm towards higher vitamin B1 accumulation. *Plant Biotechnology Journal* 2021;**19**:1253–67. https://doi.org/10.1111/pbi.13545.

[90] Mangel N, Fudge JB, Li KT, Wu TY, Tohge T, Fernie AR, et al. Enhancement of vitamin B6 levels in rice expressing Arabidopsis vitamin B6 biosynthesis de novo genes. *PlJ* 2019;**99**:1047–65. https://doi.org/10.1111/tpj.14379.

[91] Trijatmiko KR, Dueñas C, Tsakirpaloglou N, Torrizo L, Arines FM, Adeva C, et al. Biofortified indica rice attains iron and zinc nutrition dietary targets in the field. *Scientific Reports* 2016;**6**:19792. https://doi.org/10.1038/srep19792.

[92] Kawakami Y, Bhullar NK. Molecular processes in iron and zinc homeostasis and their modulation for biofortification in rice. *Journal of Integrative Plant Biology* 2018;**60**:1181–98. https://doi.org/10.1111/jipb.12751.

[93] Connorton JM, Jones ER, Rodríguez-Ramiro I, Fairweather-Tait S, Uauy C, Balk J. Wheat vacuolar iron transporter TaVIT2 transports Fe and Mn and is effective for biofortification.Plant. *Physiology* 2017;**174**:2434–44. https://doi.org/10.1104/pp.17.00672.

[94] Dinkins RD, Reddy MS, Meurer CA, Yan B, Trick H, Thibaud-Nissen F, et al. Increased sulfur amino acids in soybean plants overexpressing the maize 15 kDa zein protein. In Vitro. *Cell Dev Biol Plant* 2001;**37**:742–7. https://doi.org/10.1007/s11627-001-0123-x.

[95] Hanafy MS, Rahman SM, Nakamoto Y, Fujiwara T, Naito S, Wakasa K, et al. Differential response of methionine metabolism in two grain legumes, soybean and azuki bean,

expressing a mutated form ofArabidopsis cystathionine g-synthase.Journal of. *Plant Physiology* 2013;**170**:338–45. https://doi.org/10.1016/j.jplph.2012.10.018.

[96] Song S, Hou W, Godo I, Wu C, Yu Y, Matityahu I, et al. Soybean seeds expressing feedback-insensitive cystathionine g-synthase exhibit a higher content of methionine. *Journal of Experimental Botany* 2013;**64**:1917–26. https://doi.org/10.1093/jxb/ert053.

[97] Kim M-J, Kim JK, Kim HJ, Pak JH, Lee J-H, Kim D-H, et al. Genetic modification of the soybean to enhance the b-carotene content through seed-specific expression. *PLoS One* 2012;**7**:e48287. https://doi.org/10.1371/journal.pone.0048287.1–18.

[98] Roy S. *Genome editing for biofortification of rice: current Implications and future Aspects.* 2021. p. 297–313. https://doi.org/10.1002/9781119672425.ch17.

[99] Mourad AMI, Alomari DZ, Alqudah AM, Sallam A, Salem KFM. Recent advances in wheat (Triticum spp.) breeding. In: *Advances in plant breeding strategies: cereals*, vol. 5. Egypt: Springer International Publishing; 2019. p. 559–93. https://doi.org/10.1007/978-3-030-23108-8_15.

[100] Svitashev S, Young JK, Schwartz C, Gao H, Falco SC, Cigan AM. Targeted mutagenesis, precise gene editing, and site-specific gene insertion in maize using Cas9 and guide RNA. *Plant Physiology* 2015;**169**(2):931–45. https://doi.org/10.1104/pp.15.00793.

[101] Guo C, Wei Z, Lyu W, Geng Y. *Bioinformatics-based screening of Key genes for saponin metabolism in Quinoa.* 2021. https://doi.org/10.21203/rs.3.rs-139481/v1.

[102] Hefferon K. Nutritionally enhanced food crops; progress and perspectives. *International Journal of Molecular Sciences* 2015;**16**(2):3895–914. https://doi.org/10.3390/ijms16023895.

[103] Tong Z, Xie C, Ma L, et al. Co-expression of bacterial aspartate kinase and adenylylsulfate reductase genes substantially increases sulfur amino acid levels in transgenic alfalfa (medicago sativa L.). *PLoS ONE* 2014;**9**(2):e88310. https://doi.org/10.1371/journal.pone.0088310.

[104] Li B, Liu H, Zhang Y, et al. Constitutive expression of cell wall invertase genes increases grain yield and starch content in maize. *Plant Biotechnology Journal* 2013;**11**(9):1080–91. https://doi.org/10.1111/pbi.12102.

[105] Zhang Y, Xu X, Zhou X, et al. Overexpression of an acidic endo-β-1,3-1,4-glucanase in transgenic maize seed for direct utilization in animal feed. *PLoS ONE* 2013;**8**(12):e81993. https://doi.org/10.1371/journal.pone.0081993.

[106] Siró I, Kápolna E, Kápolna B, Lugasi A. Functional food. Product development, marketing and consumer acceptance-A review. *Appetite* 2008;**51**(3):456–67. https://doi.org/10.1016/j.appet.2008.05.060.

[107] Wilson SA, Roberts SC. Metabolic engineering approaches for production of biochemicals in food and medicinal plants. *Current Opinion in Biotechnology* 2014;**26**:174–82. https://doi.org/10.1016/j.copbio.2014.01.006.

[108] Ruiz-Lopez N, Haslam RP, Napier JA, Sayanova O. Successful high-level accumulation of fish oil omega-3 long-chain polyunsaturated fatty acids in a transgenic oilseed crop. *Plant Journal* 2014;**77**(2):198–208. https://doi.org/10.1111/tpj.12378.

[109] Ruiz-López N, Haslam RP, Venegas-Calerón M, et al. Enhancing the accumulation of omega-3 long chain polyunsaturated fatty acids in transgenic *Arabidopsis thaliana* via iterative metabolic engineering and genetic crossing. *Transgenic Research* 2012;**21**(6):1233–43. https://doi.org/10.1007/s11248-012-9596-0.

[110] Ruiz-López N, Sayanova O, Napier JA, Haslam RP. Metabolic engineering of the omega-3 long chain polyunsaturated fatty acid biosynthetic pathway into transgenic plants. *Journal of Experimental Botany* 2012;**63**(7):2397–410. https://doi.org/10.1093/jxb/err454.

[111] Adarme-Vega TC, Thomas-Hall SR, Schenk PM. Towards sustainable sources for omega-3 fatty acids production. *Current Opinion in Biotechnology* 2014;**26**:14–8. https://doi.org/10.1016/j.copbio.2013.08.003.

[112] Butelli E, Titta L, Giorgio M, et al. Enrichment of tomato fruit with health-promoting anthocyanins by expression of select transcription factors. *Nature Biotechnology* 2008;**26**(11):1301–8. https://doi.org/10.1038/nbt.1506.

[113] Giovinazzo G, D'Amico L, Paradiso A, Bollini R, Sparvoli F, DeGara L. Antioxidant metabolite profiles in tomato fruit constitutively expressing the grapevine stilbene synthase gene. *Plant Biotechnology Journal* 2005;**3**(1):57–69. https://doi.org/10.1111/j.1467-7652.2004.00099.x.

[114] D'Introno A, Paradiso A, Scoditti E, et al. Antioxidant and anti-inflammatory properties of tomato fruits synthesizing different amounts of stilbenes. *Plant Biotechnology Journal* 2009;**7**(5):422–9. https://doi.org/10.1111/j.1467-7652.2009.00409.x.

[115] Ingrosso I, Bonsegna S, De Domenico S, et al. Over-expression of a grape stilbene synthase gene in tomato induces parthenocarpy and causes abnormal pollen development. *Plant Physiology and Biochemistry* 2011;**49**(10):1092–9. https://doi.org/10.1016/j.plaphy.2011.07.012.

[116] Pandey A, Misra P, Khan MP, et al. Co-expression of Arabidopsis transcription factor, AtMYB12, and soybean isoflavone synthase, GmIFS1, genes in tobacco leads to enhanced biosynthesis of isoflavones and flavonols resulting in osteoprotective activity. *Plant Biotechnology Journal* 2014;**12**(1):69–80. https://doi.org/10.1111/pbi.12118.

[117] Lassoued R, Smyth SJ, Phillips PWB, Hesseln H. Regulatory uncertainty around new breeding techniques. *Frontiers in Plant Science* 2018;**9**. https://doi.org/10.3389/fpls.2018.01291.

[118] Rehman HM, Cooper JW, Lam HM, Yang SH. Legume biofortification is an underexploited strategy for combatting hidden hunger. *Plant Cell and Environment* 2019;**42**(1):52–70. https://doi.org/10.1111/pce.13368.

Chapter 4

# Genetically engineered plant products for health: plant-made recombinant pharmaceuticals

## 1. An overview of major accomplishments in the area of plant-based biopharmaceuticals

Over the past 3 decades, genetic engineering of plant systems has progressed tremendously. In 1986,[1] the human growth hormone gene was cloned as a chimeric construct with the nopaline synthase gene in a Ti plasmid and expressed in tobacco and sunflower. Subsequently, the mouse monoclonal antibody was shown to express and assemble successfully in tobacco.[2] Human type I collagen was generated in tobacco wherein it self-assembled into homogeneous fibrils.[3] The recombinant bovine trypsin was expressed in maize plants and marketed as TrypZean by Sigma Aldrich, and it is now used favorably in animal cell cultures as it lacks animal-derived contaminants. Rice plants have been used to produce human lactoferrrin and lysozyme. Cultured transgenic carrot and tobacco cells have been used to generate plant-based pharmaceuticals by Protalix, an Israeli company. The US Food and Drug Administration (FDA) approved the production of *Taliglucerase alfa*, a recombinant glucocerebrosidase in 2012 by Protalix and Pfizer in plant systems for the treatment of Gaucher disease. Plant cell cultures, protoplasts, and hairy root cultures have been variously employed to genetically engineer biopharmaceuticals in crops such as maize, rice, potato, peanuts, carrot, alfalfa, lettuce, tomato, tobacco, and soybeans.

### Strategies for expression of plant-based biopharmaceuticals: indirect gene delivery by agrobacterium-mediated nuclear transformation and generation of transgenic plants

Nuclear or chloroplast genome recombination has been routinely used to generate transgenic plants.[4] This has been achieved traditionally using *Agrobacterium*

**Agricultural Biotechnology. https://doi.org/10.1016/B978-0-323-88476-1.00009-6**

*tumefaciens*, which is capable of naturally infecting plants and thereby introducing foreign genes (that are cloned within the T-DNA of its Ti plasmid) into the plant genome during the infection process. This leads to the stable integration of the foreign gene into the plant nuclear genome, which then becomes a permanent vaccine generating warehouse and can hence be incorporated as a primary seed bank. Nevertheless, transgenic generation of vaccine antigens requires lengthy periods of time and is complicated by adverse effects depending on the position of its integration into the plant genome, which could result in gene silencing, injury to the host genome, or cross-hybridization at the field level with non-transgenic plants in situations where crops are grown without strict regulations. The major advantages of the use of *Agrobacterium*-mediated transformation compared to that of other methods are decrease in the copy number of the transgene and stable and intact integration of the transgene into the plant genome.

Notable examples of vaccine antigens generated by the transgenic route[4] are the VP60 polypeptide against the rabbit hemorrhagic disease virus and the cholera toxin B subunit that were successfully produced in potato leaf explants. Additionally, *Arabidopsis thaliana* has been used to transgenically express the *Helicobacter pylori* Ton B protein, while the surface antigen of HBV has been expressed in tomato plants. Other proteins such as the capsid polypeptide of the Norwalk virus, the *Escherichia coli* heat-labile enterotoxin as well as the transgenic alfalfa-generated foot and mouth disease virus, have also been expressed in transgenic plant systems. A majority of these *Agrobacterium* transformations have been performed in dicotyledonous plants.

## Direct delivery of foreign genes into plant cells by biolistic bombardment

This technique, also known as microprojectile bombardment or gene gun, does not rely on the vector and enables nuclear or chloroplast genome transformations in plant systems where it is not feasible to perform *Agrobacterium*-mediated transformation.[4] This strategy involves the coating of the transgene DNA on tungsten or gold particles used as microcarriers, which, when introduced into an equipment called the gene gun, enables these DNA-coated particles to travel at high speeds and penetrate the plant cell targets, thus leading to the stable integration of the concerned transgene into the genome of the host plant. However, the high cost of the gene gun device in addition to the requirement of trained personnel and labor to operate the equipment and the resulting damage to the plant tissues due to particle penetration preclude the widespread use of this procedure for generating transgenic plants.

## Plastid transformation

The technology of chloroplast/plastid transformation has emerged recently as a novel strategy to introduce transgenes coding for vaccine antigens into plants. In comparison to nuclear genome transformation, plastid transformation has several inherent advantages. Chloroplast genomes are much smaller than nuclear genomes and undergo maternal inheritance, thereby precluding the undesirable possibility of outcrossing via the pollen of the transgenic plant while facilitating stable expression of the candidate antigen. Further, there are several chloroplasts within a given plant cell, which result in high copy numbers of the transgenes, wherein the absence of gene silencing can lead to accumulation of increased levels of the recombinant protein up to as much as 70% of the total protein content of the leaf. Techniques for performing chloroplast transformation include the use of gene guns for biolistic bombardment or alternatively, protoplast transformation of vaccine candidate genes by PEG treatment.

Chloroplast transformation has been widely used for generating vaccines against viruses such as the rotavirus and the canine parvovirus, as well as against bacteria causing plague, lyme disease, tetanus, cholera, and anthrax, most of which have been produced in tobacco.[4] Mucosal (oral/nasal) administration of chloroplast-produced tetanus and anthrax vaccines to mice resulted in potent immune response against their respective bacterial pathogens authenticating the success of these chloroplast-based vaccines. Also, the rabies virus vaccine was produced in *Zea mays*, and the dengue virus vaccine was expressed in *Lactuca sativa* (lettuce) through chloroplast transformation. Notwithstanding, plastid transformation has its own caveats: the organs of the plant that are not photosynthetic are ineffective in expressing the recombinant protein. Moreover, plastids lack glycosylation machinery, and hence, this technology is not suitable for use in expressing foreign proteins that require complex posttranslational modifications.[4]

## Transient expression of vaccine proteins using plant virus-based expression systems

Plant virus-derived transient expression technology has recently emerged as the most popular strategy due to the inherent ability of the plant virus to replicate in large amounts within the target plant, which in turn obtains great yields of vaccines and therapeutics. Several plant viruses such as tobacco mosaic virus, tomato bushy stunt virus, potato virus X, cowpea mosaic virus, bamboo mosaic virus, alfalfa mosaic virus, papaya mosaic virus, and plum pox potyvirus have been variously employed as vaccine expression systems. Vaccines against the influenza virus and the human papilloma virus have been expressed in TMV and PVX.[4] Similarly, a norovirus vaccine has been generated in the TMV expression system.

Deconstructed second-generation viral expression vectors having minimal viral genetic elements essential for replication are popularly used. These vectors have enabled the production of higher yields of vaccine candidates when compared to those of their full virus equivalents. Plant virus-based expression systems are highly advantageous as they can express the foreign antigens in high amounts within a limited time span, while enabling the display of foreign epitopes in multiple copies on the surface of the plant virus-like particles (VLPs) and facilitating large-scale infections of these engineered viruses in plants at the field level. However, this technology requires the purification of the concerned recombinant plant virus or VLPs from the plants prior to being used as a vaccine/therapeutic.

Notable examples[4] of plant-virus based vaccine expression are the use of CPMV for expression of the mink enteritis virus and the canine parvovirus VP2 capsid proteins, the *Pseudomonas aeruginosa* outer membrane protein F and the *Staphylococcus aureus* fibronectin-binding protein D2 domain. TMV has been repurposed to express antigenic epitopes from the foot and mouth disease virus and the murine hepatitis virus in tobacco. The Geminivirus, bean yellow dwarf virus has been used to express the *Staphylococcus* enterotoxin B vaccine in addition to the generation of the Ebola virus immune complex in the leaves of *Nicotiana benthamiana*.[4]

## Production of vaccines in the geminivirus system

In comparison with a host of expression vector systems derived from plant RNA viruses, Geminivirus-based plant viral expression vectors are more advantageous for the following reasons[5]: Geminiviruses possess small, single-stranded circular RNA genomes along with twinned capsid morphology that provides several selective advantages like their high copy number, capability to establish infection in a wide range of plants and increased stability while escaping gene silencing inherent to many plant viral expression systems. Geminiviral expression vectors have become the ideal systems for producing plant-based pharmaceuticals as several proteins can be expressed tandemly from a single construct. These viruses replicate by the rolling circle mechanism driven by their replication initiator protein, Rep, which specifically initiates the replication process at a highly conserved stem-loop structure located between two major ORFs encoded within the viral genome. This mode of replication is highly efficacious and robust and results in tens of thousands of viral DNA copies within every infected cell, enabling them to become powerful, potent factories for the production of biopharmaceuticals. Also, the single-stranded viral DNA genome upon complementary DNA strand synthesis, becomes a double-stranded DNA intermediate, which associates with nucleosomes to form minichromosomes. These minichromosomes exist within the cell in a transcriptionally active, unmethylated state and drives Rep-mediated unmethylated replication that conveys resistance to the host methylation mechanism, preventing its capability to actively inhibit unwanted transcription of virus-encoded gene products within the nucleus.

Furthermore, Geminiviruses have the ability to encode suppressors of gene silencing that inhibit the host silencing machinery.[5] Hence, by using inducible or developmental promoters to selectively control Rep protein expression, virus replication and the concomitant PBP expression can be switched on or off as desired, thus dispensing with concerns of toxicity while enabling the expression of foreign proteins selectively within specific tissues such as leaves or seeds. Reportedly, some of these Geminiviral vectors have expressed foreign proteins at yields as high as 1 mg/g of fresh leaf tissue as well as at stages as early as 4 days postinoculation, thus demonstrating that Geminiviruses are a powerful, rapid, and highly efficacious platform for the production of PBPs.

## Infiltration technologies for transient expression of foreign proteins

Recently, magnifection and agroinfiltration technologies are by and large being used to transiently express proteins through the vacuum- or syringe-enabled infiltration of younger leaves of plants like *Arabidopsis* and tobacco with *Agrobacterium* harboring recombinant binary vectors or with the second-generation vectors derived from deconstructed plant viruses.[4] This agroinfiltration technology was pioneered by the Canadian biotechnology company, Medicago, which produced the Influenza virus HA antigen VLP vaccine that has now entered the stage of clinical trials. These VLP vaccines displaying the envelope proteins of animal or human viruses on their outer surface are attractive platforms for embodying virus antigenicity and eliciting protective immune response, while requiring less time span for antigen expression than stable transgenic expression systems.

Lettuce agroinfiltration was reported in lettuce.[6] Icon Genetics, a German plant biotechnology company, has used the Magni-CON technology to produce many plant-based vaccines. VLPs expressing HBV surface antigen have been reported at levels as much as 300 mg/kg of fresh leaf weight of *N. benthamiana,* and additionally, the capsid proteins of the norovirus as well as non-Hodgkin lymphoma vaccines that advanced to phase I clinical trials have reportedly been generated. The Fraunhofer USA Center for Molecular Biotechnology (CMB) has developed a launch vector gene infiltration platform that combines the deconstructed TMV vector with the molecular elements of the *A. tumefaciens* binary plasmid.[4]

In agroinfiltration, the gene of interest can be cloned directly within the *Agrobacterium* vector or can be cloned into a deconstructed plant virus vector that in turn has been integrated into the binary plasmid of *Agrobacterium*, and these can then be infiltrated into the plant tissues following Agro-transformation.[7,8] The major advantage of Agroinfiltration is its ability to generate elevated levels of the recombinant proteins, in addition to providing for inexpensive production.[9,10] In this system, the protein expression levels peak within under 7 days after infiltration, which is considerably quicker in comparison to the over 2 weeks required for the full plant virus to produce systemic infection.

Agroinfiltration has been greatly successful in generating plant-derived biologics, vaccines, and therapeutics in a safe and efficacious manner, and several of these products have advanced into clinical trials.[6] In 2009, Medicago reported the production of the hemagglutinin vaccine candidate against the H1N1 influenza virus in *N. benthamiana* within a time span of 19 days.[4] This clearly demonstrates the rapid response potential of the agroinfiltration technology in generating plant-based recombinants.

## Use of sonication and chemical (PEG) stimulation to promote DNA uptake

Delivery of recombinant DNA into plant cell protoplasts has also been attempted by sonication involving the use of sound waves to cause particle agitation in solution, which accelerates DNA delivery.[4] On the other hand, polyethylene glycol (PEG), a DNA precipitating agent has been used to enhance DNA uptake into protoplasts by endocytosis, enabling transient expression of the recombinant proteins. This affords an inexpensive option for transforming DNA into protoplasts while precluding any protoplast damage, and this technique can therefore be used in conjunction with other gene delivery strategies.

## Other augmented transfection methods

In order to prevent the probable risk of the transfer of antibiotic-coding genes into the microbiota of the gut during oral vaccine administration or when these genes are inadvertently released into the surrounding environment, plants can be genetically engineered without antibiotic selection. For instance, the cholera toxin B subunit was cloned in a T-DNA vector, which was co-transformed with another T-DNA vector coding for the hygromycin phosphotransferase, and antibiotic-free plants were selected that successfully expressed the CTB protein.[11] This was achieved by segregation and selection at the level of seed progeny. Daniell et al.,[12] successfully precluded antibiotic selection by using the betaine aldehyde dehydrogenase as a selectable marker for plastid transformation in transgenic tobacco, thus preventing antibiotic-based selection. Increased amounts of the cholera toxin B subunit and botulinum neurotoxin A were synthesized in transgenic rice seeds that were accompanied by RNAi-based knockdown of storage proteins.[4] Additionally, plants possess their own inherent N-glycosylation mechanisms, which could complicate foreign protein expression by introducing molecular heterogeneity resulting in interference of foreign antigen recognition by the immune system. Therefore, in recent studies, the native N-glycosylation genes of plants have been deleted or disabled while enzyme genes encoding mammalian N-glycosylation activities have been deliberately introduced to facilitate viable molecular recognition.[4]

## Generation of multicomponent subunit vaccines

Also, when the vaccine antigen gene is fused to genes coding for immunomodulatory molecules including mucosal adjuvants, the former's immunogenicity is enhanced several folds.[4] Common adjuvants used are cholera toxin B, bacterial enterotoxin, immunomodulators derived from bacteria and mammalian systems, as well as plant secondary metabolites. This has resulted in the production of multicomponent vaccines. A tricomponent subunit vaccine was expressed in transgenic potato wherein the cholera toxin B was fused to the rotavirus enterotoxin and the adhesion protein of *E. coli*, which afforded immunity to all the three pathogens including cholera, enterotoxigenic *E. coli*, and rotavirus.[4]

## 2. Benefits and drawbacks of using plants for the generation of biopharmaceuticals

Plants are increasingly being used for expressing several recombinant proteins free of any contamination from human or animal pathogens. The major advantage of plant-derived recombinants is that they can be easily shipped and stored at low-cost without the requirement of any cold chain equipment. Plant-based subunit vaccines can be manufactured in the seeds of cereal crops, which makes these crops ideal for long-term storage and easy transport. However, there are many caveats about plant-derived pharmaceuticals, such as the wide variability in the quantities of target antigens expressed between various species of plants or even between different parts of the same plant, which complicate the standardized manufacture of the recombinant proteins. Additionally, there is significant heterogeneity in the presentation of plant-based vaccine antigens to the immune system especially when administered by oral delivery due to degradation of the recombinant proteins in the digestive tract. Thus, more quantities of the vaccine antigen are required so as to compensate for the antigen loss. Other factors such as variability of immune responses toward candidate antigens, the resulting prospect of oral tolerance, and the ensuing risk of contamination of the food chain with plant-derived biopharmaceuticals have to be factored in while generating these products. These concerns can be circumvented by producing biopharmaceuticals in nonedible plants such as tobacco from where the candidate antigens or antibodies can be purified for delivery by parenteral administration.

Plant-based vaccines can be delivered through the mucosal route (oral/nasal) or by parenteral routes such as subcutaneous or intramuscular injection. The parenteral route is particularly advantageous for vaccines because it induces IgG expression and thereby affords potent protective immunity. Nevertheless, for parenteral administration, these vaccine antigens have to be purified before being injected. Whereas, the nasal or oral vaccines operate through the mucosal route, which facilitates the manufacturing procedures as no medical infrastructure is required for vaccine delivery while retaining the immunological and biological activities of the vaccine antigen within the gastrointestinal tract due to their encapsulation within the plant cellular organelle such as the nucleus and chloroplast. Crops like rice, maize, potato, lettuce, and carrot have been widely used in generating oral vaccines.

## 3. Difficulties encountered by plant-derived recombinant vaccines and therapeutics

Currently, most of the plant-derived recombinant pharmaceuticals are still in phase I clinical trials, while some of them are in phase II and phase III clinical trials. Notable examples include the Ebola virus vaccine generated in transgenic tobacco by Mapp Biopharmaceuticals[13], the avian influenza virus vaccine produced in transgenic tobacco by Medicago, insulin protein expressed in transgenic sunflower by SemBioSys, growth factor hormone expressed in transgenic barley (ORF Genetics) and taliglucerase alfa generated in transgenic carrot by Protalix BioTherapeutics. Nevertheless, despite all these advances in the generation of plant-based biopharmaceuticals since its inception in 1989,[13] very few of these products have been approved for human consumption. In this light, important factors to be considered include the choice of the vaccine antigen and the host plant used for its expression, consistency of dosage, and the production of these vaccines as per the FDA and GMP regulations.

## 4. Recent developments in the generation of plant-based pharmaceuticals

The Fraunhofer Center for Manufacturing Innovation at Boston University (CMI) and the Fraunhofer Center for Molecular Biology (CMB) have in collaboration constructed a completely automated factory for the first time to produce great quantities of vaccines, therapeutics, and biopharmaceuticals using genetically engineered tobacco plants and engineered plant viruses.[4] Notable among these includes ZMapp, which is a cocktail of plant-derived antibodies against the Ebola virus, which is manufactured as a drug. ZMapp is expressed in *N. benthamiana* by transient expression using a plant viral vector. This product showed more potent antibody-dependent cellular cytotoxicity compared to the anti-Ebola antibodies expressed in the mammalian system, the Chinese hamster kidney cell line. Thus, plants can serve as bioreactors for the bulk manufacture of the ZMapp antibodies to meet the surging demand for Ebola virus vaccines. Various proteins such as Ebola virus GP, NP, and VP40 have been generated using the bean yellow dwarf virus (a DNA virus) and several RNA plant viruses.[4] Lettuce plants have been used to express anti-Ebola mAbs using agroinfiltration. A yield of 0.5 mg/g of leaf weight was obtained for the anti-EBOV mAb6D8 antibody, which showed robust neutralizing and protective immunity.[4] Using transient transfection in tobacco, Lindsay et al.[14] developed a VLP vaccine against the hemagglutinin protein of influenza, which showed strong immunogenicity, both humoral and cell-mediated, in addition to showing prolonged morphological stability. These novel plant-made vaccines are now rapidly progressing through recent clinical trials.

A PVX-based vector was used to express the HPV-16 (human papilloma virus-16) E7 protein in *N. benthamiana,* which elicited CTL, Th1, and Th2 immune reactions in mice and reduced tumor development by 40%.[4] Subsequently, when this HPV-16 E7 expression was targeted to the secretory pathway of the plant, it augmented the recombinant protein expression by 5-fold while inhibiting tumor growth by as much as 80%.[15] Additionally, plant extracts derived from this study contained adjuvanting capabilities. Further, in the unicellular alga *Chlamydomonas reinhardtii,* a mutant form of HPV-16 E7 gene was fused to the chloroplast genome by homologous recombination.[16] Injection of the purified protein from this system precluded tumor development in mice by 60%. This formed an ideal platform for E7 protein production as the protein remained soluble, which rendered the protein extraction and purification cost-efficient, and the *Chlamydomonas* has inherent advantages such as the ability to grow and transform easily while adhering to GMP guidelines.

In a pioneering study, HPV pseudoviruses (PsVs) were produced in plants.[17] In recent times, plant viruses are increasingly being repurposed as viral nanoparticles (VNPs) for cancer immunotherapy. VNPs are advantageous in that they can target solid tumors wherein they can elicit strong immune reactions in a highly localized manner, in addition to being able to deliver drugs and serving as tissue imaging probes. Further, VNPs are inherently biocompatible due to their natural occurrence as compared to synthetic nanoparticles and are stable, nontoxic, amenable to genetic manipulations, easy to be manufactured on a scaled-up basis and are inexpensive. VNPs derived from the PVX nanofilament were shown to carry trastuzumab, a monoclonal antibody against breast cancer, wherein the VNPs targeted the tumor cells and induced apoptosis in breast cancer cell lines.[18]

Czapar et al.[19] showed that TMV VNPs were able to deliver drugs for chemotherapy of cancer. In particular, the drug phenanthriplatin was delivered into the tumors of mouse models having breast cancer. In another study, CPMV VNPs were shown to stimulate the immune system to mediate anticancer activity. Empty eCPMV VNPs expressed in *N. benthamiana* when vaccinated in situ into tumors were capable of cancer immunotherapy in colon, breast, ovarian, and melanoma cancer models.[20] Plant viruses are becoming increasingly popular for delivering nucleic acid therapies and are far superior than mammalian vectors due to their inability to replicate in mammalian cells and therefore increased safety of use, ability to be easily produced at high yields and at low cost while dispensing with the perils of insertional mutagenesis.

## 5. Use of plant viruses and derivatives as biopharmaceuticals

Plant viruses are increasingly being used in several biotechnological applications. Plant viruses have an innate ability to be recognized and interact with the human immune system through the pathogen associated molecular pattern

(PAMP) receptors.[21] They can elicit both humoral and cell-mediated immunity[22,23] when administered through parenteral[24] or mucosal (oral)[25] routes. Plant nanobiotechnology programs using viral nanoparticles (VNPs) and virus-like particles (VLPs) have been employed in various applications in medicine and biotechnology. Helical viruses such as PVX and TMV, which possess rod-shaped structures with high aspect ratios, are ideal systems for drug delivery. Also, these plant viruses contain single-stranded RNA genomes, which can be easily engineered to express foreign proteins. In a pioneering study, a TMV-derived deconstructed virus[9,26] called the magniCON vector was produced by Icon Genetics[9] wherein full immunoglobulin IgG molecule was expressed in high yields (4.8 g/kg tissue fresh weight) in a time span of less than 2 weeks.[27] Plant virus-derived VLPs and VNPs are not pathogenic to humans and therefore rule out undesirable factors such as contamination or side effects. They are ideal epitope display platforms for generating vaccines and provide viable scaffolds for molecular conjugation of drugs or diagnostic probes.

Several novel nanomaterials displaying a wide range of molecular characteristics have been generated from VNPs.[28] VNPs have highly symmetric molecular configurations and are monodisperse, dynamic, and polyvalent while being capable of self-assembly. Additionally, they are robust, can be produced in limited time spans, and can be used as programmable molecular scaffolds. They are also biodegradable and biocompatible, unlike synthetic nanomaterials. Using a variety of conjugation chemistries, several self-assembly strategies have been employed to encapsulate ligands such as proteins, peptides, small chemical modifiers, or even other nanoparticles into these VNPs.[29,30] These encapsulation methods include genetic engineering, chemical conjugation, and mineralization.

VLPs are noninfectious derivatives of VNPs but without the viral genome, which renders them unable to replicate, and by simulating the molecular conformations of native viruses, they afford inherent immunogenicity while ensuring safety of use.[31] VLPs are now popular as viral subunit vaccines. They process and present pathogenic antigen epitopes to the immune system and undergo easy internalization by the antigen presenting cells (APCs). They also afford antitumor immunity by their innate ability to elicit immune reactions within the tumor microenvironment and therefore prime cancer immunotherapy. VLPs are composed of multiple copies of their capsid proteins, which occur as multivalent, repetitive molecular scaffolds that enable multivalent display of foreign antigens. Thus, VLPs evince superior immunogenicity when compared to antigens in solution. Furthermore, both VLPs and VNPs derived from plant viruses are highly beneficial due to their inherent adjuvanting properties, thereby avoiding the need for administration of additional adjuvants while using these particles as vaccines.

Elucidation of TMV[32] and PVX[33] molecular structures has facilitated their use in a variety of applications such as fluorescent markers,[34] nanoparticles for biologics purification,[35] nanoparticles for in vivo imaging,[36] vaccines,[37–39] biocatalysts,[40] and assembly units for memory devices.[41] Therefore, they serve as agents for delivering genes,[42] vaccines,[43] chemotherapy,[44] immunotherapy,[45] and virus-aided sensors.[46]

TMV was characterized initially in the 19th century and has become a model system for self-assembly of virus particles as well as for the formation of VLPs and VNPs toward heterologous protein expression and for acting as a carrier for molecules of biopharmaceutical importance. Its structure and genome organization have been well-studied, which makes it readily amenable for different biomedical applications. TMV has been variously modified on the interior and exterior surface of the viral capsid by means of chemical conjugation, genetic engineering, or by both strategies. The inner cavity of TMV has been used to transport heterologous cargo, and the molecular conformation of its capsid protein is convenient for fusion of foreign peptides at both the N- and C-termini. Also, the TMV CP amino acids 59–66 form a loop, which can enable presentation of foreign peptides on the exterior surface of the intact virions or CP assemblies.[47] In particular, the TMV CP C-terminus, by virtue of being exposed on the virus surface, is the favored site for fusion of foreign peptides.

## Roles of TMV in medicine and cancer

Earlier work on plant virus-based vaccines included the foot and mouth disease virus (FMDV) 11 amino acid epitope that was fused to the C-terminus of the capsid protein of TMV. This generated nanoparticles, which provided protection against FMDV challenge in animals.[48] In another study, the murine hepatitis coronavirus (MHV) vaccine was raised by display of the immunogenic peptide of its spike protein on the surface of TMV that precluded infection against challenge by murine hepatitis virus and elicited increased anti-MHV antibody titers.[49] This was the first study wherein efficient vaccines can be produced using plant virus particles that are genetically engineered to display foreign epitopes. In yet another investigation, an immunogenic epitope of the outer membrane protein F of *Pseudomonas aeruginosa* was fused to the TMV CP C-terminus, and these engineered TMV particles were shown to confer immunity to the pathogen.[50] Similarly, the influenza virus M2e protein epitope was fused to the TMV C-terminus which generated immunity against influenza virus in murine models.[37]

TMV has been genetically engineered and successfully used for cardiovascular therapy. The thrombolytic tissue plasminogen activator (tPA) was conjugated to TMV, and these engineered virus particles functioned at an efficiency similar to free tPA and demonstrated augmented safety of use with significantly lower average bleeding times.[51] In another study, TMV was used to display the carbohydrate antigen, Tn (GalNAc-α-O-Ser/Thr) antigen,[52] which, when conjugated to TMV through the Tyr 139 amino acid of TMV, generated robust immune responses.

TMV CP has been further engineered to act as a nanocarrier for an insoluble, highly hydrophobic peptide that interacts with the transmembrane domain of the neuropilin (NRP1) receptor in cancer cells.[53] These particles were shown to have antiangiogenic effects by diminishing growth and migration of cancer cells. TMV disks, by virtue of having a flat and round morphology, possess a high aspect ratio. Finbloom et al.[54] loaded DOX onto these TMV disks, which improved survival in murine models having intracranial glioblastoma. VNPs and VLPs derived from helical viruses move more efficiently across membranes and tissues than their spherical counterparts. Therefore, the helical nature of TMV particles helps them to marginate toward the walls of blood vessels, increasing the likelihood of penetrating diseased regions of the body while accruing within tumor tissues. TMV nanotubes were used to successfully deliver DOX into tumor tissues. TMV VNPs coated with albumin were loaded with DOX, and this also showed comparable antitumor efficacy. VNPs are more efficient than VLPs as their respective RNA genome cargos serve as rulers to determine the length of the nucleoprotein-virus complex. Biomolecules present in bodily fluids such as the plasma can adsorb onto the VNP surface and influence their molecular characteristics. This phenomenon is called the protein corona which is a well-defined feature in nanotechnology[55] (Pitek et al., 2016). The VNP protein corona is typically less abundant when compared to the corona of synthetic nanoparticles. The protein corona around the TMV VNPs greatly affects their immune recognition and can augment their interactions with some cell lines such as the HeLa while having no consequences on their interactions with other cell lines. The VNP corona is constituted mainly by immunoglobulins and complement proteins. Moreover, the choice of the targeting ligand is a crucial factor controlling their interactions with the plasma proteins and therefore the VNP dispersion properties in the plasma and this affects their biodistribution as well as clearance in vivo. The protein corona needs to be factored in as an important feature during the design of VNP-based drug delivery systems as well as imaging reagents. The composition and characteristics of the corona impacts the in vivo biological fate of the VNPs and this warrants careful evaluation of the protein corona prior to moving the VNP to the stage of preclinical studies. Fig. 4.1 shows the chemical modification of surface residues of the TMV VNPs.

Increasingly, plant viral VNPs have been repurposed to deliver anticancer platinum-based drugs. The inner cavity of TMV has been loaded with drugs such as phenanthriplatin[56] and cisplatin[57] by generating stable covalent adducts or alternately by charge-based reactions. The TMV VNPs in complexation with cisplatin were efficiently absorbed by cancer cells in which cisplatin showed increased cytotoxicity when compared to free cisplatin. When these cisplatin-loaded TMV VNPs were modified on their external surface using mannose and lactose moieties, this enabled their molecular recognition by the asialoglycoprotein receptor on cell membranes,[58] and these VNPs showed increased anticancer cytotoxicity in cancer cell lines. TMV coat protein was

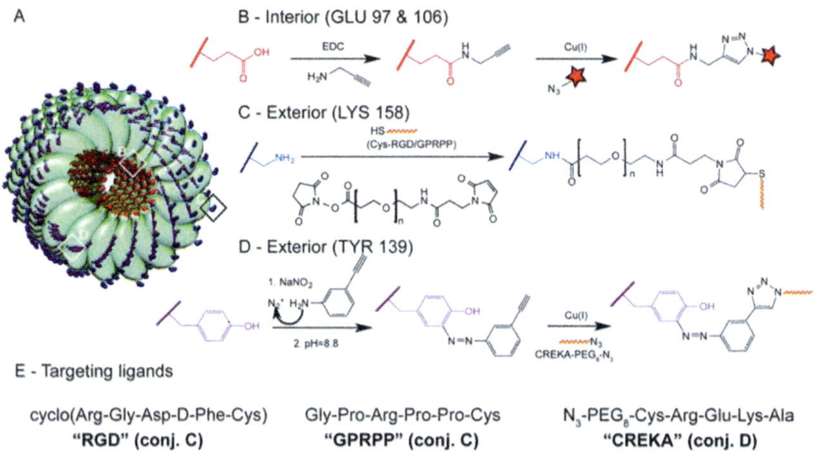


**FIGURE 4.1** Chemical modification of Tobacco Mosaic Virus. (A) Schematic illustration of TMV containing 100 coat protein subunits (the full length TMV rod contains 2130 identical coat protein unit copies) with amino acid residues exposed at the surface highlighted in color: tyrosine (purple), lysine (blue) and glutamic acid (red). (B) Internal labeling using Cy5 fluorophores through EDC conjugation of alkyne to the glutamic acid carboxyl group followed by consequent alkyne/azide-Cy5 click reaction. (C) External conjugation of cysteine tagged GPRPP and RGD peptides to lysine residues through a NHS ester−maleimide heterobifunctional PEG linker (−NH2 to −SH conjugation). (D) The CREKA peptide tagged with azide is externally conjugated to tyrosine residues through reaction with diazonium salt followed by alkyne-azide click reaction. (E) Amino acid sequences of the targeting peptide ligands. *Adapted from A.S. Pitek, A.M. Wen, S. Shukla, N.F. Steinmetz. The protein corona of plant virus nanoparticles influences their dispersion properties, cellular interactions and in vivo fates. Small. 2016 April 06; 12(13): 1758−1769.*

modified with a molecular fluorous ponytail included at specific sites, which led to the self-assembly of the TMV into spherical VNPs.[59] These VNPs showed enhanced stability of the cisplatin-VNP complexes that were generated by metal-ligated coordination.

Mitoxanthrone (MTO), an anticancer drug, was loaded onto TMV VNPs by means of a charge-driven molecular mechanism that demonstrated increased antitumor efficacy in nude mice.[60] When the antimitotic drug valine-citrulline monomethyl auristatin E was loaded onto the external surface of the TMV VNPs, this showed efficient molecular targeting and cytotoxicity in Karpas 299 non-Hodgkin's lymphoma cell line.[61] The TMV VNPs were able to enter into endolysosomal compartments, and this led to protease-mediated drug release.

TMV was used in the delivery of molecular cargo in a study by Niu and coworkers who fused the transacting activation transduction peptide (TAT) to the external surface of TMV.[62] The virus was internalized and was able to escape endo/lysosomal compartments. Moreover, the TMV-TAT fusion could successfully induce RNA silencing in hepatocellular carcinoma tumors of nude mice when delivered by intravenous and intratumoral routes. The iLOV protein was engineered to be displayed on the TMV surface, and this acted as a fluorescent probe.[63]

TMV has several inherent advantages such as a high aspect ratio and a flexuous rod-like structure that can easily target the walls of diseased blood vessels, enabling them to infiltrate into tumors in a superior manner.[64–66] Moreover, its surface characteristics can be changed using chemical or genetic approaches without compromising the structural integrity of the virus. So, different molecules such as targeting ligands, drugs, and contrast agents can be incorporated on the TMV surface in a spatially controlled fashion that yields several combinations of multifunctional systems.[67]

The interior of the TMV particles was loaded with the photosensitizer drug, Zn-EpPor (5-(4-ethynylphenyl)-10,15,20-tris(4-methylpyridin-4-ium-1-yl) porphyrin-zinc(II) triiodide), which showed high stability and prolonged shelf-life.[65] The Zn-EpPor-loaded TMV elicited augmented cell killing efficiency compared to the free drug by itself. This Zn-EpPor-loaded TMV, when conjugated to the F3 peptide in order to target it to the nucleolin shuttle protein overexpressed on HeLa cells, caused drug accumulation on cell membranes along with improved cell killing efficiency. This could be due to light-activated disruption of the cell membrane subsequent to which the drug was released followed by cellular uptake.[68]

## Applications of TMV in medical imaging and theranostics

Molecular imaging is an evolving biomedical field that enables visualization, identification, and assessment of biological systems in vivo. Notable among these imaging technologies include computed tomography (CT), magnetic resonance imaging (MRI), optical imaging, and positron emission tomography (PET), which are useful in monitoring cellular and molecular mechanisms under normal and diseased states. Ideally, a given imaging technique should provide optimal signal-to-noise ratios at the target site while at the same time reducing toxicity.[31]

VLPs are more advantageous for molecular imaging technologies compared to synthetic nanoparticles because of their low retention times and short half-life in circulation, which reduces any side effects.[28] Also, VLPs can be loaded with a variety of fluorescent labels and contrast agents and can be targeted to specific cells or tissues by means of modifications using aptamers, peptides, and antibodies.

TMV has been employed for imaging to target thrombosis and atherosclerosis.[69] TMV was conjugated with an NIR (near infra-red) dye and a peptide targeting the S100A9 molecule (a molecular marker for acute myocardial infarctions and a myeloid-related protein 14 occurring in atherosclerotic lesions). This enabled the identification of atherosclerotic lesions in ApoE−/− mice when administered intravenously, thus demonstrating that TMV can be used to detect at-risk lesions.

Luzuriaga et al. engineered a TMV-metal-organic framework (MOF) hybrid nanoparticle, which demonstrated enhanced retention of the TMV VNPs that enabled improved imaging times in murine models.[70] Similarly, the zeolitic imidazolate framework-8 was coated on the Cy5-encapsulated TMV to yield the Cy5-TMV@ZIF, which resulted in a 2.5-fold increase in fluorescence retention time compared to Cy5-TMV by itself. This combination proved to withstand harsh conditions and was nontoxic and highly stable. TMV particles were loaded with Gd-dodecane tetraacetic acid (Gd-DOTA) and altered to target the vascular cell-adhesion molecule-1 (VCAM-1), which improved the sensitivity of identification and imaging of atherosclerotic plaques in ApoE−/− mice even with low doses of the contrast agent.[71] Due to its high aspect ratio, TMV induced increased margination along the walls of the blood vessels, which resulted in enhanced signal-to-noise ratio by augmenting relaxivity and causing slower tumbling of the Gd-DOTA when complexed with TMV. This also provided higher imaging sensitivity and allowed 400 times reduction in Gd dose when compared to standard clinical doses.

In another investigation, a dysprosium (Dy3+) complex was packed within the interior cavity of TMV particles in order to improve T2 relaxivity toward MRI. Simultaneously, this facilitated better delivery of the NIR fluorescent dye, which in turn enabled dual optical-MR imaging.[31] Prostate cancer cells by virtue of the presence of integrin $\alpha 2\beta 1$ molecules were targeted by TMV whose exterior surface was labeled with an Asp-Gly-Glu-Ala peptide. According to another report, TMV particles were loaded with a metal-free paramagnetic nitroxide organic radical contrast agent (ORCA) to produce MRI and electron paramagnetic resonance probes for the detection of superoxide, which enhanced r1 and r2 relaxivities in vitro.[72] These probes functioned as both T1 and T2 contrast agents and were found to be suitable for preclinical and clinical MRI scanning. Besides, TMV with the external surface loaded with ORCA was conjugated with fluorescent dyes on its internal surface and functioned as probes that could serve as concentration markers regardless of the oxidation state of the ORCA.

An aminoxyl radical TEMPO (tetramethylpiperidin-1-oxyl, coined Compound 6) derivative was conjugated to TMV by means of a copper catalyzed azide-alkyne cyclo-addition reaction.[73] Subsequent reaction of this TMV derivative with cucurbit[8]uril (CB[8]) produced an organic radical contrast agent (ORCA) inclusion complex (designated a semirotaxane), which shielded aminoxyl radicals from reduction into hydroxylamines. The outcome of this reaction resulted in this aminoxyl-based ORCA being rendered silent for MRI. Besides, the r1 relaxivity values for this TMV-6 derivative mirrored that of Gd-DOTA, while the semirotaxane proved to be an order of magnitude more in r1 than the CB[8] and compound 6 by themselves.

TMV has also been employed in theranostics (techniques and/or drugs together used in the diagnosis and treatment of medical issues) to enable MRI and photoacoustic imaging capabilities toward photothermal therapy (PTT).

TMV in the form of nanorods was loaded with Gd followed by coating with polydopamine (PDA).[74] The PDA augmented the MRI properties and conveyed PDA contrast along with facilitating PTT. This relaxivity enhancement could be the result of trapping bulk water within the structure of the TMV by the PDA coating, which will further enhance its magnetic properties and interaction of Gd centers with water molecules. Further, the Gd-TMV-PDA viral nanoparticles exhibited strong NIR absorption *in vitro* in addition to higher photothermal conversion efficiency when compared to that of gold nanocages[75] and nanorods.[76] Additionally, these VNPs proved to be highly efficient with diminished cytotoxicity in treating PC-3 prostate and 4T1 breast cancer cells in vitro.

## PVX as an expression vector for biopharmaceuticals, theranostics, and human medicine

Mambalgin-1, a peptide that acts as a potent analgesic by inhibiting acid sensing ion channels (ASIC) in nerve cells, has been expressed using a PVX-based vector.[77] Since the ASIC plays an integral role in cancer cell growth and proliferation, Mambalgin-1 can be employed as an anticancer agent by obstruction of these channels. In this reported study, the PVX vector was used to express Mambalgin-1 in *N. benthamiana* plants, and this proved to be cytotoxic toward nervous (SH-SY5Y) cancer cells while potentiating anticancer effects through the inhibition of ASIC channels.

The scFv-TM43-E10 and scFv-Fc-TM43-E10 antibody derivatives with specificity toward the recognition of the *Salmonella typhimurium* Omp D protein[78] were expressed using a deconstructed PVX Vector in *N. benthamiana*. These antibodies showed equivalent antigen-binding specificities to those of their counterparts expressed in mammalian or microbial cell systems and were capable of successfully recognizing the *S. typhimurium* Omp D antigen. Hence, by virtue of their high sensitivity and specificity, these antibodies evince great promise for use as new diagnostic tools for detecting *S. typhimurium* infection.

PVX has been variously used in disease diagnostics, development of prophylactic vaccines/therapeutic antibodies against infectious diseases, cancer therapy, nanomedicine, and theranostics. VLPs and VNPs derived from PVX serve as ideal devices in molecular imaging, and, compared to synthetic nanoparticles, they have short half-lives in circulation in addition to decreased retention times, thus diminishing the probability of unwanted side effects. The N and M proteins of the severe acute respiratory syndrome coronavirus (SARS-CoV) have been expressed in PVX vectors.[79] The plant-derived N protein could detect the presence of antibodies in the sera of SARS-CoV patients. A universal influenza vaccine having the H1N1 influenza virus matrix protein 2 (M2e) extracellular domain epitope was expressed in a PVX vector[80] wherein it was fused to the bacterial flagellin that acted as a potent mucosal adjuvant. This PVX-derived fusion protein was shown to be expressed at high levels with yields as high as 30% of the total soluble protein and was shown to protect mouse models against influenza virus infection. The Hepatitis C virus (HCV) HVR1 epitope was synthesized as a fusion protein in a PVX-based vector, which induced IgG immune response in mice upon parenteral administration.[81] Besides, the expressed epitope recognized HCV antibodies in sera from chronically infected HCV patients. In another study, a chimeric protein generated by fusion of the antigenic epitopes of the HCV core antigen with the HBV surface antigen (HBsAg) was cloned under the control of the second capsid protein promoter of PVX.[82]

VNPs derived from PVX have been demonstrated to inhibit the growth of tumors both in animal models and in cell lines.[83] Therefore, they are being increasingly employed for immunotherapy within tumor microenvironments. In this context, PVX nanofilaments have been loaded with monoclonal antibodies of Herceptin or Trastuzumab, which proved to induce apoptosis in breast cancer cell lines.[18] A mutant form of the HPV16 E7 oncoprotein was expressed as a fusion protein with lichenase in a PVX vector. This induced a potent cytotoxic T-cell response and protected mice from tumor progression.[84] PVX VNPs when loaded with doxorubicin (DOX) greatly inhibited tumor growth in athymic mice bearing breast cancer xenografts.[85] Similarly, the PVX-DOX combination when delivered into mouse intradermal melanoma models induced prolonged survival of mice and elicited cytokine/chemokine levels that emulated that of other treatment regimes.[86]

PVX has been conjugated with fluorescent reporters, which could be of great use in in vivo imaging and theranostics.[87] Genetically engineered PVX expressing the small fluorescent iLOV protein has been demonstrated to act as a fluorescent probe that could be potentially used in vivo imaging. The tumor necrosis factor (TNF)-related apoptosis inducing ligand (TRAIL) has been displayed multivalently on PVX, which proved to increase the recruitment and induction of death receptors on cancer cell lines along with successful suppression of tumor growth in murine models bearing breast cancer.[88]

In a study by Jobsri et al.[89] a biotin/streptavidin linker was used to conjugate an idiotypic (Id) tumor-associated antigen (TAA) recombinant with PVX, which induced 7 times greater anti-Id IgG response when compared to Id by itself in a murine B-cell lymphoma model. Upon cytokine profiling of these mice, it was found to induce both IFN-$\alpha$ and IL-12. Fluorescent molecules such as GFP or mCherry were displayed on PVX VNPs by fusion to the latter's N-termini.[90] Expression of this fusion protein in *N. benthamiana* proved to enable successful in vivo particle tracking in an HT-29 mouse model and for in vitro imaging of HT-29 cells.

## Applications of CPMV and CCMV in medical biotechnology and cancer

The cowpea mosaic virus (CPMV) has been variously employed as an autonomously replicating viral vector for the expression of foreign peptides or proteins in plants. Epitopes of the *Pseudomonas aeruginosa* outer membrane (OM) protein F have been displayed on CPMV particles, which protected mice against challenges by *Pseudomonas*. Similarly, the *Staphylococcus aureus* fibronectin-binding protein (FnBP) D2 domain has been displayed on CPMV particles, which conferred protection against endocarditis in rat models.[24]

A CPMV-derived nonreplicating system was developed using a disabled version of the RNA-2 of CPMV to express large quantities of high-quality purified antibody for HIV-1 in plant[91] and influenza vaccine proteins. CPMV nanoparticles can bind vimentin, a protein present on the surface of most cells.[92] Since vimentin is overexpressed during progression of tumors, it forms a preferred target for cancer therapy. Importantly, the surface expression of vimentin correlated with CPMV uptake, which proved the capability of CPMV to detect invasive cancer cells. In a mouse model having lung cancer, CPMV nanoparticles could be readily taken up by lung cancer cells upon inhalation, and this generated an antitumor immune reaction by activating neutrophils within the tumor microenvironment.[20] Additionally, these CPMV nanoparticles showed antitumor immunity in colon, breast, and ovarian tumor mouse models.

In another investigation, mouse ovarian tumor models showed delayed tumor growth upon administration of CPMV nanoparticles along with radiotherapy.[93] This was accompanied by an enhancement in tumor infiltrating lymphocytes and showed great promise toward development as an in situ tumor vaccine.[93] A combination of CD47-blocking antibodies and CPMV nanoparticles acted synergistically to generate an antitumor response considering that CD47 is widely expressed on tumor cells, and CD47 precludes the activity of T cells and phagocytic cells[94] (Fig. 4.2). Further studies showed reduction in mouse tumors upon the use of a combination therapy comprising CPMV nanoparticles and low doses of cyclophosphamide.[95] In a study by Albakri et al.[96] CPMV upon incubation with monocytes *in vitro* augmented the release of chemokines MIP-1 1$\alpha$, CXCL10, and MIP-1$\beta$ into the supernatants of cell cultures. Similarly, activation of monocyte-derived macrophages and dendritic cells was observed upon incubation with CPMV, and this activation was associated with SYK signaling. CPMV was shown to outperform several other types of VLPs to act as a powerful immune stimulant.[97] The cowpea chlorotic mottle virus (CCMV) VLPs have been shown to deliver mRNA successfully in a study wherein the enhanced yellow fluorescent protein (EYFP) mRNA was delivered to mammalian BHK-21 cells by lipofectamine transfection.[42] This makes CCMV a promising candidate for the delivery of mRNA vaccines and a proof of concept of this finding has been demonstrated using several reporter genes.[98]

CpG ODNs (oligodeoxynucleotides) that are toll-like receptor ligands (TLR9) can be encapsulated into CCMV VLPs, and these loaded VLPs were shown to be rapidly taken up by tumor associated macrophages.[99] Additionally, they proved to inhibit tumor growth in B16F10 melanoma tumors and

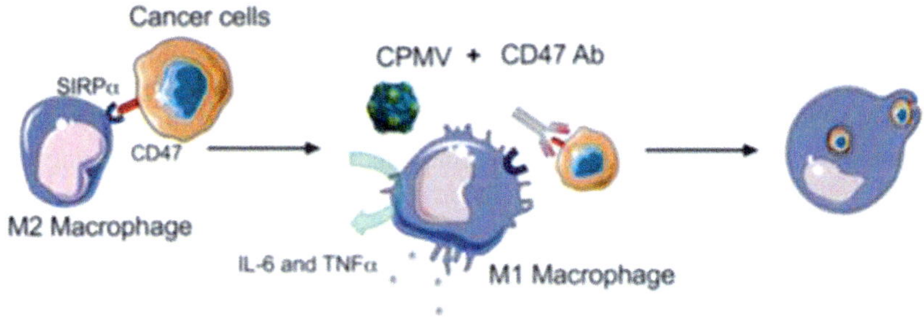


**FIGURE 4.2 A schematic representation of the synergistic antitumor action of CD47-blocking antibodies and CPMV nanoparticles.** A combination of CD47-blocking antibodies and CPMV nanoparticles acted synergistically to generate an antitumor response considering that CD47 is widely expressed on tumor cells and CD47 precludes the activity of T cells and phagocytic cells. *From Wang C, Steinmetz NF. CD47 Blockade and Cowpea Mosaic Virus Nanoparticle In Situ Vaccination Triggers Phagocytosis and Tumor Killing.* Advanced Healthcare Materials *2019;8(8). https://doi.org/10.1002/adhm.201801288.*

solid CT26 colon cancer in mice through macrophage activation. CCMV VLPs were also used to encapsulate drug-activating enzymes.[100] For instance, the bacterial cytochrome, CYPBM3, was encapsulated into CCMV, which activated chemotherapeutic prodrugs such as resveratrol and tamoxifen into their active forms. This could minimize side effects while enhancing retention and homing to the tumor site.[44,101]

## The use of geminiviruses in biotechnology and medicine

The Geminivirus, bean yellow dwarf virus (BeYDV), is popularly being used as an expression system for pharmaceutical proteins. BeYDV has been used to generate vaccines for HIV, HPV, norovirus, and HBV, as well as for producing monoclonal antibodies against the Ebola virus and West Nile Virus. Also, the lumbrokinase (PI239) (derived from the earthworm) used to dissolve blood clots, and fibrin have been expressed in BeYDV. Geminiviruses not only infect a variety of higher plants but also grow in algae and therefore have been used to express the Ebola virus vaccine in a system called Algevir.[102] Additionally, the highly immunogenic B subunit of the heat-labile enterotoxin of *E. coli* was expressed using Geminiviruses. A high yield of the recombinant proteins, 1.25 mg/g fresh biomass, was obtained within 3 days following transformation. Geminiviruses have been used to express recombinant immune complex (RIC) vaccines made in plants.[103] A bio-better vaccine against Zika virus was produced wherein the antigenic fusion site of the Zika virus envelope domain III (ZE3) was altered to include an N-terminal fusion to the IgG heavy chain (N-RIC) and this led to 40% enhancement in RIC expression. This recombinant construct elicited a potent antibody titer that corresponded with the neutralization of the Zika virus. Administration of these RICs along with the plant-made hepatitis B core (HBc) virus-like particles (VLP) displaying ZE3 resulted in 5-fold higher antibody titer that robustly neutralized the Zika virus compared to either RICs or VLPs by themselves and this high immunity was obtained just after two doses and in the absence of any adjuvant.[104] Many plant-derived human IgG1 fusion vaccine candidates were studied using the ZE3 protein domain of Zika virus. These recombinant fusion constructs were changed to make RICs and were expressed in Geminivirus vectors in plants humanized with regard to their glycosylation pathways. A yield of 1.5 mg IgG fusion/g leaf fresh weight was obtained that induced robust antibody titers against Zika virus.[105]

## 6. Conclusions and future perspectives

In summary, this chapter describes the various ways in which genetically engineered plants expressing recombinant pharmaceuticals serve to enhance human health. Various biotechnological strategies have been employed to produce vaccines and therapeutics using plant production platforms. Likewise, nutritionally improved, biofortified plants have been developed with capabilities to alleviate malnutrition as well as chronic diseases such as diabetes and cancer. Also, plants that have been utilized over the millennia in the form of herbal remedies are presently being examined for use in modern medical practices toward identifying compounds that are able to combat malaria and cancer.

In the 1980s, the plant viruses, cauliflower mosaic virus, and brome mosaic virus were used as genetically engineered DNA and RNA virus vectors respectively to express bacterial gene products.[106,107] Ever since, several plant virus-based vectors have been developed as efficacious tools for expressing recombinant proteins and for the advancement of genomic technology. Many plant viruses such as tobamoviruses,[108,109] potexviruses,[110–112] comoviruses,[113,114] geminiviruses,[115] potyviruses,[116–118] furoviruses,[119] caulimovirus,[106] and necrovirus,[120] infecting dicotyledonous plants have been employed as delivery vectors. Also, viruses including the wheat streak mosaic virus,[121] barley stripe mosaic virus,[122–124] wheat mosaic virus,[119] and the foxtail mosaic virus[125] infecting plant monocots have been used as expression vectors for medically important proteins such as therapeutic antibodies, pharmaceutical peptides and other complex proteins. Most of the above-discussed viral vectors have also been repurposed toward functional genomic purposes, such as virus-enabled genome editing, miRNA expression, virus-induced gene silencing, and the identification of virulence determinants.

The current chapter addresses the use of the important plant viruses including the PVX, TMV, CPMV, CCMV, and the Geminiviruses for medical and biotechnological applications. Significantly highlighted are the applications of VLPs and VNPs based on plant viruses for use as diagnostic, prophylactic and therapeutic agents against infectious diseases and cancer.

Additionally discussed, are their use as nanocarriers toward medical imaging and for theranostic purposes. Plant virus-derived VLPs and VNPs play important roles in the development of nanomedicines and nanotechnology. Viruses and VNPs inherently act as natural transporters of nucleic acids that protect and carry their cargo molecules, thus enabling drug delivery. By means of an assortment of molecular chemistries and attachment of a broad spectrum of functional groups, cargos such as drugs can be encapsulated, impregnated, absorbed, or conjugated to the external or internal surfaces of their capsid protein interfaces.[126] This confers molecular adaptability toward shielding cargo containing proteinaceous matrices, reversible interaction of active molecules and selective targeting of the locus of activity. VNPs are favored as natural delivery vehicles due to their uniformity of structure, biocompatibility, water solubility, functionalization flexibility and increased uptake efficiency.[127] Nanosized molecular cages are ideal platforms for drug delivery and medical imaging due to their improved stability, penetrability, cell-specific targeting and suitable pharmacokinetics. Furthermore, VNPs do not exhibit tissue tropisms, thus enabling them to be utilized in homing, recognition and interaction with cell surface receptors, transport across cell membranes and nuclear permeation.[128]

Besides, the structures of many plant viruses have been elucidated at atomic resolution, which facilitates their precise modifications according to spatial selectivity. Using genetic engineering, plant viral VLPs can be designed to acquire novel structures having definitive interactions with biological molecules and can be modified to present on their surfaces important functional groups including targeting ligands, imaging dyes, epitopes and drug payloads. VLPs, by their inherent size and shape, enable vascular transmission, active cellular intake and molecular binding while being able to tolerate extremes of environments in addition to being biocompatible. Also, even at high doses, the plant viral VLPs are by and large well-tolerated. The VLPs are easily cleared in the human biological system through proteolytic degradation, thus minimizing side effects. Hybrid VLP-based nanomaterials can be functionalized due to the capability of the VLPs to be self-assembled while being able to be modified using chemical biology techniques. VLP- and VNP-based drug delivery technologies are rapidly evolving with many candidate molecules in clinical trials that could most likely yield advanced therapeutics in the immediate future. Therefore, plant viruses will increasingly be genetically modified and repurposed for their progressive application in medicine and biotechnology. Taken together, plants and plant-based pharmaceuticals will continue to have a major role in modern medicine in the coming decades.

## References

1. Barta A, Sommergruber K, Thompson D, Hartmuth K, Matzke MA, Matzke AJM. The expression of a nopaline synthase - human growth hormone chimaeric gene in transformed

tobacco and sunflower callus tissue. *Plant Molecular Biology* 1986;**6**(5):347–57. https://doi.org/10.1007/BF00034942.
2. Hiatt A, Cafferkey R, Bowdish K. Production of antibodies in transgenic plants. *Nature* 1989;**342**(6245):76–8. https://doi.org/10.1038/342076a0.
3. Shoseyov O, Posen Y, Grynspan F. Human collagen produced in plants more than just another molecule. *Bioengineered* 2013;**5**(1). https://doi.org/10.4161/bioe.26002.
4. Venkataraman S, Makhzoum A, Shoaib E, Hefferon K. *Encyclopedia of life sciences: plant products for health.* John Wiley & Sons; 2019. https://doi.org/10.1002/9780470015902.a0023732.
5. Hefferon KL. DNA Virus Vectors for Vaccine Production in Plants: Spotlight on Geminiviruses. *Vaccines (Basel)* 2014;**2**(3):642–53. https://doi.org/10.3390/vaccines2030642.
6. Chen Q, Lai H. Gene delivery into plant cells for recombinant protein production. *BioMed Research International* 2015. https://doi.org/10.1155/2015/932161. 2015.
7. Musiychuk K, Stephenson N, Bi H, et al. A launch vector for the production of vaccine antigens in plants. *Influenza and Other Respiratory Viruses* 2007;**1**(1):19–25. https://doi.org/10.1111/j.1750-2659.2006.00005.x.
8. Peyret H, Lomonossoff GP. When plant virology met Agrobacterium: the rise of the deconstructed clones. *Plant Biotechnology Journal* 2015;**13**(8):1121–35. https://doi.org/10.1111/pbi.12412.
9. D'Aoust MA, Couture MMJ, Charland N, et al. The production of hemagglutinin-based virus-like particles in plants: a rapid, efficient and safe response to pandemic influenza. *Plant Biotechnology Journal* 2010;**8**(5):607–19. https://doi.org/10.1111/j.1467-7652.2009.00496.x.
10. Gleba Y, Klimyuk V, Marillonnet S. Magnifection - a new platform for expressing recombinant vaccines in plants. *Vaccine* 2005;**23**(17–18):2042–8. https://doi.org/10.1016/j.vaccine.2005.01.006.
11. Kashima K, Mejima M, Kurokawa S, Kuroda M, Kiyono H, Yuki Y. Comparative whole-genome analyses of selection marker-free rice-based cholera toxin B-subunit vaccine lines and wild-type lines. *BMC Genomics* 2015;**16**(1). https://doi.org/10.1186/s12864-015-1285-y.
12. Daniell H, Wiebe PO, Millan AF-S. Antibiotic-free chloroplast genetic engineering – an environmentally friendly approach. *Trends in Plant Science* 2001;**6**(6):237–9. https://doi.org/10.1016/s1360-1385(01)01949-5.
13. Rybicki EP. Plant-based vaccines against viruses. *Virology Journal* 2014;**11**(1):205. https://doi.org/10.1186/preaccept-2088894788140228.
14. Lindsay BJ, Bonar MM, Costas-Cancelas IN, et al. Morphological characterization of a plant-made virus-like particle vaccine bearing influenza virus hemagglutinins by electron microscopy. *Vaccine* 2018;**36**(16):2147–54. https://doi.org/10.1016/j.vaccine.2018.02.106.
15. Franconi R, Massa S, Illiano E, Mullar A, Cirilli A, Accardi L, Di Bonito P, Giorgi C, Venuti A. Exploiting the plant secretory pathway to improve the anticancer activity of a plant-derived HPV16 E7 vaccine. *International Journal of Immunopathology and Pharmacology* 2006;**19**(1):187–97. PMID: 16569357.
16. Demurtas OC, Massa S, Ferrante P, Venuti A, Franconi R, Giuliano G. A Chlamydomonas-derived Human Papillomavirus 16 E7 vaccine induces specific tumor protection. *PLoS One* 2013;**8**(4):e61473. https://doi.org/10.1371/journal.pone.0061473. PMID: 23626690; PMCID: PMC3634004.

17. Lamprecht RL, Kennedy P, Huddy SM, et al. Production of Human papillomavirus pseudovirions in plants and their use in pseudovirion-based neutralisation assays in mammalian cells. *Scientific Reports* 2016;**6**(1). https://doi.org/10.1038/srep20431.
18. Esfandiari N, Arzanani MK, Soleimani M, Kohi-Habibi M, Svendsen WE. A new application of plant virus nanoparticles as drug delivery in breast cancer. *Tumor Biology* 2016;**37**(1):1229–36. https://doi.org/10.1007/s13277-015-3867-3.
19. Czapar AE, Zheng YR, Riddell IA, et al. Tobacco mosaic virus delivery of phenanthriplatin for cancer therapy. *ACS Nano* 2016;**10**(4):4119–26. https://doi.org/10.1021/acsnano.5b07360.
20. Lizotte PH, Wen AM, Sheen MR, et al. In situ vaccination with cowpea mosaic virus nanoparticles suppresses metastatic cancer. *Nature Nanotechnology* 2016;**11**(3):295–303. https://doi.org/10.1038/nnano.2015.292.
21. Acosta-Ramírez E, Pérez-Flores R, Majeau N, et al. Translating innate response into long-lasting antibody response by the intrinsic antigen-adjuvant properties of papaya mosaic virus. *Immunology* 2008;**124**(2):186–97. https://doi.org/10.1111/j.1365-2567.2007.02753.x.
22. Kemnade JO, Seethammagari M, Collinson-Pautz M, Kaur H, Spencer DM, McCormick AA. Tobacco mosaic virus efficiently targets DC uptake, activation and antigen-specific T cell responses in vivo. *Vaccine* 2014;**32**(33):4228–33. https://doi.org/10.1016/j.vaccine.2014.04.051.
23. Yusibov V, Mett V, Mett V, et al. Peptide-based candidate vaccine against respiratory syncytial virus. *Vaccine* 2005;**23**(17–18):2261–5. https://doi.org/10.1016/j.vaccine.2005.01.039.
24. Brennan FR, Jones TD, Longstaff M, et al. Immunogenicity of peptides derived from a fibronectin-binding protein of *S. aureus* expressed on two different plant viruses. *Vaccine* 1999;**17**(15–16):1846–57. https://doi.org/10.1016/S0264-410X(98)00485-X.
25. Brennan FR, Bellaby T, Helliwell SM, et al. Chimeric plant virus particles administered nasally or orally induce systemic and mucosal immune responses in mice. *Journal of Virology* 1999;**73**(2):930–8. https://doi.org/10.1128/jvi.73.2.930-938.1999.
26. Lindbo JA. TRBO: a high-efficiency tobacco mosaic virus RNA-based overexpression vector. *Plant Physiology* 2007;**145**(4):1232–40. https://doi.org/10.1104/pp.107.106377.
27. Bendandi M, Marillonnet S, Kandzia R, et al. Rapid, high-yield production in plants of individualized idiotype vaccines for non-Hodgkin's lymphoma. *Annals of Oncology* 2010;**21**(12):2420–7. https://doi.org/10.1093/annonc/mdq256.
28. Steinmetz NF. Viral nanoparticles as platforms for next-generation therapeutics and imaging devices. *Nanomedicine: Nanotechnology, Biology, and Medicine* 2010;**6**(5):634–41. https://doi.org/10.1016/j.nano.2010.04.005.
29. Steinmetz NF, Evans DJ. Utilisation of plant viruses in bionanotechnology. *Organic and Biomolecular Chemistry* 2007;**5**(18):2891–902. https://doi.org/10.1039/b708175h.
30. Young M, Willits D, Uchida M, Douglas T. Plant viruses as biotemplates for materials and their use in nanotechnology. *Annual Review of Phytopathology* 2008;**46**:361–84. https://doi.org/10.1146/annurev.phyto.032508.131939.
31. Chung YH, Cai H, Steinmetz NF. Viral nanoparticles for drug delivery, imaging, immunotherapy, and theranostic applications. *Advanced Drug Delivery Reviews* 2020;**156**:214–35. https://doi.org/10.1016/j.addr.2020.06.024.
32. Namba K, Pattanayek R, Stubbs G. Visualization of protein-nucleic acid interactions in a virus. Refined structure of intact tobacco mosaic virus at 2.9 Å resolution by X-ray fiber diffraction. *Journal of Molecular Biology* 1989;**208**(2):307–25. https://doi.org/10.1016/0022-2836(89)90391-4.
33. Parker L, Kendall A, Stubbs G. Surface features of potato virus X from fiber diffraction. *Virology* 2002;**300**(2):291–5. https://doi.org/10.1006/viro.2002.1483.

34. Yi H, Nisar S, Lee SY, et al. Patterned assembly of genetically modified viral nanotemplates via nucleic acid hybridization. *Nano Letters* 2005;**5**(10):1931–6. https://doi.org/10.1021/nl051254r.
35. Werner S, Marillonnet S, Hause G, Klimyuk V, Gleba Y. Immunoabsorbent nanoparticles based on a tobamovirus displaying protein A. *Proceedings of the National Academy of Sciences of the United States of America* 2006;**103**(47):17678–83. https://doi.org/10.1073/pnas.0608869103.
36. Niehl A, Appaix F, Boscá S, et al. Fluorescent Tobacco mosaic virus-derived bio-nanoparticles for intravital two-photon imaging. *Frontiers in Plant Science* 2016;**6**(2016). https://doi.org/10.3389/fpls.2015.01244.
37. Petukhova NV, Gasanova TV, Stepanova LA, et al. Immunogenicity and protective efficacy of candidate universal influenza a nanovaccines produced in plants by Tobacco mosaic virus-based vectors. *Current Pharmaceutical Design* 2013;**19**(31):5587–600. https://doi.org/10.2174/13816128113199990337.
38. Smolenska L, Roberts IM, Learmonth D, et al. Production of a functional single chain antibody attached to the surface of a plant virus. *FEBS Letters* 1998;**441**(3):379–82. https://doi.org/10.1016/S0014-5793(98)01586-5.
39. Therien A, Bedard M, Carignan D, et al. A versatile papaya mosaic virus (PapMV) vaccine platform based on sortase-mediated antigen coupling. *Journal of Nanobiotechnology* 2017;**15**.
40. Carette N, Engelkamp H, Akpa E, et al. A virus-based biocatalyst. *Nature Nanotechnology* 2007;**2**(4):226–9. https://doi.org/10.1038/nnano.2007.76.
41. Tseng RJ, Tsai C, Ma L, Ouyang J, Ozkan CS, Yang Y. Digital memory device based on tobacco mosaic virus conjugated with nanoparticles. *Nature Nanotechnology* 2006;**1**(1):72–7. https://doi.org/10.1038/nnano.2006.55.
42. Azizgolshani O, Garmann RF, Cadena-Nava R, Knobler CM, Gelbart WM. Reconstituted plant viral capsids can release genes to mammalian cells. *Virology* 2013;**441**(1):12–7. https://doi.org/10.1016/j.virol.2013.03.001.
43. Phelps JP, Dang N, Rasochova L. Inactivation and purification of cowpea mosaic virus-like particles displaying peptide antigens from Bacillus anthracis. *Journal of Virological Methods* 2007;**141**(2):146–53. https://doi.org/10.1016/j.jviromet.2006.12.008.
44. Sánchez-Sánchez L, Cadena-Nava RD, Palomares LA, et al. Chemotherapy pro-drug activation by biocatalytic virus-like nanoparticles containing cytochrome P450. *Enzyme and Microbial Technology* 2014;**60**:24–31. https://doi.org/10.1016/j.enzmictec.2014.04.003.
45. Venuti A, Curzio G, Mariani L, Paolini F. Immunotherapy of HPV-associated cancer: DNA/plant-derived vaccines and new orthotopic mouse models. *Cancer Immunology, Immunotherapy* 2015;**64**(10):1329–38. https://doi.org/10.1007/s00262-015-1734-0.
46. Eiben S, Koch C, Altintoprak K, et al. Plant virus-based materials for biomedical applications: trends and prospects. *Advanced Drug Delivery Reviews* 2019;**145**:96–118. https://doi.org/10.1016/j.addr.2018.08.011.
47. Smith ML, Fitzmaurice WP, Turpen TH, Palmer KE. *Display of peptides on the surface of tobacco mosaic virus particles*, vol 332. Springer Science and Business Media LLC; 2009. p. 13–31. https://doi.org/10.1007/978-3-540-70868-1_2.
48. Wu L, Jiang L, Zhou Z, et al. Expression of foot-and-mouth disease virus epitopes in tobacco by a tobacco mosaic virus-based vector. *Vaccine* 2003;**21**(27–30):4390–8. https://doi.org/10.1016/S0264-410X(03)00428-6.
49. Koo M, Bendahmane M, Lettieri GA, et al. Protective immunity against murine hepatitis virus (MHV) induced by intranasal or subcutaneous administration of hybrids of tobacco mosaic

virus that carries an MHV epitope. *Proceedings of the National Academy of Sciences of the United States of America* 1999;**96**(14):7774–9. https://doi.org/10.1073/pnas.96.14.7774.

50. Staczek J, Bendahmane M, Gilleland LB, Beachy RN, Gilleland HE. Immunization with a chimeric tobacco mosaic virus containing an epitope of outer membrane protein F of *Pseudomonas aeruginosa* provides protection against challenge with *P. aeruginosa*. *Vaccine* 2000;**18**(21):2266–74. https://doi.org/10.1016/S0264-410X(99)00571-X.
51. Pitek AS, Park J, Wang Y, et al. Delivery of thrombolytic therapy using rod-shaped plant viral nanoparticles decreases the risk of hemorrhage. *Nanoscale* 2018;**10**(35):16547–55. https://doi.org/10.1039/c8nr02861c.
52. Yin Z, Nguyen HG, Chowdhury S, et al. Tobacco mosaic virus as a new carrier for tumor associated carbohydrate antigens. *Bioconjugate Chemistry* 2012;**23**(8):1694–703. https://doi.org/10.1021/bc300244a.
53. Gamper C, Spenlé C, Boscá S, et al. Functionalized tobacco mosaic virus coat protein monomers and oligomers as nanocarriers for anti-cancer peptides. *Cancers* 2019;**11**(10):1609. https://doi.org/10.3390/cancers11101609.
54. Finbloom JA, Aanei IL, Bernard J, et al. Evaluation of three morphologically distinct virus-like particles as nanocarriers for convection-enhanced drug delivery to glioblastoma. *Nanomaterials (Basel)* 2018;(12).
55. Pitek AS, Wen AM, Shukla S, Steinmetz NF. The protein corona of plant virus nanoparticles influences their dispersion properties, cellular interactions, and in vivo fates. *Small* 2016;**12**(13):1758–69. https://doi.org/10.1002/smll.201502458.
56. Vernekar AA, Berger G, Czapar AE, et al. Speciation of phenanthriplatin and its analogs in the core of tobacco mosaic virus. *Journal of the American Chemical Society* 2018;**140**(12):4279–87. https://doi.org/10.1021/jacs.7b12697.
57. Franke CE, Czapar AE, Patel RB, Steinmetz NF. Tobacco mosaic virus-delivered cisplatin restores efficacy in platinum-resistant ovarian cancer cells. *Molecular Pharmaceutics* 2018;**15**(8):2922–31. https://doi.org/10.1021/acs.molpharmaceut.7b00466.
58. Liu X, Liu B, Gao S, et al. Glyco-decorated tobacco mosaic virus as a vector for cisplatin delivery. *Journal of Materials Chemistry B* 2017;**5**(11):2078–85. https://doi.org/10.1039/c7tb00100b.
59. Gao S, Liu X, Wang Z, et al. Fluorous interaction induced self-assembly of tobacco mosaic virus coat protein for cisplatin delivery. *Nanoscale* 2018;**10**(25):11732–6. https://doi.org/10.1039/c8nr03748e.
60. Lin RD, Steinmetz NF. Tobaccco mosaic virus delivery of mitothraxone for cancer therapy. *Nanoscale* 2018;**10**(34):16307–13. https://doi.org/10.1039/c8nr04142c.
61. Kernan DL, Wen AM, Pitek AS, Steinmetz NF. Featured Article: delivery of chemotherapeutic vcMMAE using tobacco mosaic virus nanoparticles. *Experimental Biology and Medicine* 2017;**242**(14):1405–11. https://doi.org/10.1177/1535370217719222.
62. Tian Y, Zhou M, Shi H, et al. Integration of cell-penetrating peptides with rod-like bionanoparticles: virus-inspired gene-silencing technology. *Nano Letters* 2018;**18**(9):5453–60. https://doi.org/10.1021/acs.nanolett.8b01805.
63. Röder J, Fischer R, Commandeur U. Adoption of the 2A Ribosomal skip principle to tobacco mosaic virus for peptide display. *Frontiers in Plant Science* 2017;**8**. https://doi.org/10.3389/fpls.2017.01125.

64. Chariou PL, Lee KL, Pokorski JK, Saidel GM, Steinmetz NF. Diffusion and uptake of tobacco mosaic virus as therapeutic carrier in tumor tissue: effect of nanoparticle aspect ratio. *The Journal of Physical Chemistry B* 2016;**120**(26):6120–9. https://doi.org/10.1021/acs.jpcb.6b02163.
65. Sokullu E, Abyaneh HS, Gauthier MA. Plant/bacterial virus-based drug discovery, drug delivery, and therapeutics. *Pharmaceutics* 2019;**11**(5). https://doi.org/10.3390/pharmaceutics11050211.
66. Wen AM, Wang Y, Jiang K, et al. Shaping bio-inspired nanotechnologies to target thrombosis for dual optical-magnetic resonance imaging. *Journal of Materials Chemistry B* 2015;**3**(29):6037–45. https://doi.org/10.1039/c5tb00879d.
67. Rong J, Niu Z, Lee LA, Wang Q. Self-assembly of viral particles. *Current Opinion in Colloid and Interface Science* 2011;**16**(6):441–50. https://doi.org/10.1016/j.cocis.2011.09.001.
68. Chariou PL, Wang L, Desai C, et al. Let there Be light: targeted photodynamic therapy using high aspect ratio plant viral nanoparticles. *Macromolecular Bioscience* 2019;**19**(5). https://doi.org/10.1002/mabi.201800407.
69. Park J, Gao H, Wang Y, Hu H, Simon DI, Steinmetz NF. S100A9-targeted tobacco mosaic virus nanoparticles exhibit high specificity toward atherosclerotic lesions in ApoE−/− mice. *Journal of Materials Chemistry B* 2019;**7**(11):1842–6. https://doi.org/10.1039/c8tb02276c.
70. Luzuriaga MA, Welch RP, Dharmarwardana M, et al. Enhanced stability and controlled delivery of MOF-encapsulated vaccines and their immunogenic response in vivo. *ACS Applied Materials and Interfaces* 2019;**11**(10):9740–6. https://doi.org/10.1021/acsami.8b20504.
71. Bruckman MA, Jiang K, Simpson EJ, et al. Dual-modal magnetic resonance and fluorescence imaging of atherosclerotic plaques in vivo using VCAM-1 targeted tobacco mosaic virus. *Nano Letters* 2014;**14**(3):1551–8. https://doi.org/10.1021/nl404816m.
72. Dharmarwardana M, Martins AF, Chen Z, et al. Nitroxyl modified tobacco mosaic virus as a metal-free high-relaxivity MRI and EPR active superoxide sensor. *Molecular Pharmaceutics* 2018;**15**(8):2973–83. https://doi.org/10.1021/acs.molpharmaceut.8b00262.
73. Lee H, Shahrivarkevishahi A, Lumata JL, et al. Supramolecular and biomacromolecular enhancement of metal-free magnetic resonance imaging contrast agents. *Chemical Science* 2020;**11**(8):2045–50. https://doi.org/10.1039/c9sc05510j.
74. Hu H, Yang Q, Baroni S, Yang H, Aime S, Steinmetz NF. Polydopamine-decorated tobacco mosaic virus for photoacoustic/magnetic resonance bimodal imaging and photothermal cancer therapy. *Nanoscale* 2019;**11**(19):9760–8. https://doi.org/10.1039/c9nr02065a.
75. Wang Y, Black KCL, Luehmann H, et al. Comparison study of gold nanohexapods, nanorods, and nanocages for photothermal cancer treatment. *ACS Nano* 2013;**7**(3):2068–77. https://doi.org/10.1021/nn304332s.
76. Vankayala R, Huang YK, Kalluru P, Chiang CS, Hwang KC. First demonstration of gold nanorods-mediated photodynamic therapeutic destruction of tumors via near infra-red light activation. *Small* 2014;**10**(8):1612–22. https://doi.org/10.1002/smll.201302719.
77. Khezri1 G, Rouz O, Javad Davarpanah S. Heterologous expression of biologically active Mambalgin 1 peptide as a new potential anticancer, using a PVX based viral vector in Nicotiana benthamiana. *Plant Cell, Tissue and Organ Culture (PCTOC)* 2020:1–11.
78. Kopertekha L. Transient plant production of Salmonella Typhimurium diagnostic antibodies Lilya Kopertekha. *Biotechnology Reports* 2018;**20**.
79. Demurtas OC, Massa S, Illiano E, et al. Antigen production in plant to tackle infectious diseases flare up: the case of SARS. *Frontiers in Plant Science* 2016;**7**. https://doi.org/10.3389/fpls.2016.00054.

80. Mardanova ES, Kotlyarov RY, Kuprianov VV, et al. Rapid high-yield expression of a candidate influenza vaccine based on the ectodomain of M2 protein linked to flagellin in plants using viral vectors. *BMC Biotechnology* 2015;**15**(1). https://doi.org/10.1186/s12896-015-0164-6.
81. Uhde-Holzem K, Schlösser V, Viazov S, Fischer R, Commandeur U. Immunogenic properties of chimeric potato virus X particles displaying the hepatitis C virus hypervariable region I peptide R9. *Journal of Virological Methods* 2010;**166**(1–2):12–20. https://doi.org/10.1016/j.jviromet.2010.01.017.
82. Mohammadzadeh S, Roohvand F, Memarnejadian A, et al. Co-expression of hepatitis C virus polytope-HBsAg and p19-silencing suppressor protein in tobacco leaves. *Pharmaceutical Biology* 2016;**54**(3):465–73. https://doi.org/10.3109/13880209.2015.1048371.
83. Shukla S, DiFranco NA, Wen AM, Commandeur U, Steinmetz NF. To target or not to target: active vs. Passive tumor homing of filamentous nanoparticles based on potato virus X. *Cellular and Molecular Bioengineering* 2015;**8**(3):433–44. https://doi.org/10.1007/s12195-015-0388-5.
84. Demurtas OC, Massa S, Ferrante P, Venuti A, Franconi R, Giuliano G. A chlamydomonas-derived human papillomavirus 16 E7 vaccine induces specific tumor protection. *Plos One* 2013;**8**.
85. Le DHT, Lee KL, Shukla S, Commandeur U, Steinmetz NF. Potato virus X, a filamentous plant viral nanoparticle for doxorubicin delivery in cancer therapy. *Nanoscale* 2017;**9**(6):2348–57. https://doi.org/10.1039/c6nr09099k.
86. Lee KL, Murray AA, Le DHT, et al. Combination of plant virus nanoparticle-based in situ vaccination with chemotherapy potentiates antitumor response. *Nano Letters* 2017;**17**(7):4019–28. https://doi.org/10.1021/acs.nanolett.7b00107.
87. Röder J, Dickmeis C, Fischer R, Commandeur U. Systemic infection of Nicotiana benthamiana with potato virus X nanoparticles presenting a fluorescent iLOV polypeptide fused directly to the coat protein. *BioMed Research International* 2018;**2018**:1–12. https://doi.org/10.1155/2018/9328671.
88. Le DHT, Commandeur U, Steinmetz NF. Presentation and delivery of tumor necrosis factor-related apoptosis-inducing ligand via elongated plant viral nanoparticle enhances antitumor efficacy. *ACS Nano* 2019;**13**(2):2501–10. https://doi.org/10.1021/acsnano.8b09462.
89. Jobsri J, Allen A, Rajagopal D, et al. Plant virus particles carrying tumour antigen activate TLR7 and induce high levels of protective antibody. *PLoS One* 2015;**10**(2). https://doi.org/10.1371/journal.pone.0118096.
90. Shukla S, Dickmeis C, Fischer R, Commandeur U, Steinmetz NF. In planta production of fluorescent filamentous plant virus-based nanoparticles. In: *Methods in molecular biology*, vol 1776. United States: Humana Press Inc.; 2018. p. 61–84. https://doi.org/10.1007/978-1-4939-7808-3_5.
91. Sainsbury F, Sack M, Stadlmann J, et al. Rapid transient production in plants by replicating and non-replicating vectors yields high quality functional anti-HIV antibody. *PLoS ONE* 2010;**5**(11):e13976. https://doi.org/10.1371/journal.pone.0013976.
92. Steinmetz NF, Cho CF, Ablack A, Lewis JD, Manchester M. Cowpea mosaic virus nanoparticles target surface vimentin on cancer cells. *Nanomedicine* 2011;**6**(2):351–64. https://doi.org/10.2217/nnm.10.136.
93. Patel R, Czapar AE, Fiering S, Oleinick NL, Steinmetz NF. Radiation therapy combined with cowpea mosaic virus nanoparticle in situ vaccination initiates immune-mediated tumor regression. *ACS Omega* 2018;**3**(4):3702–7. https://doi.org/10.1021/acsomega.8b00227.

94. Wang C, Steinmetz NF. CD47 blockade and cowpea mosaic virus nanoparticle in situ vaccination Triggers Phagocytosis and tumor killing. *Advanced Healthcare Materials* 2019;**8**(8). https://doi.org/10.1002/adhm.201801288.
95. Cai H, Wang C, Shukla S, Steinmetz NF. Cowpea mosaic virus immunotherapy combined with cyclophosphamide reduces breast cancer tumor burden and inhibits lung metastasis. *Advanced Science* 2019;**6**(16):1802281. https://doi.org/10.1002/advs.201802281.
96. Albakri MM, Veliz FA, Fiering SN, Steinmetz NF, Sieg SF. Endosomal toll-like receptors play a key role in activation of primary human monocytes by cowpea mosaic virus. *Immunology* 2020;**159**(2):183–92. https://doi.org/10.1111/imm.13135.
97. Shukla S, Wang C, Beiss V, et al. The unique potency of Cowpea mosaic virus (CPMV): in situ cancer vaccine. *Biomaterials Science* 2020;**8**(19):5489–503. https://doi.org/10.1039/d0bm01219j.
98. Biddlecome A, Habte HH, McGrath KM, et al. Delivery of self-amplifying RNA vaccines in in vitro reconstituted virus-like particles. *PLoS ONE* 2019;**14**(6). https://doi.org/10.1371/journal.pone.0215031.
99. Cai H, Shukla S, Steinmetz NF. The antitumor efficacy of CpG oligonucleotides is improved by encapsulation in plant virus-like particles. *Advanced Functional Materials* 2020;**30**(15):1908743. https://doi.org/10.1002/adfm.201908743.
100. Koyani R, Pérez-Robles J, Cadena-Nava RD, Vazquez-Duhalt R. Biomaterial-based nanoreactors, an alternative for enzyme delivery. *Nanotechnology Reviews* 2017;**6**(5):405–19. https://doi.org/10.1515/ntrev-2016-0071.
101. Strable E, Finn MG. Chemical modification of viruses and virus-like particles. *Current Topics in Microbiology and Immunology* 2009;**327**:1–21. https://doi.org/10.1007/978-3-540-69379-6_1.
102. Bañuelos-Hernández B, Monreal-Escalante E, González-Ortega O, Angulo C, Rosales-Mendoza S, Algevir. An expression system for microalgae based on viral vectors. *Frontiers in Microbiology* 2017;**8**.
103. Kim M-Y, Reljic R, Kilbourne J, et al. Novel vaccination approach for dengue infection based on recombinant immune complex universal platform. *Vaccine* 2015;**33**(15):1830–8. https://doi.org/10.1016/j.vaccine.2015.02.036.
104. Diamos AG, Pardhe MD, Sun H, et al. Codelivery of improved immune complex and virus-like particle vaccines containing Zika virus envelope domain III synergistically enhances immunogenicity. *Vaccine* 2020;**38**(18):3455–63. https://doi.org/10.1016/j.vaccine.2020.02.089.
105. Diamos AG, Hunter JGL, Pardhe MD, et al. High level production of monoclonal antibodies using an optimized plant expression system. *Frontiers in Bioengineering and Biotechnology* 2020;**7**. https://doi.org/10.3389/fbioe.2019.00472.
106. Brisson N, Paszkowski J, Penswick JR, Gronenborn B, Potrykus I, Hohn T. Expression of a bacterial gene in plants by using a viral vector. *Nature* 1984;**310**(5977):511–4. https://doi.org/10.1038/310511a0.
107. French R, Janda M, Ahlquist P. Bacterial gene inserted in an engineered RNA virus: efficient expression in monocotyledonous plant cells. *Science* 1986;**231**(4743):1294–7. https://doi.org/10.1126/science.231.4743.1294.
108. Takamatsu N, Ishikawa M, Meshi T, Okada Y. Expression of bacterial chloramphenicol acetyltransferase gene in tobacco plants mediated by TMV-RNA. *The EMBO Journal* 1987;**6**(2):307–11. https://doi.org/10.1002/j.1460-2075.1987.tb04755.x.

109. Yusibov V, Shivprasad S, Turpen TH, Dawson W, Koprowski H. *Plant viral vectors based on tobamoviruses*, vol 240. Springer Science and Business Media LLC; 2000. p. 81–94. https://doi.org/10.1007/978-3-642-60234-4_4.
110. Baulcombe DC, Chapman S, Cruz S. Jellyfish green fluorescent protein as a reporter for virus infections. *The Plant Journal* 1995;**7**(6):1045–53. https://doi.org/10.1046/j.1365-313x.1995.07061045.x.
111. Palmer K, Gleba Y. Plant viral vectors. *Current Topics in Microbiology and Immunology* 2014;**375**:1–194. https://doi.org/10.1007/978-3-642-40829-8.
112. Zhang H, Wang L, Hunter D, Voogd C, Joyce N, Davies K. A Narcissus mosaic viral vector system for protein expression and flavonoid production. *Plant Methods* 2013;**9**(1):28. https://doi.org/10.1186/1746-4811-9-28.
113. Sainsbury F, Lavoie PO, D'Aoust MA, Vézina LP, Lomonossoff GP. Expression of multiple proteins using full-length and deleted versions of cowpea mosaic virus RNA-2. *Plant Biotechnology Journal* 2008;**6**(1):82–92. https://doi.org/10.1111/j.1467-7652.2007.00303.x.
114. Zhang C, Bradshaw JD, Whitham SA, Hill JH. The development of an efficient multipurpose bean pod mottle virus viral vector set for foreign gene expression and RNA silencing. *Plant Physiology* 2010;**153**(1):52–65. https://doi.org/10.1104/pp.109.151639.
115. Stanley J. Geminiviruses: plant viral vectors. *Current Opinion in Genetics and Development* 1993;**3**(1):91–6. https://doi.org/10.1016/S0959-437X(05)80347-8.
116. Lellis AD, Kasschau KD, Whitham SA, Carrington JC. Loss-of-susceptibility mutants of *Arabidopsis thaliana* reveal an essential role for eIF(iso)4E during potyvirus infection. *Current Biology* 2002;**12**(12):1046–51. https://doi.org/10.1016/S0960-9822(02)00898-9.
117. Majer E, Navarro JA, Daròs JA. A potyvirus vector efficiently targets recombinant proteins to chloroplasts, mitochondria and nuclei in plant cells when expressed at the amino terminus of the polyprotein. *Biotechnology Journal* 2015;**10**(11):1792–802. https://doi.org/10.1002/biot.201500042.
118. Seo JK, Choi HS, Kim KH. Engineering of soybean mosaic virus as a versatile tool for studying protein-protein interactions in soybean. *Scientific Reports* 2016;**6**. https://doi.org/10.1038/srep22436.
119. Jarugula S, Charlesworth SR, Qu F, Stewart LR. Soil-borne wheat mosaic virus infectious clone and manipulation for gene-carrying capacity. *Archives of Virology* 2016;**161**(8):2291–7. https://doi.org/10.1007/s00705-016-2863-2.
120. Zhou B, Zhang Y, Wang X, et al. Oral administration of plant-based rotavirus VP6 induces antigen-specific IgAs, IgGs and passive protection in mice. *Vaccine* 2010;**28**(37):6021–7. https://doi.org/10.1016/j.vaccine.2010.06.094.
121. Choi IR, Stenger DC, Jack Morris T, French R. A plant virus vector for systemic expression of foreign genes in cereals. *Plant Journal* 2000;**23**(4):547–55. https://doi.org/10.1046/j.1365-313X.2000.00820.x.
122. Cheuk A, Houde M. A new barley stripe mosaic virus allows large protein overexpression for rapid function analysis. *Plant Physiology* 2018;**176**(3):1919–31. https://doi.org/10.1104/pp.17.01412.
123. Lee WS, Hammond-Kosack KE, Kanyuka K. Barley stripe mosaic virus-mediated tools for investigating gene function in cereal plants and their pathogens: virus-induced gene silencing, host-mediated gene silencing, and virus-mediated overexpression of heterologous protein. *Plant Physiology* 2012;**160**(2):582–90. https://doi.org/10.1104/pp.112.203489.

124. Yuan C, Li C, Yan L, et al. A high throughput barley stripe mosaic virus vector for virus induced gene silencing in monocots and dicots. *PLoS ONE* 2011;**6**(10):e26468. https://doi.org/10.1371/journal.pone.0026468.
125. Bouton C, King RC, Chen H, et al. Foxtail mosaic virus: a viral vector for protein expression in cereals. *Plant Physiology* 2018;**177**(4):1352–67. https://doi.org/10.1104/pp.17.01679.
126. Elzoghby AO, Samy WM, Elgindy NA. Protein-based nanocarriers as promising drug and gene delivery systems. *Journal of Controlled Release* 2012;**161**(1):38–49. https://doi.org/10.1016/j.jconrel.2012.04.036.
127. Ma Y, Nolte RJM, Cornelissen JJLM. Virus-based nanocarriers for drug delivery. *Advanced Drug Delivery Reviews* 2012;**64**(9):811–25. https://doi.org/10.1016/j.addr.2012.01.005.
128. Ferrer-Miralles N, Rodríguez-Carmona E, Corchero JL, García-Fruitós E, Vázquez E, Villaverde A. Engineering protein self-assembling in protein-based nanomedicines for drug delivery and gene therapy. *Critical Reviews in Biotechnology* 2015;**35**(2):209–21. https://doi.org/10.3109/07388551.2013.833163.

Chapter 5

# Genome-edited crops

## 1. An overview of genome editing and its action mechanism

Of critical concern is the inordinate imbalance between cultivable lands necessary for food production and the steady escalation in urban development required to accommodate the rise in population at the expense of agricultural lands. With the accelerating emergence of new sequencing technologies, the complete genome sequences of several important crops have been characterized. There is a compelling need to use this genomic information to develop better crops to define gene functions, determine significant agronomic traits, and generate cultivars that are high-yielding and with beneficial characteristics. To define gene function, mutants need to be generated by induction of physical or chemical mutagenesis or by using insertional mutagenesis through T-DNA technology[1] or by RNA interference-based gene silencing.[2] While the above methods have yielded the development of some elite cultivars, these mutations are random, time-consuming, with poor accuracy and incomplete abolition of gene function, necessitating laborious screening and identification of the mutants. Considering that a vast amount of genome information still requires to be characterized, this calls for the generation of customizable, precise, superior, and high-efficiency tools to target and edit plant genes within a short period of time. The gene editing technology affords a vastly controlled and quicker approach to introduce specific changes in the target-gene functions at precise positions within the genome for enabling gene silencing or increasing gene expression.

Changes of a few nucleotides at a genome site would constitute genome editing, whereas if a large sequence obtained from some other species is inserted, it would constitute a genetically modified (GM) crop. Examples of crops subjected to gene editing include crops such as sorghum, rice,[3] maize,[4] *Brassica oleracea* and barley,[5] poplar,[6] and *Arabidopsis*.[7–9] The clustered regularly interspaced short palindromic repeats/CRISPR-associated protein 9 (CRISPR/Cas9) system was introduced into wheat (*Triticum aestivum*), rice (*Oryza sativa*), *Arabidopsis thaliana,* and *Nicotiana benthamiana* by three independent research groups in 2013.[8,10,11] This paved the way for plant breeders to determine the specific introduction of variations in gene sequences toward quick and facile enhancement of agricultural crops.

**Agricultural Biotechnology. https://doi.org/10.1016/B978-0-323-88476-1.00001-1**

Efforts have been undertaken to apply artificial sequence-specific nucleases (SSNs) that can elicit double strand breaks (DSBs) at targeted sequences in the genomes.[12] Thereupon, the DSBs induced are repaired through the existing DNA repair mechanisms resulting in targeted mutagenesis by the removal, addition, or replacement of DNA nucleotides, a process currently designated as "genome editing". Fig. 5.1a illustrates the CRISPR-Cas9 genetic scissors used to edit the genome; Fig. 5.1b shows the pathways of double-strand break repair involved in genome editing.

The DNA repair process can take place either by the error-prone nonhomologous end joining (NHEJ) or, on the other hand, by the more precise homologous recombination (HR)[15,16] of which, the former occurs more frequently in somatic cells. The reason for this is that HR involves a more difficult and complex process necessitating the introduction of both the donor DNA and the expression cassette harboring the SSNs.[15–17] Presently, the most popular technologies for genome editing include the ZFNs,[18] the transcription

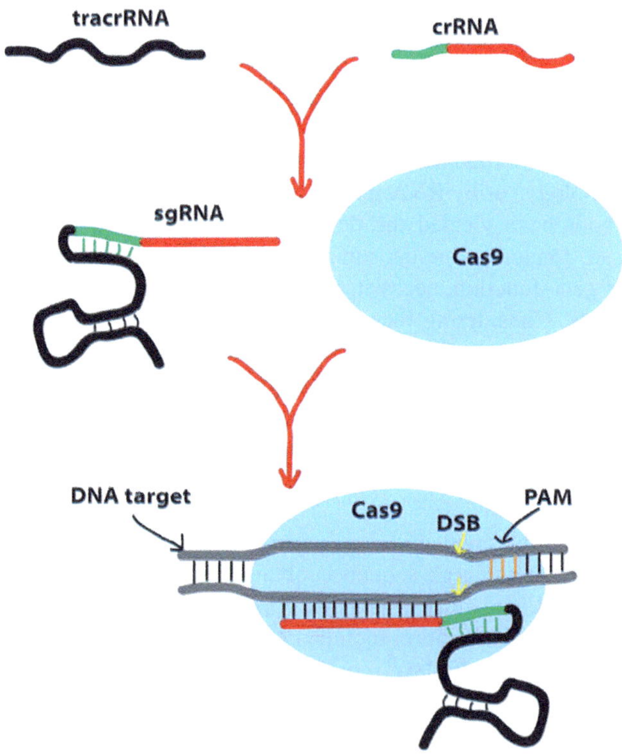


**FIGURE 5.1A** Molecular mechanism of CRISPR/Cas9 genome editing. sgRNA is formed by joining the crRNA (green and red) and tracrRNA (black). PAM (mostly NGG) is depicted as three yellow base paring lines on the target DNA (shown in grey). The location of double stranded breaks is shown by a pair of yellow arrows.[13]

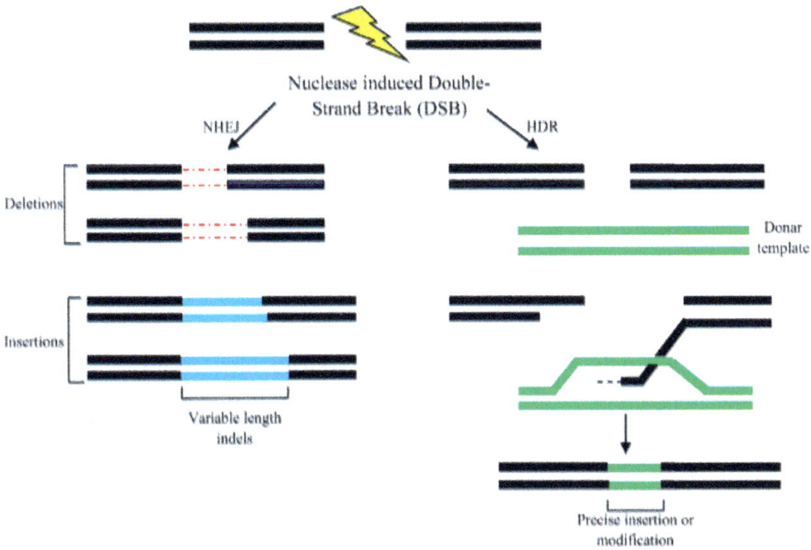


**FIGURE 5.1B** Pathways of DSB (double-strand break) repair. DSBs generated in the process of genome editing can be repaired by either of two pathways, namely, non-homologous end-joining (NHEJ) and homology-directed DNA repair (HDR). At the sites of the DSBs, NHEJ includes short deletions and/or insertions (termed as indels) whereas HDR includes accurate mutations or insertions using a donor template.[14]

activator-like effector nucleases (TALENs),[19] or the recently generated type II CRISPR/Cas, called the CRISPR/Cas9 system.[20] Of all of these above technologies, the CRISPR/Cas9 system holds the highest potential for use in genome editing due to its facile design and construction. The ZFNs and TALENs are both based on similar designs containing a couple of sequence-specific DNA-binding proteins (two zinc-finger motifs/TALEN motifs) that flank a target sequence, while the C-terminus of each zinc-finger motif/TALEN motif is followed by a FokI endonuclease, cleaving the target DNA in the form of a dimer (Fig. 5.1).[18,19] Nevertheless, the ZFNs and TALENs require complex and laborious protein engineering for their construction and therefore their applications to genome editing of plants has been circumscribed.[20,21] On the other hand, the construction of the CRISPR/Cas9 system is far easier as it contains just the Cas9 protein and the synthetic sgRNA (single-guide RNA)[22] wherein only the sgRNA has to be designed to pair with the sequence of the target DNA. By simply altering the synthetic sgRNA sequence, new targeting specificities can be generated,[22] which enables easy modifications/editing of the genome sequences.

## 2. CRISPR/Cas9 complex: mechanism of action

The CRISPR molecule contains a fragment of DNA in which noncontiguous short DNA repeats are interspaced by bits of variable sequences named spacers. There are motifs associated with the precursors of spacers (protospacers), and these are called protospacer adjacent motifs (PAMs). PAMs are composed of short stretches of di- or trinucleotides that occur immediately next to or one position after the protospacers and play a significant role in the sequence-specific recognition of protospacers and in directing the spacer orientation that is integrated into the arrays of repeats.[23] The CRISPR RNA precursor (pre-cRNA) that is transcribed from a CRISPR locus is cleaved within the repeats by a complex of Cas proteins to yield mature RNA molecules called the crRNA. The resulting crRNA contains a spacer flanked by short repeats, and this serves as a small-guide RNA enabling the Cas proteins to mediate genome editing.[24]

Makarova et al.[25] described that CRISPR/Cas9-based immunity occurs in three stages, namely, adaptation, expression, and interference. In the first adaptation stage, short foreign plasmid or virus-derived sequences undergo integration into the CRISPR array as new spacers. In the second, namely, the expression stage, transcription and maturation of the crRNAs take place, and, in the final, interference stage, the crRNAs guides Cas proteins to destroy the complementary foreign target sequences, which match with the spacer sequences. Further, Makarova et al.[25] also classified the CRISPR/Cas systems into three different types: I, II, and III. Each type possesses its own signature protein(s) wherein it depends on this protein(s) to complete the genome editing process. Of these, types I and III utilize a complex of several Cas proteins that function as signature proteins, whereas type II contains a single large Cas9 signature protein that is multifunctional and produces crRNA as well as cleaves the target DNA.[25] In contrast to the type I and III systems, the type II system has a relatively facile construction and is more amenable to being engineered as a genome editing tool.

At the type II system expression stage, a trans-noncoding small RNA (the transactivating CRISPR RNA, tracrRNA) found in *Streptococcus pyogenes* guides the crRNA maturation by means of base pairing with the repeat sequences of the pre-crRNA transcripts, with the help of the Cas9 protein and an RNase III.[26] Jinek et al.[22] synthesized an sgRNA by fusion of the crRNA and tracrRNA wherein the 5′ 20 nucleotide sequence of the sgRNA, that is the DNA target interaction sequence, could be changed into any concerned sequence of interest. Thus, the *S. pyogenes* Cas9 nuclease can be programmed with an sgRNA to create DSBs at a position three nucleotide pairs upstream of the PAM.[22,27] The position and sequence of the PAMs vary in different CRISPR/Cas systems.[23] The canonical SpCas9 PAM is 5′-NGG-3′ (N stands for any of the four nucleotides) and a 20 bp target DNA sequence followed by a PAM is (N)20 NGG.[22,23] The Cas9 nuclease possesses two domains, namely, the HNH and RuvC-like[25] wherein the former cleaves the DNA strand complementary to the guide RNA, while the latter cleaves the noncomplementary sequence with respect to the guide RNA sequence.[22] This system is versatile, highly efficient, and programmable and has developed into the model CRISPR/Cas9 system, which is most often used to perform genome editing using NHEJ/HR-mediated mutations.

Most of the Cas9 systems to date have been developed to target DNA.[28] On the other hand, the class 2 type VI systems such as LwaCas13a[29] and PspCas13b,[30] driven by the single effector protein C2c2 (now called Cas13) have been engineered to edit RNA. The C2c2 protein is guided by a single crRNA and can specifically act on and cleave single-stranded target RNA possessing complementary protospacers through its HEPN domains. Thus, C2c2 can be applied toward RNA knockdown and to investigate RNA biology.[30]

## 3. Delivery of CRISPR/CAS components into plants

The transmission of gene editing reagents into plant cells and the performance of the editing events are vital steps in genome editing.[28] Technologies such as protoplast transfection and particle bombardment or *Agrobacterium*-mediated transfer DNA (T-DNA) transformation have been used to deliver the CRISPR-mediated editing components including RNA, DNA, and ribonucleoproteins (RNPs). Among these, protoplast transfection is predominantly used for transient expression, while particle bombardment and Agrobacterium-based transformation are being used for generating edited plants. In general, continued activity of the Cas9 nuclease is not required. Nevertheless, the gene editing activity will be diluted out as the cells keep dividing. Whereas for Agrobacterium-introduced genes, they would segregate away from the position of nuclease activity to which they are mostly not linked.[31]

## 4. Precision plant breeding applications using CRISPR/Cas9 technology

Agriculture is facing constant threats worldwide due to abiotic stresses such as heat, frost, drought, and salinity, as well as biotic stresses such as bacteria, fungi, and viruses.[32] Due to the exploding increase in the global population, food supply shortages, and environmental deterioration, there is an avid interest among crop breeders to increase crop yields, quality, and crop resistance/stress tolerance.[33] In this context, genome editing mediated by the CRISPR/Cas9 technology can be utilized not only for research on functional genomics but also for enhancing crop varieties.

### Crop trait enhancement by gene knockout

Removal of negative elements is a useful strategy to be adopted toward crop improvement. Hence, knock out of genes conferring unwanted traits is the most facile and common purpose of using CRISPR/Cas9.[28]

### Improvement of crop traits using gene knock-in and replacement

Several agronomic traits are determined by single-nucleotide substitutions, changes in gene expression, or inclusion of new gene functions. Targeted and precise modifications of genes using knock-ins or base replacements enable breeding by introducing novel alleles without any linkage drag or creating allelic variants that do not occur naturally.[34] Also, knock-ins can be employed to change multiple elite traits by accumulating genes in a single variety. However, due to the fact that the HDR DNA repair pathway is infrequent, these techniques are not routinely employed; and therefore, their application is rather limited. In a study by Shi et al.[35] maize drought tolerance was augmented using CRISPR/Cas9-based gene editing. The gene *ARGOS8* codes for a negative regulator of ethylene expression and is synthesized at diminished levels in most of the inbred maize lines. HDR was used to knock-in the *GOS2* promoter to substitute the native *ARGOS8* promoter and this enhanced *ARGOS8* expression, which led to higher crop yields under drought conditions. In yet another study,[36] a tomato line was created with extended shelf life by means of T317A replacement into the *ALC* (*alcobaca*) gene using CRISPR/Cas9-mediated gene editing.

Augmented efficiency of gene targeting was reported in rice,[37] cassava,[38] wheat,[39] tomato,[40,41] and potato[42] using a Geminivirus-derived DNA replicon to rise the number of repair templates toward improved HDR efficiency. Purple tomato with high anthocyanin content was produced by,[41] using Geminivirus replicon-based gene editing to cause a ten-fold higher insertion frequency of the cauliflower mosaic virus 35S promoter into the site upstream of the *ANT1* (anthocyanin mutant 1) gene that led to constitutive *ANT1* expression. Moreover, in tomatoes, a 25% efficiency in the repair of a fast-neutron-induced *crtiso* (*carotenoid isomerase*) allele having a 281 base pair deletion was reported by Dahan-Meir et al.[40] using a high-efficacy reporter-free and selection-free gene targeting system.

The conserved domains of the 5-enolpyruvylshikimate-3-phosphate synthase (*EPSPS*) and the endogenous acetolactate synthase (*ALS*) genes having substitutions of key amino acids could cause resistance to glyphosate or sulfonylurea-based herbicides. Base changes in the ALS gene induced by HDR-based gene editing machinery conferred sulfonylurea-based herbicide-resistance in soybean,[43] rice,[44–46] and maize.[4,47] Also, HDR-mediated glyphosate resistance was reported in cassava[38] and flax (*Linum usitatissimum*)[48] harboring double amino acid substitutions (T102I/P106S (TIPS) and T102I/P106A) in the *EPSPS* gene. Li et al.[49] used an intron-targeting method performed through NHEJ-driven gene replacement and insertion to generate TIPS ESPS glyphosate-resistant rice. This higher editing frequency strategy is a good alternative to the low-frequency HDR-driven gene editing.

## Gene modifications for augmenting yield traits

In several crop plants, the process of flowering is regulated by seasonal variations in day-length, which can circumscribe the range of geographical cultivation of these crops.[50] The time of flowering can be regulated by altering the flowering genes and their molecular interactions.[51] The flowering times of crops such as tomato[52] and soybean (*Glycine max*)[53] have been modified by mutating flowering genes such as *SELFPRUNING 5G* (SP5G)[52] and *FLOWER LOCUS T* (FT)[53] using the CRISPR/Cas9 system.

Hybrid rice varieties are important for augmenting rice yields and the two-line hybrid breeding system has been developed using the popular male sterile line namely, the thermos-sensitive genetic male sterility (*TGMS*) line.[54] 11 new *TGMS* lines have been generated in 1 year using CRISPR/Cas9-based mutations in the *TGMS* line, which proved the ability of the CRISPR/Cas9 system to expedite hybrid rice breeding. The rice cultivar Zhinghua 11 has been engineered by the CRISPR/Cas9 system specifically to mutate four of the yield-related genes, namely, *DEP1, Gn1a, IPA1,* and *GS3*. The resulting *dep1, gn1a,* and *gs3* mutants possessed denser erect panicles, enhanced grain number, and greater grain size, respectively,[55] proving the ability of the CRISPR/Cas9 system to augment significant agronomic traits such as crop yield. In maize (*Zea mays*), compared to the conventional leaf structure, an upright leaf structure can capture more light and thereby grow at greater field densities.[56] The dominant gene *LIGULELESS1* (*LG1*) codes for the former and its recessive counterpart in the homozygous form generates upright leaf structure.[56] The sgRNA/Cas9 targeting the *ZmLG-1* was used to create a desired-target mutator (DTM) in T0 transgenic maize plants, which was then introduced into six distinct recipient maize lines through hybridization with transgenic plants.[57] Thus, genome editing of *ZmLG-1* was accomplished in the recipient maize lines that all contained significantly smaller leaf angles compared to those of wild type maize plants.[57] The above study employs an important strategy to perform genome editing but not via delivery of sgRNA/Cas9 expression cassettes into target crops by traditional transformation. This is a significant finding considering that for several major crops, procedures for transformation and plant regeneration are either unachievable or inefficient, in addition to requiring meticulous and time-taking tissue culture steps.

Plant architecture can be determined by a novel type of plant hormone called the strigolactones, which function by inhibiting branching of the shoot (tillering) and by modulating root growth.[58] In a study by Butt et al.[59] the rice gene *CAROTENOID CLEAVAGE DIOXYGENASE 7* (*CCD7*) that controls a crucial step in strigolactone biosynthesis was knocked out using the CRISPR/Cas9 system. The resulting ccd7 rice mutants had ideal phenotypes such as high-tillering and a dwarf habit.[59] Moreover, CRISPR/Cas9 has been used to construct various loss of function mutants with phenotypes such as grain dormancy (*HvPM19*) in barley (*Hordeum vulgare*),[5] valve-margin development (*BnALC*) in polyploid oilseed rape (*Brassica napus*),[60] dwarfing and reduced fruit dehiscence (*BolC.GA4.a*) in *Brassica oleracea*[5] and chloroplast development (*GhCLA1*) in cotton (*Gossypium hirsutum*).[61]

## Gene modifications for enhanced produce quality

Zhang et al.[62] used the CRISPR/Cas9 system to knock out the dominant Waxy gene that controls the amylose content in rice, and the resultant mutants were shown to possess diminished amylose contents and augmented glutinosity, thus demonstrating the potential of CRISPR/Cas9 system to enhance rice quality. In yet another investigation,[63] the gene encoding a granule-bound starch synthase (*GBSS*) was edited by the CRISPR/Cas9 system in tetraploid potato (*Solanum tuberosum*), which resulted in loss of GBSS enzyme activity, and consequently, the tubers showed diminished amylose content with a simultaneous enhancement in amylopectin/amylose ratio. The capability of CRISPR/Cas9 to induce mutagenesis of multiple targets simultaneously affords a convenient and powerful tool for improvement of polyploid crops, wherein several functionally related genes or multiple genes in a single-family control the same genetic trait. Rice containing an excess of cadmium can be hazardous to humans and animals who feed on the contaminated rice.[64] To circumvent this contamination, novel Indica rice varieties containing decreased cadmium content have been developed using the CRISPR/Cas9 technology to disrupt the metal transporter gene *OsNramp5*, thus enabling the generation of safe rice varieties in paddy fields contaminated with cadmium.

Traits involved in determining crop quality vary based on the concerned breeding requirements. Quality enhancement through genome editing has focused on specific parameters such as fragrance, starch content, nutritional value, and storage ability of the respective crops. CRISPR-Cas9-mediated knockout of the *Waxy* gene generated rice with diminished amylose content that led to rice with better cooking quality, as well as enhanced eating quality.[62,65] A similar venture on the *Waxy* gene of corn produced by DuPont Pioneer generated high crop yields for enhanced commercial use.[66] CRISPR/Cas9-mediated mutation of the starch branching enzyme SBEIIb yielded rice with higher amylose and resistant starch content, wherein the high amylose content would benefit those with diet-associated noninfectious chronic diseases.[67] Fragrance is an important parameter for determining rice quality, and rice varieties possessing desirable fragrances upon cooking have augmented market value. The main fragrance compound in fragrant rice is 2-acetyl-1-pyrroline that is synthesized by a mutation in the betaine aldehyde dehydrogenase 2 (*BADH2*) gene. Shan et al.[68] used TALEN-based disruption of the *OsBADH2* gene to create a fragrant rice variety with equivalent 2-acetyl-1pyrroline content (0.35–0.75 mg/kg) to that of the fragrant arising by natural mutation.

The cereal crop gluten proteins cause celiac disease in over 7% of the population in Western countries. The main gluten-encoding α-gliadin gene family in wheat contains almost 100 genes or pseudogenes. Sanchez-Leon et al.[69] generated low-gluten wheat by the simultaneous CRISPR/Cas9-mediated knockout of the most conserved domains of the α-gliadin family members. CRISPR/Cas9-based editing has also been used to produce other high-quality crops such as tomatoes with prolonged shelf-life,[70,71] *Brassica napus*[72] and *Camelina sativa*[73,74] with high oleic acid content in their seeds, high-quality tomatoes with increased γ-aminobutyric acid[75,76] or lycopene content,[77] as well as potatoes with diminished levels of toxic steroidal glycoalkaloids.[78]

## Genome editing using CRISPR/Cas9 technology for abiotic stress tolerance

The CRISPR/Cas9 system has been used to mutate the mitogen-activated protein kinases3 (*SlMAPK3*) gene, whereupon the involvement of *SlMAPK3* in plant defense responses against drought was identified in tomatoes.[79] Also, the regulatory mechanism of stress tolerance induced by the stress/ABA-activated protein kinase2 (*SAPK2*) gene was shown by generating mutants using CRISPR/Cas9 in rice plants.[80] In *Arabidopsis*, the C-repeat binding factors (CBFs) have been implicated in cold stress response using CRISPR/Cas9 to mutate the wild-type genes *CBF1, CBF2,* and *CBF3* into double and triple mutants.[81] Thus, with an insight into gene function, plant breeders can use candidate genes to produce crops with improved stress tolerance.

Maize varieties with drought tolerance have been developed by Shi et al.[35] In maize, the *ARGOS8* gene functions as a negative modulator of ethylene responses. However, the *ARGOS8* expression is poor in maize. Toward enhancing *ARGOS8* expression, CRISPR/Cas9 was utilized to insert the sequence of the *GOS2* promoter into the 5′-untranslated region of the native *ARGOS8* gene or to modify the *ARGOS8* native promoter into the *GOS2* promoter by homologous recombination. These *ARGOS8* maize variants generated augmented grain yields under drought conditions in the fields.[35] From the above examples, it is well evident that CRISPR/Cas9-based genome editing can be used to induce tolerance to abiotic stresses caused by extreme drought, temperature, salinity, deficiency of nutrients,[82] and exceeding levels of heavy metals.[83]

Also, contamination of cultivable land has necessitated measures to preclude toxic heavy metal accumulation in crops. Using knockout of the genes *OsHAK1*, *OsNramp5,* and *OsARM1,* rice strains with decreased levels of cadmium, radioactive cesium, and arsenic, respectively, have been generated.[83–85] Rice plants with enhanced grain yield, improved tolerance to high temperatures and decreased preharvest sprouting were produced by CRISPR/Cas9-based *pyl1/4/6* (pyrabactin resistance 1-like) triple knockout of *OsPYL* abscisic acid receptor gene family.[86]

## CRISPR/Cas9 genome editing toward biotic stress resistance

The versatility of applications of the GenEd technology also extends to plant disease control. In many cases, resistant and susceptible alleles of a given gene vary only by a few nucleotides.[87] GenEd technology can be easily applied to mutate the susceptible alleles to resistant ones directly rather than using the conventional and time-consuming crosses and back crosses. Also, GenEd can be employed to generate mutants of polyploid plants, in which mutants are often challenging to isolate.

In *Arabidopsis*, the gene-enhanced disease resistance 1 (*EDR1*) has been shown to be a negative regulator of powdery mildew resistance.[88] In a study by

Zhang et al.[89] three homologs of the *EDR1* gene were knocked out simultaneously in wheat (*Triticum aestivum*), which enabled the resultant Taedr1 plants to exhibit increased resistance to powdery mildew. The gene *MILDEWRESISTANCE LOCUS (MLO)* encodes a protein that causes susceptibility to powdery mildew in plants.[90] Augmented tolerance to powdery mildew was demonstrated in tomato[91] and wheat,[92] wherein the *MLO* gene was mutated using the RNA-guided Cas9 system. It is tough to find resistance to powdery mildew of wheat (*Blumeria graminis* f. sp. tritici). Naturally occurring recessive mutations within the *MLO* gene that confer a high degree of enduring disease resistance have earlier been obtained in the diploid barley plants,[93] but not in wheat. The TALENS GenEd complex was used to mutate all three homologous alleles of the wheat *MLO* locus that naturally encodes a suppressor of a defense response. These triply mutated lines showed resistance to the fungus.[92] Similar studies were also carried out using the CRISPR/Cas9 system.[92] Ethylene responsive factors (ERFs) are implicated in the ethylene response pathway and can control multiple stress tolerance.[94] The *OsERF922* gene was mutated by CRISPR/Cas9, which led to increased resistance to blast fungus in rice plants.[95] Resistance to *Botrytis cinerea* was demonstrated in grape (*Vitis vinifera*) mutants, wherein CRISPR/Cas9 was shown to ablate the function of the *VvWRKY52* gene.[96]

*Xanthomonas citri* subsp. *citri,* a bacterial pathogen, causes citrus canker, wherein it stimulates the expression of the *CsLOB1* canker susceptibility gene.[97] Citrus varieties with resistance to canker were demonstrated by the disruption of the *CsLOB1* gene using the CRISPR/Cas9 system in the grapefruit Duncan (*Citrus paradisi* Macf.)[98] In several different host plants, the *eIF* genes coding for the eukaryotic translation initiation factors have been found to act as recessive resistance genes.[99] *Xanthomonas oryzae* pv. *Oryzae* causes bacterial blight in rice crops. Zhou et al.[100] created rice plants with resistance to this disease by deletion of the *OsSWEET13* promoter. The CRISPR/Cas9 technology has shown great promise for suppression of viral infections for which there are no good control methods. Pathogenic viruses use the plant host cell machinery for their replication, and therefore, they are vulnerable to an introduced Cas9 nuclease that specifically recognizes their nucleotide sequences and destroys them. Virus-resistant *Arabidopsis*[101] and cucumber (*Cucumis sativus*)[102] plants were generated by mutations of the *eIF(iso)4E* and *eIF4E* genes, respectively, using the CRISPR/Cas9 technology. CRISPR/Cas9 has also been used to induce virus resistance in crops to create cotton leaf curl disease−resistant *clcud* cotton,[103] broad potyvirus−resistant *eif4e* cucumber,[102] and tungro disease-resistant *eif4g* rice.[104] In rice plants, the *eIF4G* gene negatively regulates the recessive Rice tungro spherical virus (RTSV) resistance trait that is responsible for causing rice tungro disease. Macovei et al.[104] used the CRISPR/Cas9 system to eliminate the *eIF4G* gene, which resulted in RTSV-resistant rice varieties. Ali et al.[105] identified highly conserved sequences in Geminiviruses that have been targeted by CRISPR/

Cas9. Further, Cas-like nucleases such as from *Francisella novicida* specifically act on RNA instead of DNA, and this offers the promise of being able to delete key nucleotide sequences of RNA viruses when they start their infectious process within the plant cell.[106] However, for this purpose, the concerned plant should carry the respective nuclease and its guide sequences within its genome. In other words, it should be a GM plant adapted to committing GenEd.

Disruption of *OsCYP71A1* gene inhibited serotonin biosynthesis while greatly enhancing salicylic acid levels, which conferred resistance to stem borers and plant hoppers, the two most detrimental pests of rice.[107] The above reports provide compelling evidence of the use of the CRISPR/Cas9 system in generating crops with augmented resistance to a host of different diseases. However, besides building disease resistance by editing the plant genome, this system has also been used to mutate the viral genome.[32] In a recent study, the Cas9 nuclease from *Francisella novicida* (FnCas9) along with the sgRNA with specificities toward cucumber mosaic virus (CMV) and tobacco mosaic virus (TMV) have been used to edit RNA virus genomes in Arabidopsis and tobacco plants, which provided immunity against these RNA viruses.[108] Tables 5.1 and 5.2 show some representative examples of genome editing in crops.

**TABLE 5.1 CRISPR/Cas9 technology for genome editing in fruit crops.**

| Species | Target gene(s) | Target traits | References |
|---|---|---|---|
| Tomato (*solanum lycopersicum*) | RIN | Regulating fruit ripening | Ito et al.[70] |
| Tomato (*solanum lycopersicum*) | ORRM4 | RNA editing factors involving in fruit ripening | Yang et al.[109] |
| Tomato (*solanum lycopersicum*) | ALC | Ripening of fruit | Yu et al.[36] |
| Tomato (*solanum lycopersicum*) | SlWUS, SlCLV3 | Fruit size | Rodriguez-leal et al.[110] |
| Apple (*malus prunifolia*) | DIPM-1, DIPM-2, DIPM-4 | Resistance to fire blight disease | Malnoy et al.[111], Osakabe et al.[112] |
| Grapevine (*Vitis vinifera*) | *MLO-7* | Susceptible gene to powdery mildew | Malnoy et al.[111], Osakabe et al.[112] |
| Grapevine (*Vitis vinifera*) | *IdnDH* | Biosynthesis of tartaric acid (TA) | Ren et al.[113] |

*Continued*

**TABLE 5.1 CRISPR/Cas9 technology for genome editing in fruit crops.—cont'd**

| Species | Target gene(s) | Target traits | References |
|---|---|---|---|
| Banana (banana cv. Rasthali) | *PDS* | Albino phenotype and decrease of carotenoid content | Kaur et al.[114] |
| Cucumber (*Cucumis sativus* L.) | *eIF4E* | Resistance against cucumber vein yellowing virus | Chandrasekaran et al.[102] |
| Wild strawberry (*Fragaria vesca*) | *TAA1 ARF8* | Auxin biosynthesis and signaling | Zhou et al.[115] |

Some examples of the use of CRISPR/Cas9 technology for genome editing in fruit crops.
From Zhou J, Li D, Wang G, Wang F, Kunjal M, Joldersma D, Liu, Z. Application and future perspective of CRISPR/Cas9 genome editing in fruit crops. Journal of Integrative Plant Biology 2020;**62**(3):269–286.

**TABLE 5.2 Gene editing in cereal crops.**

| Species | Target gene(s) | Gene function | References |
|---|---|---|---|
| Maize | *ALS2* | A key enzyme for the biosynthesis of branched-chain amino acids (major targets for herbicides) | Svitashev et al.[4] |
| Maize | *ARGOS8* | Increased grain yield under drought stress | Shi et al.[35] |
| Wheat | *TaVIT2* | Fe content | Connorton et al.[116] |
| Wheat | *Alphagliadin* | Low-gluten | Sánchez-león et al.[69] |
| Rice | *osPPKL1* | Increases length and yield | Zhang et al.[117] |
| Rice | *OsMPK5* | Various abiotic stress tolerance and disease resistance | Xie and Yang[118] |
| Rice | *OsHAK-1* | Low cesium accumulation | Cordones et al.[84] |
| Rice | *eIF4G* | Resistance to rice tungro spherical virus | Macovei et al.[104] |
| Rice | *OsRR22* | Salinity tolerance | Zhang et al.[119] |

Some examples of gene editing investigations in cereal crops.
From Ansari WA, Chandanshive SU, Bhatt V, Nadaf AB, Vats S, Katara JL, Sonah H, Deshmukh R. Genome editing in cereals: approaches, applications and challenges. International Journal of Molecular Sciences 2020;**21**(11):4040. https://doi.org/10.3390/ijms21114040.

## 5. CRISPR/Cas9-based control of gene transcription or translation

There are many levels at which gene expression can be coordinated, such as, for instance, at transcriptional and posttranscriptional levels. Therefore, regulating gene expression in a targeted and precise fashion to enable phenotypic diversity is essential for plant breeding.[120] Earlier work has indicated that several mutations driving domestication, breeding, and evolution of crop species mostly occurred within the cis-regulatory promoter regions of genes.[121–123]

In recent times, CRISPR/Cas9 has been employed to inactivate the promoters of three target genes controlling plant architecture, branching of the influorescence, and fruit size in tomatoes. This created diverse cis-regulatory alleles with some of the promoter variants exhibiting continuous variability in all the three traits.[110] Therefore, even if CRISPR/Cas9 has been mainly used to disrupt the coding sequences of several genes to generate null alleles, the above investigation[110] showed that CRISPR/Cas9 can also be used to mutate regulatory elements driving levels and patterns of gene expression, from which crop enhancement can be achieved by creation of novel alleles and favorable variations of quantitative traits.

CRISPR/Cas9 has also been used to mutate the upstream ORF of *LsGGP2* (GDP-I-galactose phosphorylase) encoding a vital enzyme essential for vitamin C biosynthesis in *Lactuca sativa* (lettuce). This generated lettuce plants containing augmented ascorbate contents and superior tolerance to oxidative stresses.[124] This study demonstrated the enormous potential of the CRISPR/Cas9 technology in modulating both transcriptional and translational levels of gene regulation, which in turn can be applicable in investigating biological mechanisms and also in enhancing crop performance. Functional genomics investigations associated with interrogation of gene function can be executed in a far superior and facile manner using the CRISPR/Cas9 system, particularly for several genes that are only transcribed but are not translated.[125]

## 6. Use of CRISPR/Cas9 technology in mutant library construction

The precise functional analysis of all the genes contained in the genome of plant species with a known sequence is a significant challenge. Toward addressing this challenge, it is imperative to generate a saturated mutant library[126,127] and herein comes the important role of the CRISPR/Cas9 system in generating mutants as its targeting specificity can be readily modified by simply changing the 20-bp target-binding sequence within the sgRNA. It is therefore appropriate and advantageous to employ CRISPR/Cas9 to execute genome-scale mutagenesis as well as forward genetic screening studies. Pooled sgRNA libraries were transformed into tomato plants, and mutant collections were produced.[128] In this study, the leucine-rich repeat subfamily XII genes associated with immunity and a homolog of a gene coding for a boron efflux transporter in *Arabidopsis* were discovered as genes targeted by the sgRNA using extensive genetic screening and genome sequencing.[128]

Mutant libraries created using the CRISPR/Cas9 system have also been generated in rice, wherein several loss-of-function mutants were produced by transforming synthetic sgRNA libraries.[126,129] Upon cultivation of these mutant rice plants, it was shown that the resulting plants showed phenotypic changes such as spotted leaves, altered leaf color, decreased plant height, enhanced tiller angle, as well as sterility and lethality. Additionally, the interaction between the phenotype and genotype in these CRISPR/Cas9 mutant libraries could be identified in a facile manner using sgRNAs. This strategy therefore affords an easy way to discover mutant alleles.[126,129] Overall, the studies presented above show the great potential of the CRISPR/Cas9 mutant libraries in applications such as crop improvement and research on functional genomics. Thus, mutagenesis can be performed genome-wide by means of an efficacious crop transformation system.

## 7. Use of base editors in plant breeding

Several economically important agricultural traits are determined by single-nucleotide polymorphisms within coding or noncoding regions of the genome, and therefore, base editing is of great use in crop enhancement and plant breeding. For instance, the herbicide resistance trait is conferred by base editing within coding regions. The *ALS* gene has been targeted by a plant cytidine base editor in wheat,[130] rice,[131] watermelon,[132] and Arabidopsis[28] to confer sulfonylurea- or imidazolinone-resistance. Similarly, the acetyl-coenzyme A carboxylase *ACCase* gene has been targeted by a plant adenine base editor to generate rice varieties with haloxyfop-R-methyl resistance.[133] Gene editing can be used to target splicing sites mediating alternative splicing, and this can lead to mis-splicing. For example, disruption of the splicing acceptor site on an intron of the *AtPDS* mRNA through conversion of A to G led to altered splicin.[134] Similarly, conversion of G to A at the splice donor site resulted in the constitutive retention of an *AtHAB* mRNA intron, which conferred abscisic acid hypersensitivity.[135]

Various variations have been adopted to modify the properties of the CRISPR/Cas9 machinery. By increasing or decreasing the length of the RNA:DNA guide, the precision of the nucleotide match can be changed. In some instances, nuclease activity for one of the strands has been abolished, leaving a nickase action.[136] Larger deletions or insertions can be generated by two complementary nickases, each of which recognizes a 20 bp sequence, resulting in lengthy single-stranded DNA stretches, which will have a strong propensity to look for homologous binding.[9,136] Other modifications include the application of a disarmed nuclease bereft of nuclease activity, that will exhibit competitive binding to a defined site, thus blocking the access of other molecules like transcription factors and thereby downregulate or shut off gene expression[137] (Fig. 5.1b).

The CRISPR/Cas9 complex can hold transcription activators near a site.[125] or retain a specific repair enzyme at this site, resulting in inducing site-specific mutation even without cutting the DNA backbone.[138]

## 8. Accelerating hybrid breeding

Another powerful strategy to enhance crop productivity is hybrid breeding. An essential condition for generating high-quality hybrid varieties is to develop a maternal line that is male-sterile. Great advances have been made to yield male-sterile lines using CRISPR/Cas9-driven gene knockout, and from this technique, male-sterile thermosensitive *tms5* lines have been developed in maize[139] and *csa* rice.[140] Additionally, photosensitive genic male-sterile *ms45* wheat[141] and csa rice[142] have also been generated using the above strategy. One of the major obstacles to exploiting heterosis in plant breeding is hybrid sterility. To circumvent reproductive barriers in hybrids of *japonica-indica*, the *OgTPR1* at the *S1* locus[143] and the *SaF/SaM* at the sterility locus *Sa*[144] were disrupted. Also, one or two copies of the *Sc* gene were knocked out in the *indica* allele *Sc−I,* which rescued male fertility in hybrids of *japonica-indica*.[145] A similar knockout of the *ORF2* toxin gene that is known to cause the recently identified selfish-gene suicide in rice, augmented the fertility of the *japonica-indica* hybrids. Knockout of three vital meiotic genes, *PAIR1, REC8,* and *OSD1,* in rice using genome editing substituted mitosis for meiosis.[107] Concurrent activation of *BBM1* in the egg cell.[146] or knockout of *MTL*[147] generated asexual propagation lines that facilitated a fixation of hybrid heterozygosity through seed propagation. Further, genome editing is also a highly efficient tool for improving several other traits, such as reducing growth times,[148] augmenting haploid breeding,[149,150] and surmounting self-incompatibility in potato diploids.[151] Fig. 5.2 depicts the distribution of traits changed by genome editing obtained in a recent survey.

## 9. Biological, political challenges, and regulatory decisions on GenEd crops

As described above, CRISPR/Cas has great potential to augment food security world-wide toward sustainable agriculture. Regulatory enforcement to date concerning GenEd crops has by and large followed the standards applied for mutation breeding, i.e., product-based classification. For instance, GenEd mushrooms (a fungus and not a plant) generated with a diminished tendency to brown using CRISPR-derived mutation of the gene for polyphenol oxidase, were not deemed to require GM regulation.[152] Moreover, the USDA has also considered that plants developed using this technology would not be

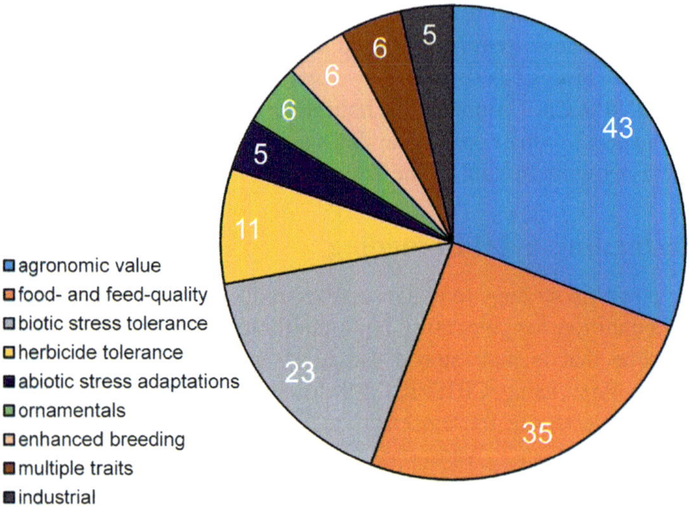


**FIGURE 5.2 Genome editing trait distribution.** Distribution of genome editing traits by applications. *From Menz J, Modrzejewski D, Hartung F, Wilhelm R, Sprink T. Genome edited crops touch the market: a view on the global development and regulatory environment.* Frontiers in Plant Science *2020;11. https://doi.org/10.3389/fpls.2020.586027.*

categorized as GM.[66] Whereas in Canada, GenEd crops are covered under the existing plants expressing novel traits (PNT) regulation, that is, product or phenotype-based.[153] It is unreasonable to consider that genetic changes in plants caused by GM or GenEd technologies are more hazardous compared to those developed using conventional breeding or mutation-induced breeding measures that may themselves lead to inadvertent mutations or gene combinations. In all the respective cases, several cycles of field trials and analysis are required to exclude the most undesirable or less than effective outcomes. Nevertheless, the plant lines developed by classical breeding in the present environment undergo less thorough scrutiny compared to GM or GedEd crops. If the same standards enforced on GM/GenEd crops were to be applied to those developed by classical breeding, few of the new cultivars would survive the investigation.

## 10. Conclusions and future perspectives

In summary, the genome editing of plants using the CRISPR/Cas technology affords a facile, efficacious, and resourceful tool for the investigation of precision plant breeding. Base-editing methodologies allow the direct conversion of one given DNA nucleotide into another, a process that is independent of the formation of double-strand breaks. Genome editing through the use of CRISPR/Cas RNPs in plants affords a DNA-free editing method with low frequency of off-target effects. The editing of cis-regulatory elements including promoter regions and the upstream open reading frames using the CRISPR/Cas system facilitates tweaking of gene expression, thereby resulting in quantitative trait variation. Gene knockouts, replacement, and insertion using the CRISPR/Cas system are currently being expeditiously applied toward enhancing crop quality, yield, stress tolerance, and disease resistance in addition to improving hybrid breeding as well as crop domestication.

However, substantial impediments still remain to be surmounted for wider application of this technology in a precise and efficacious manner across different plant species. Important among the hurdles to CRISPR/Cas-mediated enhancements in plants is the successful delivery of nuclease reagents and improved homology-directed repair frequency that still have to be addressed. The remarkable capability of genome editing to elicit targeted, sequence-specific genetic diversity throughout the plant genome has led to formidable advancements in crop breeding. The robustness, simplicity, and adaptability of the CRISPR/Cas systems renders genome editing to become an efficient tool toward precise crop enhancements through gene knock-in, knockout, point mutations, gene replacement, fine-tuning of gene expression, as well as other alterations at any given locus in crops.

This technology is also of use in the construction of high-throughput mutant libraries and antiviral breeding. Nevertheless, efficiently translating these technologies from the level of the laboratory to the field necessitates rapid identification of the basis of significant genetic traits, improved efficiency of gene targeting, effectual delivery of the components of the CRISPR/Cas system into plant cells, and the subsequent regeneration of plants with or without the requirement of tissue culture as well as the availability of base editors with enhanced targeting scope and frequency. Mobilizing the concepts of plant systems biology and synthetic biology in addition to advancements in functional genomics along with the advent of genome-editing technology, high-throughput next generation sequencing, and other associated techniques will enable the engineering of modern crops with highly enhanced qualities.

Additionally, it is important that there be public dialogue regarding genome editing and financing of technologies useful in providing consumer-friendly genetic traits such as healthier crop products with extended shelf-lives. Given what has been already achieved in tomato plants, designer fruits and other plant products as well as their de novo domestication are not far from reality. Despite the availability of these advanced genome editing technologies, there remains restriction of their widespread use for enhancing agricultural crops due to global regulatory requirements. The productivity of novel, healthier food can be enhanced only by overcoming these restrictions. Crop plants cultivated using CRISPR/Cas genome editing could be classified as nongenetically modified organisms in order to enable their quick acceptance at the field level. It is anticipated that implementation of this technology in agriculture will significantly remodel crop productivity leading to another green revolution that will ensure the food and nutritional requirements of the ever-increasing global populations.

## References


1. Jeon J-S, Lee S, Jung K-H, et al. T-DNA insertional mutagenesis for functional genomics in rice. *The Plant Journal* 2000;**22**(6):561–70. https://doi.org/10.1046/j.1365-313x.2000.00767.x.
2. Wang L, Zheng J, Luo Y, et al. Construction of a genome wide RNAi mutant library in rice. *Plant Biotechnology Journal* 2013;**11**(8):997–1005. https://doi.org/10.1111/pbi.12093.
3. Jiang W, Zhou H, Bi H, Fromm M, Yang B, Weeks DP. Demonstration of CRISPR/Cas9/sgRNA-mediated targeted gene modification in Arabidopsis, tobacco, sorghum and rice. *Nucleic Acids Research* 2013;**41**(20):e188. https://doi.org/10.1093/nar/gkt780.
4. Svitashev S, Young JK, Schwartz C, Gao H, Falco SC, Cigan AM. Targeted mutagenesis, precise gene editing, and site-specific gene insertion in maize using Cas9 and guide RNA. *Plant Physiology* 2015;**169**(2):931–45. https://doi.org/10.1104/pp.15.00793.
5. Lawrenson T, Shorinola O, Stacey N, et al. Induction of targeted, heritable mutations in barley and *Brassica oleracea* using RNA-guided Cas9 nuclease. *Genome Biology* 2015;**16**(1). https://doi.org/10.1186/s13059-015-0826-7.
6. Fan D, Liu T, Li C, et al. Efficient CRISPR/Cas9-mediated targeted mutagenesis in populus in the first generation. *Scientific Reports* 2015;**5**. https://doi.org/10.1038/srep12217.
7. Feng Z, Zhang B, Ding W, et al. Efficient genome editing in plants using a CRISPR/Cas system. *Cell Research* 2013;**23**(10):1229–32. https://doi.org/10.1038/cr.2013.114.
8. Li JF, Norville JE, Aach J, et al. Multiplex and homologous recombination-mediated genome editing in *Arabidopsis* and *Nicotiana benthamiana* using guide RNA and Cas9. *Nature Biotechnology* 2013;**31**(8):688–91. https://doi.org/10.1038/nbt.2654.
9. Schiml S, Fauser F, Puchta H. The CRISPR/Cas system can be used as nuclease for in planta gene targeting and as paired nickases for directed mutagenesis in *Arabidopsis* resulting in heritable progeny. *The Plant Journal* 2014;**80**(6):1139–50. https://doi.org/10.1111/tpj.12704.
10. Nekrasov V, Staskawicz B, Weigel D, Jones JDG, Kamoun S. Targeted mutagenesis in the model plant *Nicotiana benthamiana* using Cas9 RNA-guided endonuclease. *Nature Biotechnology* 2013;**31**(8):691–3. https://doi.org/10.1038/nbt.2655.
11. Shan Q, Wang Y, Li J, Zhang Y, Chen K, Kunling C, et al. CRISPR/Cas genome editing and precision plant breeding in agriculture. *Annual Review of Plant Biology* 2013;**31**:667–97.
12. Bao A, Burritt DJ, Chen H, Zhou X, Cao D, Tran L-SP. The CRISPR/Cas9 system and its applications in crop genome editing. *Critical Reviews in Biotechnology* 2019;**39**(3):321–36. https://doi.org/10.1080/07388551.2018.1554621.
13. Zhou J, Li D, Wang G, Wang F, Kunjal M, Joldersma D, Liu Z. Application and future perspective of CRISPR/Cas9 genome editing in fruit crops. *Journal of Integral Plant Biology* 2020;**62**(3):269–86.
14. Barman A, Deb B, Chakraborty S. A glance at genome editing with CRISPR-Cas9 technology. *Current Genetics* 2020;**66**(3):447–62.
15. Bleuyard JY, Gallego ME, White CI. Recent advances in understanding of the DNA double-strand break repair machinery of plants. *DNA Repair* 2006;**5**(1):1–12. https://doi.org/10.1016/j.dnarep.2005.08.017.
16. Puchta H, Dujon B, Hohn B. Two different but related mechanisms are used in plants for the repair of genomic double-strand breaks by homologous recombination. *Proceedings of the National Academy of Sciences of the United States of America* 1996;**93**(10):5055–60. https://doi.org/10.1073/pnas.93.10.5055.
17. Puchta H. The repair of double-strand breaks in plants: mechanisms and consequences for genome evolution. *Journal of Experimental Botany* 2005;**56**(409):1–14. https://doi.org/10.1093/jxb/eri025.

18. Osakabe K, Osakabe Y, Toki S. Site-directed mutagenesis in *Arabidopsis* using custom-designed zinc finger nucleases. *Proceedings of the National Academy of Sciences of the United States of America* 2010;**107**(26):12034–9. https://doi.org/10.1073/pnas.1000234107.
19. Miller JC, Tan S, Qiao G, et al. A TALE nuclease architecture for efficient genome editing. *Nature Biotechnology* 2011;**29**(2):143–50. https://doi.org/10.1038/nbt.1755.
20. Kumar V, Jain M. The CRISPR-Cas system for plant genome editing: advances and opportunities. *Journal of Experimental Botany* 2015;**66**(1):47–57. https://doi.org/10.1093/jxb/eru429.
21. Soda N, Verma L, Giri J. CRISPR-Cas9 based plant genome editing: significance, opportunities and recent advances. *Plant Physiology and Biochemistry* 2018;**131**:2–11. https://doi.org/10.1016/j.plaphy.2017.10.024.
22. Jinek M, Chylinski K, Fonfara I, Hauer M, Doudna JA, Charpentier E. A programmable dual-RNA-guided DNA endonuclease in adaptive bacterial immunity. *Science* 2012;**337**(6096):816–21. https://doi.org/10.1126/science.1225829.
23. Mojica FJM, Díez-Villaseñor C, García-Martínez J, Almendros C. Short motif sequences determine the targets of the prokaryotic CRISPR defence system. *Microbiology* 2009;**155**(3):733–40. https://doi.org/10.1099/mic.0.023960-0.
24. Brouns SJJ, Jore MM, Lundgren M, et al. Small CRISPR RNAs guide antiviral defense in prokaryotes. *Science* 2008;**321**(5891):960–4. https://doi.org/10.1126/science.1159689.
25. Makarova KS, Haft DH, Barrangou R, et al. Evolution and classification of the CRISPR-Cas systems. *Nature Reviews Microbiology* 2011;**9**(6):467–77. https://doi.org/10.1038/nrmicro2577.
26. Deltcheva E, Chylinski K, Sharma CM, et al. CRISPR RNA maturation by trans-encoded small RNA and host factor RNase III. *Nature* 2011;**471**(7340):602–7. https://doi.org/10.1038/nature09886.
27. Garneau JE, Dupuis ME, Villion M, et al. The CRISPR/cas bacterial immune system cleaves bacteriophage and plasmid DNA. *Nature* 2010;**468**(7320):67–71. https://doi.org/10.1038/nature09523.
28. Chen Y, Wang Z, Ni H, Xu Y, Chen Q, Jiang L. CRISPR/Cas9-mediated base-editing system efficiently generates gain-of-function mutations in *Arabidopsis*. *Science China Life Sciences* 2017;**60**(5):520–3. https://doi.org/10.1007/s11427-017-9021-5.
29. Abudayyeh OO, Gootenberg JS, Essletzbichler P, et al. RNA targeting with CRISPR-Cas13. *Nature* 2017;**550**(7675):280–4. https://doi.org/10.1038/nature24049.
30. Cox DBT, Gootenberg JS, Abudayyeh OO, et al. RNA editing with CRISPR-Cas13. *Science* 2017;**358**(6366):1019–27. https://doi.org/10.1126/science.aaq0180.
31. Georges F, Ray H. Genome editing of crops: a renewed opportunity for food security. *GM Crops & Food* 2017;**8**(1):1–12. https://doi.org/10.1080/21645698.2016.1270489.
32. Zaidi SSA, Tashkandi M, Mansoor S, Mahfouz MM. Engineering plant immunity: using CRISPR/Cas9 to generate virus resistance. *Frontiers in Plant Science* 2016;**7**(2016). https://doi.org/10.3389/fpls.2016.01673.
33. Zhang Q. Strategies for developing green super rice. *Proceedings of the National Academy of Sciences of the United States of America* 2007;**104**(42):16402–9. https://doi.org/10.1073/pnas.0708013104.
34. Luo M, Gilbert B, Ayliffe M. Applications of CRISPR/Cas9 technology for targeted mutagenesis, gene replacement and stacking of genes in higher plants. *Plant Cell Reports* 2016;**35**(7):1439–50. https://doi.org/10.1007/s00299-016-1989-8.
35. Shi J, Gao H, Wang H, et al. ARGOS8 variants generated by CRISPR-Cas9 improve maize grain yield under field drought stress conditions. *Plant Biotechnology Journal* 2017;**15**(2):207–16. https://doi.org/10.1111/pbi.12603.

36. Yu QH, Wang B, Li N, et al. CRISPR/Cas9-induced targeted mutagenesis and gene replacement to generate long-shelf life tomato lines. *Scientific Reports* 2017;**7**(1). https://doi.org/10.1038/s41598-017-12262-1.
37. Wang M, Lu Y, Botella JR, Mao Y, Hua K, Zhu JK. Gene targeting by homology-directed repair in rice using a geminivirus-based CRISPR/Cas9 system. *Molecular Plant* 2017;**10**(7):1007–10. https://doi.org/10.1016/j.molp.2017.03.002.
38. Hummel AW, Chauhan RD, Cermak T, et al. Allele exchange at the EPSPS locus confers glyphosate tolerance in cassava. *Plant Biotechnology Journal* 2018;**16**(7):1275–82. https://doi.org/10.1111/pbi.12868.
39. Gil-Humanes J, Wang Y, Liang Z, et al. High-efficiency gene targeting in hexaploid wheat using DNA replicons and CRISPR/Cas9. *Plant Journal* 2017;**89**(6):1251–62. https://doi.org/10.1111/tpj.13446.
40. Dahan-Meir T, Filler-Hayut S, Melamed-Bessudo C, et al. Efficient in planta gene targeting in tomato using geminiviral replicons and the CRISPR/Cas9 system. *The Plant Journal* 2018;**95**(1):5–16. https://doi.org/10.1111/tpj.13932.
41. Čermák T, Baltes NJ, Čegan R, Zhang Y, Voytas DF. High-frequency, precise modification of the tomato genome. *Genome Biology* 2015;**16**(1). https://doi.org/10.1186/s13059-015-0796-9.
42. Butler NM, Baltes NJ, Voytas DF, Douches DS. Geminivirus-mediated genome editing in potato (Solanum tuberosum l.) using sequence-specific nucleases. *Frontiers in Plant Science* 2016;**7**(2016). https://doi.org/10.3389/fpls.2016.01045.
43. Li Z, Liu ZB, Xing A, et al. Cas9-guide RNA directed genome editing in soybean. *Plant Physiology* 2015;**169**(2):960–70. https://doi.org/10.1104/pp.15.00783.
44. Butt H, Eid A, Ali Z, et al. Efficient CRISPR/Cas9-mediated genome editing using a chimeric single-guide RNA molecule. *Frontiers in Plant Science* 2017;**8**. https://doi.org/10.3389/fpls.2017.01441.
45. Endo M, Mikami M, Toki S. Biallelic gene targeting in rice. *Plant Physiology* 2016;**170**(2):667–77. https://doi.org/10.1104/pp.15.01663.
46. Sun Y, Zhang X, Wu C, et al. Engineering herbicide-resistant rice plants through CRISPR/Cas9-mediated homologous recombination of acetolactate synthase. *Molecular Plant* 2016;**9**(4):628–31. https://doi.org/10.1016/j.molp.2016.01.001.
47. Svitashev S, Schwartz C, Lenderts B, Young JK, Mark Cigan A. Genome editing in maize directed by CRISPR-Cas9 ribonucleoprotein complexes. *Nature Communications* 2016;**7**. https://doi.org/10.1038/ncomms13274.
48. Sauer NJ, Narváez-Vásquez J, Mozoruk J, et al. Oligonucleotide-mediated genome editing provides precision and function to engineered nucleases and antibiotics in plants. *Plant Physiology* 2016;**170**(4):1917–28. https://doi.org/10.1104/pp.15.01696.
49. Li J, Meng X, Zong Y, et al. Gene replacements and insertions in rice by intron targeting using CRISPR-Cas9. *Nature Plants* 2016;**2**(10). https://doi.org/10.1038/nplants.2016.139.
50. Xu M, Xu Z, Liu B, et al. Genetic variation in four maturity genes affects photoperiod insensitivity and PHYA-regulated post-flowering responses of soybean. *BMC Plant Biology* 2013;**13**(1). https://doi.org/10.1186/1471-2229-13-91.
51. Subudhi PK, De Leon TB, Tapia R, et al. Genetic interaction involving photoperiod-responsive Hd1 promotes early flowering under long-day conditions in rice. *Scientific Reports* 2018;**8**(1). https://doi.org/10.1038/s41598-018-20324-1.
52. Soyk S, Müller NA, Park SJ, et al. Variation in the flowering gene SELF PRUNING 5G promotes day-neutrality and early yield in tomato. *Nature Genetics* 2017;**49**(1):162–8. https://doi.org/10.1038/ng.3733.

53. Cai Y, Chen L, Liu X, et al. CRISPR/Cas9-mediated targeted mutagenesis of GmFT2a delays flowering time in soya bean. *Plant Biotechnology Journal* 2018;**16**(1):176–85. https://doi.org/10.1111/pbi.12758.
54. Zhou H, Liu Q, Li J, et al. Photoperiod- and thermo-sensitive genic male sterility in rice are caused by a point mutation in a novel noncoding RNA that produces a small RNA. *Cell Research* 2012;**22**(4):649–60. https://doi.org/10.1038/cr.2012.28.
55. Li M, Li X, Zhou Z, et al. Reassessment of the four yield-related genes Gn1a, DEP1, GS3, and IPA1 in rice using a CRISPR/Cas9 system. *Frontiers in Plant Science* 2016;**7**(2016). https://doi.org/10.3389/fpls.2016.00377.
56. Johnston R, Wang M, Sun Q, Sylvester AW, Hake S, Scanlon MJ. Transcriptomic analyses indicate that maize ligule development recapitulates gene expression patterns that occur during lateral organ initiation. *Plant Cell* 2014;**26**(12):4718–32. https://doi.org/10.1105/tpc.114.132688.
57. Li C, Liu C, Qi X, et al. RNA-guided Cas9 as an in vivo desired-target mutator in maize. *Plant Biotechnology Journal* 2017;**15**(12):1566–76. https://doi.org/10.1111/pbi.12739.
58. Waldie T, McCulloch H, Leyser O. Strigolactones and the control of plant development: lessons from shoot branching. *Plant Journal* 2014;**79**(4):607–22. https://doi.org/10.1111/tpj.12488.
59. Butt H, Jamil M, Wang JY, Al-Babili S, Mahfouz M. Engineering plant architecture via CRISPR/Cas9-mediated alteration of strigolactone biosynthesis. *BMC Plant Biology* 2018;**18**(1). https://doi.org/10.1186/s12870-018-1387-1.
60. Braatz J, Harloff HJ, Mascher M, Stein N, Himmelbach A, Jung C. CRISPR-Cas9 targeted mutagenesis leads to simultaneous modification of different homoeologous gene copies in polyploid oilseed rape (*Brassica napus*). *Plant Physiology* 2017;**174**(2):935–42. https://doi.org/10.1104/pp.17.00426.
61. Wang P, Zhang J, Sun L, et al. High efficient multisites genome editing in allotetraploid cotton (*Gossypium hirsutum*) using CRISPR/Cas9 system. *Plant Biotechnology Journal* 2018;**16**(1):137–50. https://doi.org/10.1111/pbi.12755.
62. Zhang J, Zhang H, Botella JR, Zhu JK. Generation of new glutinous rice by CRISPR/Cas9-targeted mutagenesis of the Waxy gene in elite rice varieties. *Journal of Integrative Plant Biology* 2018;**60**(5):369–75. https://doi.org/10.1111/jipb.12620.
63. Andersson M, Turesson H, Nicolia A, Fält AS, Samuelsson M, Hofvander P. Efficient targeted multiallelic mutagenesis in tetraploid potato (*Solanum tuberosum*) by transient CRISPR-Cas9 expression in protoplasts. *Plant Cell Reports* 2017;**36**(1):117–28. https://doi.org/10.1007/s00299-016-2062-3.
64. Hu Y, Cheng H, Tao S. The challenges and solutions for cadmium-contaminated rice in China: a critical review. *Environment International* 2016;**92–93**:515–32. https://doi.org/10.1016/j.envint.2016.04.042.
65. Ma X, Zhang Q, Zhu Q, et al. A robust CRISPR/Cas9 system for convenient, high-efficiency multiplex genome editing in monocot and dicot plants. *Molecular Plant* 2015;**8**(8):1274–84. https://doi.org/10.1016/j.molp.2015.04.007.
66. Waltz E. CRISPR-edited crops free to enter market, skip regulation. *Nature Biotechnology* 2016;**34**(6). https://doi.org/10.1038/nbt0616-582. 582-582.
67. Sun Y, Jiao G, Liu Z, et al. Generation of high-amylose rice through CRISPR/Cas9-mediated targeted mutagenesis of starch branching enzymes. *Frontiers in Plant Science* 2017;**8**. https://doi.org/10.3389/fpls.2017.00298.
68. Shan Q, Zhang Y, Chen K, Zhang K, Gao C. Creation of fragrant rice by targeted knockout of the OsBADH2 gene using TALEN technology. *Plant Biotechnology Journal* 2015;**13**(6):791–800. https://doi.org/10.1111/pbi.12312.

69. Sánchez-León S, Gil-Humanes J, Ozuna CV, et al. Low-gluten, nontransgenic wheat engineered with CRISPR/Cas9. *Plant Biotechnology Journal* 2018;**16**(4):902–10. https://doi.org/10.1111/pbi.12837.
70. Ito Y, Nishizawa-Yokoi A, Endo M, Mikami M, Toki S. CRISPR/Cas9-mediated mutagenesis of the RIN locus that regulates tomato fruit ripening. *Biochemical and Biophysical Research Communications* 2015;**467**(1):76–82. https://doi.org/10.1016/j.bbrc.2015.09.117.
71. Li R, Fu D, Zhu B, Luo Y, Zhu H. CRISPR/Cas9-mediated mutagenesis of lncRNA1459 alters tomato fruit ripening. *The Plant Journal* 2018;**94**(3):513–24. https://doi.org/10.1111/tpj.13872.
72. Okuzaki A, Ogawa T, Koizuka C, et al. CRISPR/Cas9-mediated genome editing of the fatty acid desaturase 2 gene in *Brassica napus*. *Plant Physiology and Biochemistry* 2018;**131**:63–9. https://doi.org/10.1016/j.plaphy.2018.04.025.
73. Jiang WZ, Henry IM, Lynagh PG, Comai L, Cahoon EB, Weeks DP. Significant enhancement of fatty acid composition in seeds of the allohexaploid, *Camelina sativa*, using CRISPR/Cas9 gene editing. *Plant Biotechnology Journal* 2017;**15**(5):648–57. https://doi.org/10.1111/pbi.12663.
74. Morineau C, Bellec Y, Tellier F, et al. Selective gene dosage by CRISPR-Cas9 genome editing in hexaploid *Camelina sativa*. *Plant Biotechnology Journal* 2017;**15**(6):729–39. https://doi.org/10.1111/pbi.12671.
75. Li R, Li R, Li X, et al. Multiplexed CRISPR/Cas9-mediated metabolic engineering of γ-aminobutyric acid levels in Solanum lycopersicum. *Plant Biotechnology Journal* 2018;**16**(2):415–27. https://doi.org/10.1111/pbi.12781.
76. Nonaka S, Arai C, Takayama M, Matsukura C, Ezura H. Efficient increase of Γ-aminobutyric acid (GABA) content in tomato fruits by targeted mutagenesis. *Scientific Reports* 2017;**7**(1). https://doi.org/10.1038/s41598-017-06400-y.
77. Li X, Wang Y, Chen S, et al. Lycopene is enriched in tomato fruit by CRISPR/Cas9-mediated multiplex genome editing. *Frontiers in Plant Science* 2018;**9**. https://doi.org/10.3389/fpls.2018.00559.
78. Nakayasu M, Akiyama R, Lee HJ, et al. Generation of α-solanine-free hairy roots of potato by CRISPR/Cas9 mediated genome editing of the St16DOX gene. *Plant Physiology and Biochemistry* 2018;**131**:70–7. https://doi.org/10.1016/j.plaphy.2018.04.026.
79. Wang L, Chen L, Li R, et al. Reduced drought tolerance by CRISPR/Cas9-Mediated SlMAPK3 mutagenesis in tomato plants. *Journal of Agricultural and Food Chemistry* 2017;**65**(39):8674–82. https://doi.org/10.1021/acs.jafc.7b02745.
80. Lou D, Wang H, Liang G, Yu D. OsSAPK2 confers abscisic acid sensitivity and tolerance to drought stress in rice. *Frontiers in Plant Science* 2017;**8**. https://doi.org/10.3389/fpls.2017.00993.
81. Jia Y, Ding Y, Shi Y, Zhang X, Gong Z, Yang S. The cbfs triple mutants reveal the essential functions of CBF s in cold acclimation and allow the definition of CBF regulons in *Arabidopsis*. *New Phytologist* 2016;**212**(2):345–53. https://doi.org/10.1111/nph.14088.
82. Abdelrahman M, Al-Sadi AM, Pour-Aboughadareh A, Burritt DJ, Tran LSP. Genome editing using CRISPR/Cas9–targeted mutagenesis: an opportunity for yield improvements of crop plants grown under environmental stresses. *Plant Physiology and Biochemistry* 2018;**131**:31–6. https://doi.org/10.1016/j.plaphy.2018.03.012.
83. Tang L, Mao B, Li Y, et al. Knockout of OsNramp5 using the CRISPR/Cas9 system produces low Cd-accumulating indica rice without compromising yield. *Scientific Reports* 2017;**7**(1). https://doi.org/10.1038/s41598-017-14832-9.

84. Nieves-Cordones M, Mohamed S, Tanoi K, et al. Production of low-Cs+ rice plants by inactivation of the K+ transporter OsHAK1 with the CRISPR-Cas system. *Plant Journal* 2017;**92**(1):43–56. https://doi.org/10.1111/tpj.13632.
85. Wang FZ, Chen MX, Yu LJ, et al. OsARM1, an R2R3 MYB Transcription factor, is involved in regulation of the response to arsenic stress in rice. *Frontiers in Plant Science* 2017;**8**. https://doi.org/10.3389/fpls.2017.01868.
86. Miao C, Xiao L, Hua K, et al. Mutations in a subfamily of abscisic acid recepto genes promote rice growth and productivity. *Proceedings of the National Academy of Sciences of the United States of America* 2018;**115**(23):6058–63. https://doi.org/10.1073/pnas.1804774115.
87. Bryan GT, Wu KS, Farrall L, et al. A single amino acid difference distinguishes resistant and susceptible alleles of the rice blast resistance gene Pi-ta. *Plant Cell* 2000;**12**(11):2033–45. https://doi.org/10.1105/tpc.12.11.2033.
88. Frye CA, Tang D, Innes RW. Negative regulation of defense responses in plants by a conserved MAPKK kinase. *Proceedings of the National Academy of Sciences of the United States of America* 2001;**98**(1):373–8. https://doi.org/10.1073/pnas.98.1.373.
89. Zhang Y, Bai Y, Wu G, et al. Simultaneous modification of three homoeologs of TaEDR1 by genome editing enhances powdery mildew resistance in wheat. *The Plant Journal* 2017;**91**(4):714–24. https://doi.org/10.1111/tpj.13599.
90. Humphry M, Consonni C, Panstruga R. mlo-based powdery mildew immunity: silver bullet or simply non-host resistance? *Molecular Plant Pathology* 2006;**7**(6):605–10. https://doi.org/10.1111/j.1364-3703.2006.00362.x.
91. Nekrasov V, Wang C, Win J, Lanz C, Weigel D, Kamoun S. Rapid generation of a transgene-free powdery mildew resistant tomato by genome deletion. *Scientific Reports* 2017;**7**(1). https://doi.org/10.1038/s41598-017-00578-x.
92. Wang Y, Cheng X, Shan Q, et al. Simultaneous editing of three homoeoalleles in hexaploid bread wheat confers heritable resistance to powdery mildew. *Nature Biotechnology* 2014;**32**(9):947–51. https://doi.org/10.1038/nbt.2969.
93. Piffanelli P, Ramsay L, Waugh R, et al. A barley cultivation-associated polymorphism conveys resistance to powdery mildew. *Nature* 2004;**430**(7002):887–91. https://doi.org/10.1038/nature02781.
94. Jung J, Won SY, Suh SC, et al. The barley ERF-type transcription factor HvRAF confers enhanced pathogen resistance and salt tolerance in *Arabidopsis*. *Planta* 2007;**225**(3):575–88. https://doi.org/10.1007/s00425-006-0373-2.
95. Wang F, Wang C, Liu P, et al. Enhanced rice blast resistance by CRISPR/Cas9-targeted mutagenesis of the ERF transcription factor gene OsERF922. *PLOS one* 2016;**11**(4):e0154027. https://doi.org/10.1371/journal.pone.0154027.
96. Wang X, Tu M, Wang D, et al. CRISPR/Cas9-mediated efficient targeted mutagenesis in grape in the first generation. *Plant Biotechnology Journal* 2018;**16**(4):844–55. https://doi.org/10.1111/pbi.12832.
97. Hu Y, Zhang J, Jia H, et al. Lateral organ boundaries 1 is a disease susceptibility gene for citrus bacterial canker disease. *Proceedings of the National Academy of Sciences of the United States of America* 2014;**111**(4):E521–9. https://doi.org/10.1073/pnas.1313271111.
98. Jia H, Zhang Y, Orbović V, et al. Genome editing of the disease susceptibility gene CsLOB1 in citrus confers resistance to citrus canker. *Plant Biotechnology Journal* 2017;**15**(7):817–23. https://doi.org/10.1111/pbi.12677.
99. Lellis AD, Kasschau KD, Whitham SA, Carrington JC. Loss-of-susceptibility mutants of *Arabidopsis thaliana* reveal an essential role for eIF(iso)4E during potyvirus infection. *Current Biology* 2002;**12**(12):1046–51. https://doi.org/10.1016/S0960-9822(02)00898-9.

100. Zhou J, Peng Z, Long J, et al. Gene targeting by the TAL effector PthXo2 reveals cryptic resistance gene for bacterial blight of rice. *The Plant Journal* 2015;**82**(4):632–43. https://doi.org/10.1111/tpj.12838.
101. Pyott DE, Sheehan E, Molnar A. Engineering of CRISPR/Cas9-mediated potyvirus resistance in transgene-free *Arabidopsis* plants. *Molecular plant pathology* 2016;**17**(8):1276–88. https://doi.org/10.1111/mpp.12417.
102. Chandrasekaran J, Brumin M, Wolf D, et al. Development of broad virus resistance in non-transgenic cucumber using CRISPR/Cas9 technology. *Molecular plant pathology* 2016;**17**(7):1140–53. https://doi.org/10.1111/mpp.12375.
103. Iqbal Z, Sattar MN, Shafiq M. CRISPR/Cas9: a tool to circumscribe cotton leaf curl disease. *Frontiers in Plant Science* 2016;**7**(2016). https://doi.org/10.3389/fpls.2016.00475.
104. Macovei A, Sevilla NR, Cantos C, et al. Novel alleles of rice eIF4G generated by CRISPR/Cas9-targeted mutagenesis confer resistance to Rice tungro spherical virus. *Plant Biotechnology Journal* 2018;**16**(11):1918–27. https://doi.org/10.1111/pbi.12927.
105. Ali Z, Ali S, Tashkandi M, Zaidi SSEA, Mahfouz MM. CRISPR/Cas9-Mediated immunity to geminiviruses: differential interference and evasion. *Scientific Reports* 2016;**6**. https://doi.org/10.1038/srep26912.
106. Fondong VN, Nagalakshmi U, Dinesh-Kumar SP. Novel functional genomics approaches: a promising future in the combat against plant viruses. *Phytopathology* 2016;**106**(10):1231–9. https://doi.org/10.1094/PHYTO-03-16-0145-FI.
107. Chen K, Wang Y, Zhang R, Zhang H, Gao C. CRISPR/Cas genome editing and precision plant breeding in agriculture. *Annual Review of Plant Biology* 2019;**70**:667–97.
108. Abdullah E, Idris A, Saparon A. Establishing RNA virus resistance in plants by harnessing CRISPR immune system. *ARPN Journal of Engineering and Applied Science* 2017;**12**:3218–21.
109. Yang YF, Zhu GN, Li R, Yan SJ, Fu DQ, Zhu BZ, Tian HQ, Luo YB, Zhu HL. The RNA editing factor SlORRM4 is required for normal fruit ripening in tomato. *Plant Physiology* 2017;**175**:1690–702.
110. Rodríguez-Leal D, Lemmon ZH, Man J, Bartlett ME, Lippman ZB. Engineering quantitative trait variation for crop improvement by genome editing. *Cell* 2017;**171**(2):470–80. https://doi.org/10.1016/j.cell.2017.08.030. e8.
111. Malnoy M, Viola R, Jung MH, Koo OJ, Kim S, Kim JS, Velasco R, Nagamangala Kanchiswamy C. DNA-Free Genetically Edited Grapevine and Apple Protoplast Using CRISPR/Cas9 Ribonucleoproteins. *Frontiers in Plant Science* 2016;**7**:1904.
112. Osakabe Y, Liang Z, Ren C, Nishitani C, Osakabe K, Wada M, Komori S, Malnoy M, Velasco R, Poli M, Jung MH, Koo OJ, Viola R, Nagamangala Kanchiswamy C. CRISPR-Cas9-mediated genome editing in apple and grapevine. *Nature Protocols* 2018;**13**:2844–63.
113. Ren C, Liu X, Zhang Z, Wang Y, Duan W, Li S, Liang Z. CRISPR/Cas9-mediated efficient targeted mutagenesis in Chardonnay (*Vitis vinifera* L.). *Scientific Reports* 2016;**6**:32289.
114. Kaur N, Alok A, Shivani Kaur N, Pandey P, Awasthi P, Tiwari S. CRISPR/Cas9-mediated efficient editing in phytoene desaturase (PDS) demonstrates precise manipulation in banana cv. Rasthali genome. *Functional Integral Genomics* 2018;**18**:89–99.
115. Zhou J, Wang G, Liu Z. Efficient genome-editing of wild strawberry genes, vector development, and validation. *Plant Biotechnology Journal* 2018;**16**:1868–77.
116. Connorton JM, Jones ER, Rodríguez-Ramiro I, Fairweather-Tait S, Uauy C, Balk J. Wheat vacuolar iron transporter TaVIT2 transports Fe and Mn and is effective for biofortification. *Plant Physiology* 2017;**174**(4):2434–44.

117. Zhang X, Wang J, Huang J, Lan H, Wang C, Yin C, Wu Y, Tang H, Qian Q, Li J, Zhang H. Rare allele of OsPPKL1 associated with grain length causes extra-large grain and a significant yield increase in rice. *Proceedings of the National Academy of Sciences United States of America* 2012;**109**(52):21534−9.
118. Xie KB, Yang YN. RNA-guided genome editing in plants using a CRISPRCas system. *Molecular plant* 2013;**6**:1975−83.
119. Zhang A, Liu Y, Wang F, Li T, Chen Z, Kong D, Bi J, Zhang F, Luo X, Wang J, Tang J, Yu X, Liu G, Luo L. Enhanced rice salinity tolerance via CRISPR/Cas9-targeted mutagenesis of the OsRR22 gene. *Molecular Breed* 2019;**39**:47.
120. Mahas A, Neal Stewart C, Mahfouz MM. Harnessing CRISPR/Cas systems for programmable transcriptional and post-transcriptional regulation. *Biotechnology Advances* 2018;**36**(1):295−310. https://doi.org/10.1016/j.biotechadv.2017.11.008.
121. Meyer RS, Purugganan MD. Evolution of crop species: genetics of domestication and diversification. *Nature Reviews Genetics* 2013;**14**(12):840−52. https://doi.org/10.1038/nrg3605.
122. Olsen KM, Wendel JF. A bountiful harvest: genomic insights into crop domestication phenotypes. *Annual Review of Plant Biology* 2013;**64**:47−70. https://doi.org/10.1146/annurev-arplant-050312-120048.
123. Wittkopp PJ, Kalay G. Cis-regulatory elements: molecular mechanisms and evolutionary processes underlying divergence. *Nature Reviews Genetics* 2012;**13**(1):59−69. https://doi.org/10.1038/nrg3095.
124. Zhang H, Si X, Ji X, Fan R, Liu J, Chen K, Wang D, Gao C. Genome editing of upstream open reading frames enables translational control in plants. *Nat Biotechnol.* 2018;**36**(9):894−8.
125. Konermann S, Brigham MD, Trevino AE, et al. Genome-scale transcriptional activation by an engineered CRISPR-Cas9 complex. *Nature* 2015;**517**(7536):583−8. https://doi.org/10.1038/nature14136.
126. Lu Y, Ye X, Guo R, et al. Genome-wide targeted mutagenesis in rice using the CRISPR/Cas9 system. *Molecular Plant* 2017;**10**(9):1242−5. https://doi.org/10.1016/j.molp.2017.06.007.
127. Shalem O, Sanjana NE, Zhang F. High-throughput functional genomics using CRISPR-Cas9. *Nature Reviews Genetics* 2015;**16**(5):299−311. https://doi.org/10.1038/nrg3899.
128. Jacobs TB, Zhang N, Patel D, Martin GB. Generation of a collection of mutant tomato lines using pooled CRISPR libraries. *Plant Physiology* 2017;**174**(4):2023−37. https://doi.org/10.1104/pp.17.00489.
129. Meng X, Yu H, Zhang Y, et al. Construction of a genome-wide mutant library in rice using CRISPR/Cas9. *Molecular Plant* 2017;**10**(9):1238−41. https://doi.org/10.1016/j.molp.2017.06.006.
130. Zong Y, Song Q, Li C, et al. Efficient c-to-t base editing in plants using a fusion of ncas9 and human apobec3a. *Nature Biotechnology* 2018;**36**(10):950. https://doi.org/10.1038/nbt.4261.
131. Shimatani Z, Kashojiya S, Takayama M, et al. Targeted base editing in rice and tomato using a CRISPR-Cas9 cytidine deaminase fusion. *Nature Biotechnology* 2017;**35**(5):441−3. https://doi.org/10.1038/nbt.3833.
132. Tian S, Jiang L, Cui X, et al. Engineering herbicide-resistant watermelon variety through CRISPR/Cas9-mediated base-editing. *Plant Cell Reports* 2018;**37**(9):1353−6. https://doi.org/10.1007/s00299-018-2299-0.

133. Li C, Zong Y, Wang Y, et al. Expanded base editing in rice and wheat using a Cas9-adenosine deaminase fusion. *Genome Biology* 2018;**19**(1). https://doi.org/10.1186/s13059-018-1443-z.
134. Kang BC, Yun JY, Kim ST, et al. Precision genome engineering through adenine base editing in plants. *Nature Plants* 2018;**4**(7):427–31. https://doi.org/10.1038/s41477-018-0178-x.
135. Xue C, Zhang H, Lin Q, Fan R, Gao C. Manipulating mRNA splicing by base editing in plants. *Science China Life Sciences* 2018;**61**(11):1293–300. https://doi.org/10.1007/s11427-018-9392-7.
136. Cong L, Ran FA, Cox D, et al. Multiplex genome engineering using CRISPR/Cas systems. *Science* 2013;**339**(6121):819–23. https://doi.org/10.1126/science.1231143.
137. Gilbert LA, Larson MH, Morsut L, et al. XCRISPR-mediated modular RNA-guided regulation of transcription in eukaryotes. *Cell* 2013;**154**(2):442. https://doi.org/10.1016/j.cell.2013.06.044.
138. Nishida K, Arazoe T, Yachie N, et al. Targeted nucleotide editing using hybrid prokaryotic and vertebrate adaptive immune systems. *Science* 2016;**353**(6305). https://doi.org/10.1126/science.aaf8729.
139. Li J, Zhang H, Si X, et al. Generation of thermosensitive male-sterile maize by targeted knockout of the ZmTMS5 gene. *Journal of Genetics and Genomics* 2017;**44**(9):465–8. https://doi.org/10.1016/j.jgg.2017.02.002.
140. Zhou H, He M, Li J, et al. Development of commercial thermo-sensitive genic male sterile rice accelerates hybrid rice breeding using the CRISPR/Cas9-mediated TMS5 editing system. *Scientific Reports* 2016;**6**. https://doi.org/10.1038/srep37395.
141. Singh M, Kumar M, Albertsen MC, Young JK, Cigan AM. Concurrent modifications in the three homeologs of Ms45 gene with CRISPR-Cas9 lead to rapid generation of male sterile bread wheat (*Triticum aestivum* L.). *Plant Molecular Biology* 2018;**97**(4–5):371–83. https://doi.org/10.1007/s11103-018-0749-2.
142. Li Q, Zhang D, Chen M, et al. Development of japonica photo-sensitive genic male sterile rice lines by editing carbon starved anther using CRISPR/Cas9. *Journal of Genetics and Genomics* 2016;**43**(6):415–9. https://doi.org/10.1016/j.jgg.2016.04.011.
143. Xie Y, Xu P, Huang J, et al. Interspecific hybrid sterility in rice is mediated by OgTPR1 at the S1 locus encoding a peptidase-like protein. *Molecular Plant* 2017;**10**(8):1137–40. https://doi.org/10.1016/j.molp.2017.05.005.
144. Xie Y, Niu B, Long Y, et al. Suppression or knockout of SaF/SaM overcomes the Sa -mediated hybrid male sterility in rice. *Journal of Integrative Plant Biology* 2017;**59**(9):669–79. https://doi.org/10.1111/jipb.12564.
145. Shen R, Wang L, Liu X, et al. Genomic structural variation-mediated allelic suppression causes hybrid male sterility in rice. *Nature Communications* 2017;**8**(1). https://doi.org/10.1038/s41467-017-01400-y.
146. Khanday I, Skinner D, Yang B, Mercier R, Sundaresan V. A male-expressed rice embryogenic trigger redirected for asexual propagation through seeds. *Nature* 2019;**565**(7737):91–5. https://doi.org/10.1038/s41586-018-0785-8.
147. Wang C, Liu Q, Shen Y, et al. Clonal seeds from hybrid rice by simultaneous genome engineering of meiosis and fertilization genes. *Nature Biotechnology* 2019;**37**(3):283–6. https://doi.org/10.1038/s41587-018-0003-0.
148. Li X, Zhou W, Ren Y, et al. High-efficiency breeding of early-maturing rice cultivars via CRISPR/Cas9-mediated genome editing. *Journal of Genetics and Genomics* 2017;**44**(3):175–8. https://doi.org/10.1016/j.jgg.2017.02.001.

149. Dong L, Li L, Liu C, Liu C, Shuaifeng G, et al. Genome editing and double fluorescence proteins enable robust maternal haploid induction and identification in maize. *Molecular Plant* 2018;**11**:1214–7.
150. Yao L, Zhang Y, Liu C, Liu Y, Wang Y, et al. OsMATL mutation induces haploid seed formation in indica rice. *Nature Plants* 2018;**4**:530–3.
151. Ye M, Peng Z, Tang D, et al. Generation of self-compatible diploid potato by knockout of S-RNase. *Nature Plants* 2018;**4**(9):651–4. https://doi.org/10.1038/s41477-018-0218-6.
152. Waltz E. Gene-edited CRISPR mushroom escapes US regulation. *Nature* 2016;**532**(7599):293. https://doi.org/10.1038/nature.2016.19754.
153. Wolt JD, Wang K, Yang B. The regulatory status of genome-edited crops. *Plant Biotechnology Journal* 2016;**14**(2):510–8. https://doi.org/10.1111/pbi.12444.

Chapter 6

# Gene drives focusing on agriculture

## 1. Introduction

The world's most important crop cultivation challenges such as herbicide tolerance and insect resistance have been addressed by the introduction of transgenic plant technology.[1] These days, over 90% of all the cultivated transgenic crops worldwide are tolerant to herbicides.[2] This means that these plants can be sprayed with herbicides without damaging the crop itself while at the same time retarding the growth of the neighboring weeds. Another attribute of transgenic crops is insect resistance. The *Bacillus thuringiensis* Bt insecticidal protein is expressed in transgenic plants the world over to preclude insect infestation. When insects ingest these transgenic crops expressing the Bt toxin, they are killed specifically, and nontarget insects in the vicinity of these crops, but are not pests, are left unharmed.[3]

On the other hand, cisgenic crops do not possess a transgene originating from another species but instead contain a gene from a sexually compatible version of the same plant[4]; for instance, from a Chinese chestnut that is blight resistant to the American chestnut that is vulnerable to blight. Thus, cisgenesis is used to create plants expressing genes from plants that are closely related and are developed to retrieve useful genes, which have been lost during several years of cultivation using conventional breeding methods. A good example of cisgenesis application includes the Wheat Stem Rust initiative that has been developed to engineer wheat crops harboring multiple resistance genes against the fungal pathogen Ugg99.[5]

Additionally, the RNAi-driven gene silencing technology has been increasingly employed in genetic engineering of commercially important crop species. In this technology, crops are engineered to synthesize the antisense RNA version of a given gene within the plant genome or is present as part of the genome of the invading pathogen, such as a virus. Consequently, the target gene expression is blocked by the gene silencing mechanism. Papaya crops resistant to the Papaya ringspot virus (RSV) have been developed using antisense RNA expression targeted against the viral genome. Gonsalves, 1998[6] reports that this technology has successfully revived the papaya industry in

**Agricultural Biotechnology. https://doi.org/10.1016/B978-0-323-88476-1.00002-3**

Hawaii. Further, China has employed this technology to save its small papaya crop sector. Taken together, although the RSV-resistant papaya has not proved to be used in agriculture in many countries owing to political resistance, agriculturists elsewhere have disseminated this technology informally and have found it efficient in combating the fatal RSV virus.[7]

The technologies elaborated as above have been adopted worldwide, and yet, they pose several caveats, such as protracted delays from field trials lasting several years, as well as legal challenges, while being invariably complicated. A notable example of this is the golden rice technology developed to address vitamin A deficiency, wherein it has faced several challenges in breeding.[4] Presently, crops addressing protection against biotic stresses such as pests and weeds have met with the most success in cultivation. However, the recent advent of genome editing has enabled agriculturists to look into a wider, more rewarding frontier.

Hamilton (1967)[8] first used the term "gene drive," which denotes the concept of the spread of genetic elements through populations in a non-Mendelian fashion (over 50% rate) to the next generation.[9] Thus far, selfish DNA elements or natural gene drives such as meiotic drive elements, transposable elements, B chromosomes, endonucleases, *Wolbachia* endosymbionts, and homing endonuclease genes have been defined.[9–13] The great potential of gene drives lies in the idea that they are responsible for the spread of genetic elements across populations even if they are harmful to the organism.

Vast developments in molecular biology have enabled such genetic systems to be engineered in unique and novel ways. Methods like the application of endonucleases, chromosome inversions, and lately, the use of CRISPR/Cas9-based strategies have facilitated the manipulation of genomes toward biasing inheritance away from the usual Mendelian patterns. This has the capability to develop novel gene drive tools in different organisms.[14–16] The CRISPR/Cas9 synthetic gene drive approaches could be engineered to develop suppression and modification drives.[14,17] Suppression drives serve to diminish or eliminate populations of target organisms, while the modification or replacement drives enable the spread of genetic modifications through populations.

Gene drives generate the most basic changes in organisms, augmenting both probable advantages and risks. For instance, to stop the spread of diseases such as dengue, malaria, and Zika vectored by mosquitos, gene drives driven through the genome editing technology have been adopted. The gene drive concept was first introduced by an entomologist in the 1960s who proposed that breeding programs could be engineered such that male offspring of mosquitoes could be favorably produced due to the presence of a male-generating factor expressed in the genomes of some of the male mosquitoes. Therefore, when such male mosquitoes carrying this male generating factor are released into the natural environment, this could change the sex ratio of the mosquito population, resulting in the reduction in the number of females to a level below that necessary for efficacious disease transmission.[18,19] Nevertheless, the introduction of genome editing technologies using CRISPR/Cas9 mechanisms has offered unprecedented contingencies to decrease mosquito populations.[20]

Gene drives operate by establishing a program of biased inheritance such that the capability of a gene or genetic element to transmit from parent to offspring via sexual reproduction becomes increased. Therefore, the existence of this genetic element enhances in frequency and disseminates from a given generation to the next until the predominant majority or all members of the wild population in the respective species harbor the same genetic element. Contrary to classical Mendelian inheritance wherein every offspring has a 50% probability of inheriting a given gene from one of their parents, the precept of gene drive technology dictates that a majority or all of the offspring would inherit a given genetic trait under the control of the drive. For developing the genome-edited mosquitoes discussed above, genes conferring a recessive female sterility characteristic were edited. In this instance, CRISPR/Cas9-edited gene drive constructs targeted and edited each sterility gene as well as its homolog such that they were incorporated into the female sterility gene locus. The outcome of this approach was a great increase in the number of sterile females. Modeling studies on mosquito populations demonstrated that this gene drive could be utilized to efficiently target female reproduction (since only female mosquitoes bite humans).[21] This gene editing technology was further used to edit mosquito genomes such that they are unable to spread infectious diseases.[22]

## 2. Gene drives and genome editing

The use of gene drive mechanisms could inhibit or eliminate several invasive species that pose a threat to biodiversity, eradicate weeds, or change pathogens capable of damaging crops or transmitting diseases. The gene drive technologies could introduce novel traits into extant populations and could save plant species that are endangered or revive extinct species. As an example, toward protection of the native plant species' biodiversity in the United States, gene drives are being formulated to inhibit the dissemination of Centaurea maculosa, a nonindigenous spotted knapweed. This weed has been known to damage natural ecosystems and cause soil erosion. The unbridled spread of knapweed could be controlled using gene drives directed toward suppression of the gene determining the sex of the plants in a strategy similar to the gene drives controlling mosquito populations employed in protecting humans from mosquito-borne diseases. Such a measure could create an imbalance in the sex ratio of the plants, thus leading to a population crash.[23] Nevertheless, in comparison to mosquitoes, knapweed exhibits slow growth, and it is not yet clear how parameters such as the distance and rate of pollen spread in the wild environment would impact the gene drive process.[24]

Another application of gene drives in the plant system would be to eradicate pigweed from cultivable fields. The pigweed reproduces at a rapid rate and has developed resistance against the widely used herbicide, glyphosate.

This resistant trait could be reversed by employing the gene drive strategy, thus making the pigweed susceptible to glyphosate. On the other hand, a suppression gene drive that generates a biased sex ratio could be devised to result in the collapse of the pigweed population.[24]

Genome editing strategies can be adopted in conjunction with gene drive measures to eliminate weed species and additionally to control and eliminate pest populations. Gene drive crops that do not serve as hosts for insects and microbial populations such as viruses, bacteria, and fungi could be created. As researchers gain further insights into the nature of specific proteins determining pathogen–host interactions, the use of gene drives to destroy these interactions could facilitate the growth of future generations of agricultural crops that would no longer support the growth of such pathogens.

However, despite these developments in the gene drive technology, gene drive measures will not function on invasive species of plants that do not reproduce sexually or those with a very slow rate of reproduction. Additionally, these gene drives may possibly have to be applied repeatedly over time as plant species undergo natural selection and subsequently lose the genetically introduced trait.[25] Also, a few individual plants in a particular population could develop resistance to the gene drive, leading to the re-emergence of a plant population that is recalcitrant to its further use. Gene drive measures can permanently alter entire animal or plant communities within a comparably short time period, for outcomes better or worse. Unforeseen context that could lead to irreversible consequences involving destabilization of current ecosystems brings forth pause to the idea of using gene drives in the absence of mandatory social contracts with all the relevant stakeholders across the world.

## 3. Pest management through genetic control

Application of synthetic pesticides has become a dominant practice to control pest species such as weeds, pathogens, as well as disease-vectoring and herbivorous insects. This has been by and large successful in diminishing losses in crop yield. However, the widespread emergence of resistance to pesticides[26–28] has evolved, and this has resulted in growing concerns about the impacts of pesticide use on human health in addition to off-target environmental effects. Therefore, there is a compelling need to adopt alternative strategies for pest control. In this light, genetic control of pest species through manipulation of their genomes to diminish their fitness becomes an appealing and increasingly achievable goal.

### Control of agricultural pests through indirect genetic engineering measures

Practices such as plant breeding, cultivar selection, or genetic engineering of the crop germplasm have been variously employed to control agricultural pests. Several crop species and their wild counterparts contain genetic variation for resistance to pathogens[29] and insect pests[30] Crop varieties could also differ in their combative[31] and allelopathic potential[32] against weeds. Where genetic traits which confer improved tolerance or even resistance to pathogens and pest species have been bred out of varieties of modern crops, these traits could be re-established via marker-assisted breeding, provided they do not have harmful nutritional and agricultural impacts.

Modern agricultural technologies such as genome editing have been driven by the evolution of transgenic crops expressing novel pest resistance and herbicide resistance traits, and these crops have become increasingly commercialized.[33] Transgenic crops harboring herbicide resistance have transformed weed management by facilitating the application of nonselective, broad-spectrum herbicides for controlling weeds in crops.[34] Also, several agricultural crops have been engineered to express insecticidal proteins toward control of insect pests, thus removing or diminishing the necessity for insecticide applications. Nevertheless, despite some successes, these crops face notable difficulties in proliferating the engineered genes through widely disseminated natural populations.

### Control of insect pests through direct genetic engineering measures

For the control of insect pests through direct genetic manipulation, self-limiting and self-sustaining approaches have been adopted.[35] The self-limiting technology relies on the introduction of inundative amounts of modified insects that mate with existing local populations, causing consequent suppression of the insect populations through lethal matings. On the other hand, self-sustaining approaches depend on strategies driving genetic alterations in the insect populations. Mostly, these "driven" genetic modifications diminish the fitness and/or status of the pests and enhance them in frequency through non-Mendalian inheritance over multiple generations with the potential for population replacement. Fig. 6.1 shows the mechanism of weed control using the CRISPR/Cas9 technology.

## 4. Controlling weeds through the use of CRISPR/Cas9 gene drives

The tenet for suppression of populations of weed species using CRISPR/Cas9 rests with the presumption that these gene drives can be applied to introduce and disseminate a fitness load capable of limiting the establishment, plenitude, dispersion, endurance, and impact of weed populations (Fig. 6.1). This could be accomplished through genetic manipulations targeting distinct traits of the weeds determining competitiveness, persistence, seed dormancy, morphology, and phenology.[35] However, this is limited by our insufficient comprehension of the molecular biology of weed traits. Despite this, it is possible that the Rht-1

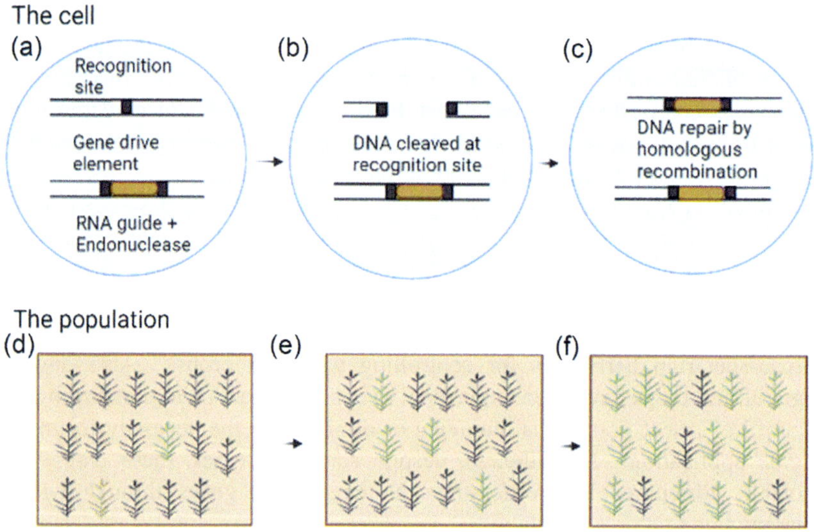


**FIGURE 6.1 Agricultural weed control through the CRISPR-Cas9 gene drive system.** Following an efficacious mating between the wild-type weeds and the genetically modified introduced weed species, plants carrying the gene drive modification would be generated in its heterozygous state (A). The Cas9 nuclease directed by the guide RNA will then cleave the DNA on the wild type chromosome at the region of the recognition site (B). Subsequently, the drive chromosome would be used as a template to repair the cut via homologous recombination. This would lead to the generation of plants homozygous for the state of the gene drive (C). At the population level, the weeds possessing the engineered gene drive (blue plants) would be introduced into the wild weed population (green plants) (D), and this would spread over generations (E, F) till weeds harboring the drive allele would dominate the weed population. *From Neve P. Gene drive systems: do they have a place in agricultural weed management?* Pest Management Science *2018;**74**(12):2671–2679. https://doi.org/10.1002/ps.5137.*

dwarfing gene homologs used in wheat breeding[36] could be found in related grass weed species. If these genes could be steered through the weed populations, they may diminish the competence of the weeds.

Likewise, investigations in weedy rice have brought into light the genetic basis of the developed traits such as seed dormancy and seed-shattering that play roles in the endurance and transmission of wild rice, thus providing yet another molecular target for attempts to diminish weed fitness in agroecosystems.[37] Also, meiotic drive systems biasing the sex ratio of dioecious weedy species like Amaranthus spp. could be adopted. Other targets to reduce plant reproduction and fertility could focus on genetic engineering measures interfering with gametogenesis to curb production of pollen or ovules, leading to curtailment and/or biases in gamete generation. Weed genomics, an emerging field in genetic engineering, has identified the value of molecular genetics research into the "weediness" traits to decipher the evolution of those traits, and this could serve as a probable novel source of gene variation toward crop improvement.[38] Given the promise of weed curtailment through direct genetic control, these measures afford more impetus if gene drive systems can be used to modify or knockout weedy traits in the wild.

For sensitizing drives, population suppression is not dependent on the drive element itself, but instead on consequent management, the applicability of which can be regulated in space and time.[35] By virtue of the above advantages, sensitizing drives can be constructed to disseminate genetically engineered genes lacking any inherent negative impacts on fitness of the organism, enabling their spread, but significantly, making sure that suppression of the population is subject to specific management measures. Obviously, the major use of sensitizing drives for management of weed populations would be to revert populations that are herbicide-resistant back to herbicide sensitivity. Of note is the fact that most of the herbicide targets are well-defined and several mutations conferring resistance have been elucidated,[27] due to which this reversal is technically feasible.

Driving the susceptibility traits through populations of weeds to the extent at which efficient control of formerly resistant populations would be feasible but would take many generations, and at this time, the target herbicide cannot be applied, since it will eliminate plants possessing the very gene drive elements propagating this reversion. Resistant weed plants, even when occurring at low frequencies in weed populations following an effective gene drive event, could lead to the rapid emergence of resistance when herbicide use is recommenced.

The ensuing management of new sensitive weed populations would have to consider carefully constructed resistance management methodologies. Nevertheless, this approach is complicated by the recognition that several of the herbicide resistance mechanisms are determined by complex, likely polygenic systems whose genetic architecture has yet to be fully characterized.[39] Using a combination of synthetic chemistry as well as structural and synthetic biology, sensitizing gene drives could be applied to alter essential, highly conserved plant genes, thereby sensitizing them to specific herbicide molecules.[15] Such edited genes would have to sustain their regular enzymatic activity to efficiently spread across populations. The above strategies would help design specifically targeted chemical interventions while limiting any nontarget and environmental impacts.

In recent times, technologies for weed control by targeting and destroying mature weed seeds that circumscribe the replenishment of seed banks have been engineered.[35] Such strategies depend on the retention of the weed seeds on mature weed plants during the time of crop harvest, so as to facilitate their culling and destruction. However, for *Alopecurus myosuroides,* the main annual grass weed in north-west Europe, early seed shattering circumscribes the promise of the above approaches. Targeting the seed-shattering gene loci using gene drives leading to seed retention on parent weed plants until the crop harvest could be integrated with harvest weed control strategies to afford a novel control method for this and other recalcitrant weed species. This is conceptually probable since single nucleotide polymorphisms within target genes have determined the loss of the shattering phenotypic trait during rice domestication.[40]

## Weed gene drives

As elaborated above, numerous routes exist for agricultural weed management, in particular, sensitizing gene drives. However, barriers do exist in the generation of transformation systems for weed control, specifically those compatible with CRISPR/Cas9-driven genome editing and recognition of good molecular targets. Therefore, weed control gene drives should be able to overcome constraints in the molecular genetics of weed systems as well as barriers posed by population biology and ecological determinants. The successful construction and execution of gene drives is very much dependent on successful collaborations between weed ecologists, molecular biologists, and evolutionary biologists.

For success in weed management gene drives, the edited genes should disperse through populations at frequencies high enough to lead to satisfactory population control within an equitable time frame. Based on modeling studies, about 10–20 generations are required to spread the edited genes in order to enable fixation in natural weed populations, subject to consequences in fitness of the edited genes (selection coefficient) and conversion efficacy of the CRISPR/Cas9 targeting and editing reactions.[41] In this light, there is an imperative need to perform more modeling studies to identify the potential of weed control gene drives for agricultural weed management.

Even as genome engineering has been largely used as a molecular tool for genetic advancement of economically or culturally important crop species, synthetic tools for genetic regulation of wild plant populations have received comparatively little attention.[35,42,33,34] Gene drives are constituted of selfish genes that can bias segregation ratios at the time of meiosis or during development of gametes.[35] Thus, they are able to disperse through populations, even while imposing a cost on the fitness of their host plants.[11] In principle, these gene drives could be engineered to introduce desirable genetics in wild populations. The conceptual potential of gene drives has long been foreseen, but the molecular genetic tools necessary to engineer them for specific purposes and species did not exist until recently.

However, presently new insights into genome editing technologies such as the CRISPR/Cas9 system have emerged as strategies for genetic modification of plants.[35,36–38] For the gene drive to take place, there has to be synergy between intervention measures occurring at the molecular, genetic, cellular, organismal, and population levels. This is particularly significant when evaluating the use of gene drives as a means of managing wild plant populations, considering that plants possess a diverse array of life histories, cytogenetic arrangements, and DNA repair mechanisms.

Control of weeds is a complex and evolving problem associated with the interplay of natural-human systems wherein it is difficult to discern simple solutions to control invasive agricultural weeds. One of the quickly developing and increasingly critical issues is the widespread emergence of herbicide resistance.[39] Another compelling issue is associated with the growth of invasive alien weeds that are challenging to control, since they often infest sensitive, remote, and rugged regions where traditional management measures may be difficult to implement.[40] The fundamental precept of gene drives operates in two different ways: population suppression and population sensitization.[35] Toward population suppression, key traits such as fertility, establishment, and persistence are directly targeted by suppression drives that have high potential to facilitate low-input control of alien, invasive weeds in agricultural and natural environments.

In contrast to suppression drives, sensitizing gene drives can be employed to subject a target population to becoming susceptible to specific weed management intervention measures. For instance, weeds have by and large acquired herbicide resistance.[41] An endonuclease-derived gene drive could disrupt the herbicide-resistant alleles at target sites and instead substitute them for the wild-type susceptibility alleles, thus controlling the weed population.[35] Table 6.1 below summarizes the gene drive potential for some of the major agricultural weeds prevailing globally.

The capability of gene drive systems toward weed population control via direct genetic interventions is considered with regard to major genetic and biological characteristics of weeds, their global distribution and complexity of control (which are together referred to here as herbicide-resistance risk). The weed species mating system determines the intrinsic capability of gene drives, since drives would successfully disseminate only the edited genes by means of outcrossing species. Species here are categorized as being predominantly selfing or mostly outcrossing. Species having high fecundity will most probably have the largest populations, which will therefore prolong the time taken for the establishment of the gene drives in order to result in substantial population suppression. Likewise, a high degree of seed bank endurance will diminish the rate at which these genes proliferate across populations. Species having evolved resistance to over five modes of actions are referred to as carrying a very high resistance risk, while species that have generated resistance to over two or fewer modes of actions are referred to as intermediate risk. For those species containing large, unsequenced genomes as well as for those which are polyploid, the recognition and targeting of genes would be more complicated. Species possessing $<500$ Mb genome size are classified as having small genomes, those having a genome sized 500 Mb to 2 Gb are classified as medium-sized, while those having $>2$ Gb genome size are considered as large.

**TABLE 6.1 Summary of gene drive potential for some of the main, global agricultural weeds.**

| Species | Mating system | Fecundity | Seed bank persistence | Resistance risk | Ploidy | Genome | Distribution |
|---|---|---|---|---|---|---|---|
| *Amaranthus palmeri* | Outcrossing (dioecious) | High | Low | Very high | Diploid | Medium, sequenced | North and South America |
| *Amaranthus tuberculatus* | Outcrossing (dioecious) | High | Medium | Very high | Diploid | Medium, congener sequenced | North America |
| *Alopecurus myosuroides* | Outcrossing | Medium | Medium | Very high | Diploid | Large, no sequence | Europe, Asia |
| *Lolium rigidum* | Outcrossing | Medium | Medium | Very high | Diploid | Large, congener sequenced | Australia, Europe, South America, Africa |
| *Avena fatua* | Selfing | Low | Medium | Very high | Hexaploid | Large, no sequence | North and South America, Australia, Europe, Africa, Asia |
| *Echinochloa crus-galli* | Selfing | Medium | Medium | Very high | Hexaploid | Medium, sequenced | North and South America, Australia, Europe, Africa Australia, Europe, Africa, Asia |

From Neve P. Gene drive systems: do they have a place in agricultural weed management? Pest Management Science 2018;74(12):2671e2679. https://doi.org/10.1002/ps.5137.

In addition to the above, gene drives could potentially be adopted as molecular genetic tools for the conservation of endangered plant populations. Genetic rescue of small, declining plant species is conventionally implemented through transplantation of small numbers of individual genetically discrete immigrant plants into small populations. The major purpose of this is to introduce favorable genetic variations into threatened populations from a few immigrant plants that are better adapted to the existing conditions. This results in the enhancement of adaptive potential, while locally, this adaptive molecular variation is not swamped.[42,43]

As an example, rapidly driven new, adaptive variations in plant populations could enhance population fitness by augmenting resilience to invasive parasites and environmental fluctuations.[44] Such adaptive measures would increase the rate of distribution of the beneficial genes as well as the ability to future-proof these populations by inducing the adaptive traits even before the arrival of the selective threat. Nevertheless, the major problem in this approach is the difficulty in identifying genes underlying these adaptive traits.

Therefore, genetic traits that are clearly deciphered could be good targets for initial trials at genetic engineering. Traits determining pathogen resistance are comparatively easily phenotyped and are also well defined at molecular and genetic levels. Several species and communities of plants are endangered by the threat of invasive pathogens. The fungal pathogen, *Puccinia psidii* has recently endangered several widespread and common species of plants in the family Myrtaceae (inclusive of Eucalyptus spp.)[45] in Australia. Resistance to this pathogen has been detected,[46] and investigations into the identification of the genetic basis of this resistance are currently under way.[45] Therefore, the subsequent cloning of the resistance genes would enable the utilization of gene drives to engineer resistance to *P. psidii* in susceptible species and populations.

## 5. Genetic modification of polyploids

Synthetic drives could also be constructed to function at much smaller genomic levels. Genetic engineering of polyploid plant species is often greatly challenging due to the occurrence of genome duplication and the existence of multiple homologous alleles. Therefore, even as CRISPR-based genome editing has been engineered in many polyploid plant species such as strawberries, potato, wheat, and oilseed rape, the results from such genetic manipulations are to a large extent variable, ranging from the editing of only one allele[47,48] through varying combinations of homologous edited alleles[49] to complete editing of all the alleles in a genome.[47–50] This poses a major caveat as gene edits across all of the alleles may be essential to engineering a desired modified trait in the polyploid species. In such cases, endonuclease-based gene drives could be a really useful tool for engineering polyploid species, provided they can be constructed to drive the gene edits across all gene copies and homologous alleles within their genomes.

The majority of the present focus regarding synthetic gene drives has involved the design and construction of CRISPR/Cas9 systems or synthetic toxin-antidote strategies (e.g,[35,38,44]). Nevertheless, there are several selfish genetic elements occurring naturally with the ability to skew the segregation ratios elucidated in plants.[11,51] Discernment of the mechanisms facilitating the drive of these elements as well as the factors influencing the frequency of drives within populations can likely offer insights into ideal ways to design and apply these synthetic drive systems. An overall strategy to mechanistically organize different varieties of drive elements[51] is to identify whether they disrupt the transmission ratios before or after gamete formation.

## 6. Pregametic gene drives

Meiotic or pregametic gene drives disrupt the transmission ratios at the time of meiosis, such that gametes bearing the drive allele have a greater probability of being generated.[11] In plants, the meiotic drive systems delineated have evolved strategies that skew chromosome segregation to the plant cell that finally evolves into the female gamete (i.e., "female meiotic drive"[11]). This is feasible since for most flowering plant species, only one product derived from female meiosis forms a valid megaspore that will form the female gametocyte possessing a fertile egg cell.

The other three products of meiosis go through programmed cell death.[52] This genetic asymmetry during meiosis enables the generation of selfish genetic elements, as it competes with homologous chromosomes for existence in the single surviving egg cell. The abnormal (Ab10) meiotic gene drive system in maize (Zea mays)[53] has been well-characterized, wherein Ab10 occurs as a selfish version of the "normal" chromosome 10 (N10). Compared to N10, Ab10 possesses large segments of repetitive DNA addressed as knobs on the long arm of the chromosome. Even as the mechanisms facilitating biased transmission are multifaceted,[54] these knob structures skew transmission by aiding in the active migration of Ab10 along meiotic spindles to the fertile egg cell before the N10 chromosomes. The outcome of this process is strong biased segregation wherein the Ab10 is conveyed to 60%–80% of the progeny.[55]

Another well-defined instance of female meiotic drive in plants is *Mimulus guttatus* (monkey flower) centromere-associated distorter (D) locus.[56,57] Originally identified through investigations of non-Mendelian inheritance in hybrids resulting from the cross of *M. guttatus* and its closely related *M. nasutus*, this D locus is analogous to Ab10 in that it causes the preferential transport of chromosomes harboring the drive element to the egg cell at the time of female meiosis. This leads to strong bias in transmission, wherein 98% of the resultant progeny from heterospecific crosses carry the D locus, and 58% result from conspecific crosses (with the variation likely to be due to the presence of suppressor elements in *M. guttatus*.[58]

B chromosomes happen to be the most well-defined class of selfish genetic elements in plant systems. These B chromosomes are supernumerary elements accumulating in a non-Mendelian manner. They generally cause deleterious consequences in their host plants and have been found in over 1000 flowering plant species.[35,59] The B chromosome drive is facilitated by a host of irregular mitotic and meiotic processes that enable them to bias transmission ratios and accrue selfishly in the germline.[59,60] Notwithstanding, similar to the above described meiotic drives, B chromosomes frequently accomplish drives by causing genetic asymmetries at the time of gametogenesis (mainly pollen mitosis).

In the case of plant systems with available data, premeiotic gene drives are generally maintained at intermediary frequencies. However, in spite of their transmission advantage, there are costs to fitness associated with transporting the drive element that could account for the maintenance of polymorphisms and the reason why drives do not lead to fixation. Theoretically, the Ab10 maize system could achieve a transmission rate of 83% in heterozygotes, while the empirical transmission rates occur between 60% and 80%.[55] Nevertheless, in wild teosinte communities, Ab10 occurs at only low-intermediate frequencies,[61] likely due to decreases in the viabilities of pollen and seeds as well as the size of seeds in Ab10 homozygotes.[55]

Likewise, the Mimulus can attain D locus showing a roughly 100% rate of transmission in interspecific crossing investigations, but when occurring in natural communities, it segregates at frequencies between 30% and 40%,[58] showing considerable (20%) costs in pollen fertility[56] and diminished seed size.[58]

Similarly, the B chromosome population dynamics could reflect the opposing influences of segregation bias and harmful effects on the host. The B chromosomes at low copy numbers have only a few deleterious effects, but when accumulated in plants, they diminish the latter's growth and fecundity.[59] Such accumulative deleterious outcomes perhaps contribute to limiting B chromosome accumulation and stabilizing their frequencies within plant communities.[62] Therefore, even as the exact mechanisms accounting for fitness costs are yet to be determined, from a physiological context, deleterious effects might simply reflect the costs incurred due to the replication and maintenance of nonessential DNA involving the accumulation of B chromosomes.[63]

## 7. Postgametic gene drives

In postgametic drives, segregation distortion is accomplished by mechanisms rendering gametes inviable after the completion of meiosis. To achieve this, postgametic drives should be able to perform the following functions: (a) the drive allele should have the capability to differentiate competing meiotic products (i.e., distinction of self from nonself) and (b) the drive allele should render inviable the competing meiotic product.[51]

In plant systems, most of the postgametic drives have been identified through observations of hybrid sterility, after crosses between species of crops and their wild relatives. Consequent genetic analysis of sterility traits has revealed loci biasing the transmission of alleles by functioning as "gamete killers."[64] While the various postgametic drives elucidated thus far are accomplished through a broad range of molecular phenomena, several of them segregate as a single locus comprised of multiple tightly linked genes that facilitate direct obstruction of gametes transmitting nonself alleles, or implement the production of a toxin and an antidote.[65] For instance, the *Oryza sativa* qHMS7 locus through a poison-antidote type strategy functions on male gametes.[66] In backcrossed hybrids of sativa/meridionalis, the pollen inheriting the meridionalis qHMS7 haplotype becomes sterile. Here, the sativa qHMS7 locus possesses the closely linked ORF2 gene encoding a toxin that kills pollen and the ORF3 gene encoding an antidote to this ORF2 toxin. On the other hand, the meridionalis qHMS7 locus codes for a nontoxic version of ORF2 while lacking ORF3.

The dynamics of gamete killers in plant populations still remains by and large obscure. In all the elucidated systems, postgametic drive loci have been established within a species, wherein the phenotype is observable only through interspecific crosses. This could reflect the ability of these loci to drive through populations at high efficiency. This potent transmission advantage conveyed by the killing of competing gametes means that, theoretically, they could become established in populations rather quickly.

Nevertheless, the advantage of transmission could be counteracted by costs, the most obvious for which is diminished fertility, which has a powerful potential to affect population dynamics of pollen-destroying post-gametic drives.[67] How these limitations could be overcome in the synthetic gene drive context warrants further studies involving the role of evolution of selfish elements in driving their emergence.[65] Gene drives are composed of genetic elements that are transmitted efficiently between sexual reproducing species despite their ability to diminish their fitness.[15,18] Therefore, gene drives can transmit traits with greater efficiency than that expected by Mendelian inheritance.[15,68] Gene drives for accomplishing pest control have been under serious consideration for more than a decade.[13,69–76] However, in practice, there are challenges to its implementation. The advent of the CRISPR/Cas9 gene editing technology has changed that scenario.

In order to work, the gene drives should be administered only in populations that undergo sexual reproduction and possess short generation times. Also, factors such as the target pest breeding structure, geographical distribution, and the rate of gene flow determining the efficacy of gene drives have to be considered.[77] Therefore, not all species of pests would be appropriate for population control using the gene drive technology. Moreover, hazards involved in the deployment of the gene drives have to be considered.

## 8. Use of genetic engineering measures to control insect vector transmission of plant disease pathogens

Thrips, aphids, plant hoppers, leafhoppers, trehoppers, and whiteflies are significant vectors of plant pathogens such as viruses and bacteria.[78–81] The destruction from vectored pathogens can be diminished by generating plants resistant to the respective pathogen using breeding and transgenic technologies.[82,83] Additionally, pathogen suppression can be accomplished by producing plants resistant to the insect vector.

## 9. Nascent and forthcoming genetic engineering technologies and gene drives

A remarkably distinct approach for inhibiting plant pathogen vector populations or for blocking their ability to transmit plant pathogens involves the use of insect transgenesis via gene drives or selfish genetic elements that create super-Mendelian inheritance of traits harbored by these insects that block insect reproduction or inhibit pathogen replication.[13,94] Such gene drive measures have advanced quickly over the past few years.[95]

However, gene drive approaches are not possible in the absence of obligate sexual reproduction. Therefore, most aphids do not qualify for the application of gene drives and additionally, some of the pathogen-harboring thrips could also reproduce by parthenogenesis.[96] On the hand, *Bemisia tabaci* whiteflies exhibit haplodiploidy[97] and hence could be targeted via the meiotic stages of their females. However, this would not be as efficient as it would be in the naturally sexual, diploid leafhoppers and planthoppers[98] The Asian citrus psyllid is a target for gene drives as it is responsible for the transmission of pathogens causing citrus greening.[99] Equivalent measures of containing agricultural weed species using the CRISPR/Cas9 technology show great promise although circumventing the enormous seed production capability and the self-fertilizing life cycle of several weed species renders this very challenging. Gene drive technologies are currently in their infancy but would in all likelihood develop into beneficial species-specific, recallable tools to manage pests subject to geographical containment.[100]

## 10. Sterile insect methodology

The sterile insect methodology (SIT) (Fig. 6.2) concerns the repetitive release of sterilized insects (commonly males) that are mass produced.[101,102] The target insect species is reared en masse, and where practically the sexes are segregated, the males are sterilized and released subsequently. Offspring resulting from wild females, which have mated with the sterilized released males, are inviable, wherein the most usual strategy of sterilization involves ionizing radiation due to which dominant lethal mutations are generated in the sperm.[103]

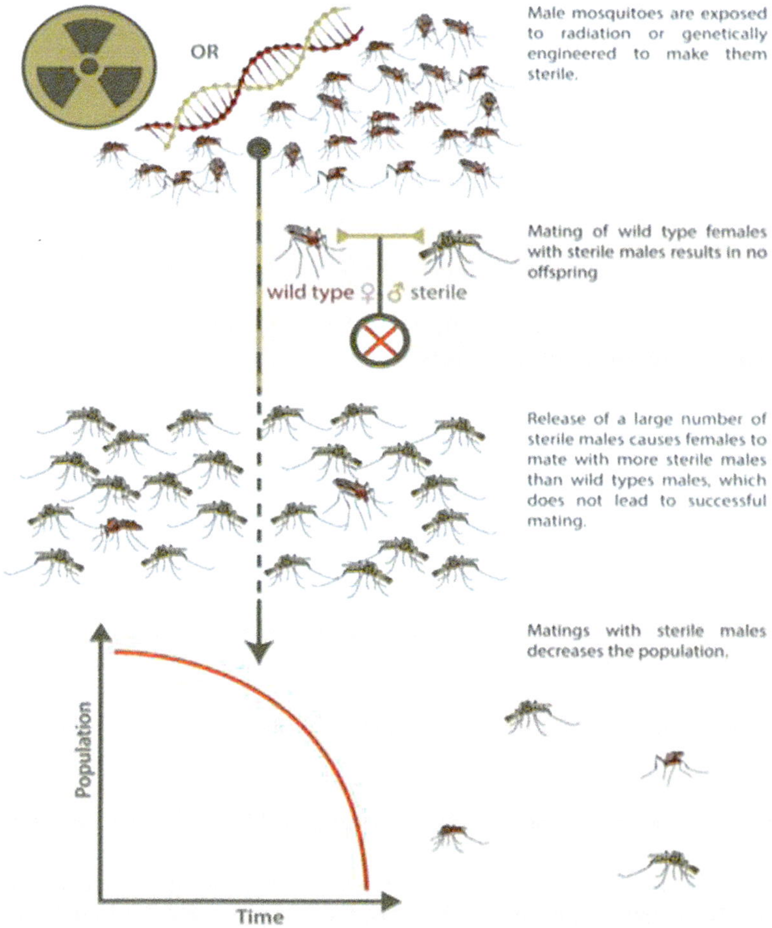


**FIGURE 6.2 Sterile insect technology (SIT).** Populations of invasive insect species are controlled through the release of sterilized males using the SIT. For a start, this entails maintenance of a seed colony from which groups of eggs are obtained and subsequently amplified over several generations to generate sizable release cohorts. In the traditional SIT technique, males are subjected to radiation to cause sterility. However, recent technologies have made it feasible to genetically engineer sterile male insects in order to offset costs to fitness and mating efficacy accompanying the use of radiation. Therefore, matings that take place between the above sterile males and the wild type females do not generate offspring. Obviously, these sterilized males should be released into the wild insect population in great numbers to increase the likelihood of matings between wild type females and the sterile males instead of with the wild type males so as to decrease the population of the invasive insect species over time. *From Teem et al. Genetic biocontrol for invasive species.* Front Bioengineering Biotechnology *2020;8:452. https://doi.org/10.3389/fbioe.2020.00452.*

In Lepidopterans, the deleterious effects induced by lower doses of radiation lead to sterility that is inherited by the F1 generation. Sterility can also be induced by chemical sterilants or by genetic engineering[17,104] or by incompatible insect strategies such as that induced by Wolbachia infections[105] or using the method of hybrid sterility.[106]

## 11. Insects infected with Wolbachia

Wolbachia are maternally inherited, intracellular endosymbionts that modulate reproduction of their host organisms in several ways to serve their own inherent maternal transmission.[107] They are predicted to exist in 40%–60% of the insect species.[108–110] Apart from vertical transmission, they are also able to spread horizontally to cause infection in new host species,[107,111] which nevertheless seems to be a mostly rare event.[13] Wolbachia strains are considered to be selfish genetic elements employed in incompatible insect technique (IIT) regimens.[17]

The most prevalent Wolbachia drive system is cytoplasmic incompatibility (CI). In other words, sperm–egg incompatibility.[107,112] Unidirectional CI takes place between populations of insects infected and uninfected with Wolbachia. The outcome of this is that viable offspring are generated only when uninfected females mate with uninfected males. Instead, when they mate with Wolbachia-infected males, there is no viable offspring produced. On the other hand, infected females can mate successfully with both infected and uninfected males and therefore have a reproductive advantage.

As a consequence of the above, this Wolbachia infection will disseminate across the population.[17,112] Bidirectional CI occurs in populations infected with different, incompatible strains of Wolbachia wherein viable offspring are produced only from matings between males and females infected with the same Wolbachia strain. This Wolbachia-induced CI can be used to control or alter pest populations. Release of males infected with Wolbachia that are incompatible with the uninfected females could cause population suppression (as if the males have been sterilized). Since males serve as dead-end hosts for Wolbachia, the strain that is introduced does not stay in the target population.

## 12. Genetically modified insects

Genetic modification of insects is undertaken in order to introduce novel traits such as those that suppress target populations of insects.[113] Currently, GM insect applications are confined to insects possessing a self-limiting trait that generates nonviable offspring due to the establishment of a repressible lethal genetic program.[104,114] Similar to SIT, this strategy involves the repetitive en masse release of males harboring the lethal trait followed by mating with females, producing offspring that die before they reach adulthood.[115] The next generation of GM insects was accomplished such that male offspring survive, therefore enabling further mating cycles to diminish the pest population.[113]

Such self-limiting technique has also been developed in order to control pest arthropod species like the diamondback moth (*Plutella xylostella*), the fall army worm (*Spodotera frugiperda*), the spotted wing drosophila (*Drosophila suzukii*), the olive fly (*Bactrocera olea*), and the Mediterranean fruit fly (*Ceratitis capitata*) (www.oxitec.com).

Field trials were conducted on a smaller scale in Geneva (NY, USA) in 2017 using the OX4319—L diamondback moth to investigate the dispersal, suppression efficacy, and survival of the regional diamondback moth population spreading on cotton.[116] The only release of GM agricultural pests on a larger scale was accomplished with pink bollworms sterilized with radiation following their genetic modification to express DsRed2 fluorescent marker. Nearly, 20 million of these insects were released, which showed field performance equivalent to that of the standard strain.[117,118]

## 13. Synthetic gene drives and their constructs

Over the past 30 years, synthetic gene constructs have been vastly developed, and our capability to engineer genes and genomes affords novel potential for controlling insect pest populations. Toxin-antidote programs involving maternal-effect selfish gene elements first elucidated in flour beetles (*Tribolium castaneum*, Coleoptera: Tenebrionidae,[119] survive by causing maternally driven lethality of all of those offspring that do not inherit the specific selfish gene(s) from the paternal and/or maternal genome. These selfish genetic elements have evolved for numerous insect systems.[120–122] This combination of the toxin-antidote scheme determines the gene drives effecting synthetically engineered maternal genetic factors.

The miRNA toxin, which impedes embryonic development, is one such combination expressed at the time of formation of eggs in females and is connected to an antidote that manifests during development. If the toxin is conveyed to offspring from the miRNA-bearing female, then development is abrogated. On the other hand, if the offspring additionally inherits the antidote comprising an miRNA resistance gene that is expressed only at the time of development, normal growth and development is restored. Recent studies have shown that such toxin-antidote systems could enable the control of *Drosophila suzukii*.[120]

Yet another synthetic drive program is founded on nucleases. Nucleases have been engineered to distort sex ratios (via X-chromosome shredding), diminish fertility, change resistance to pathogens, or are linked to specific lethality-inducing genes. Nucleases skew the inheritance pattern through “homing” wherein alleles carrying a synthetic homing gene, an effector that decreases fitness will cleave susceptible alleles inducing repair of the damaged DNA. By means of patterns of base sequence on the undamaged allele that harbors the nuclease, this gene is copied into the cleaved site. Therefore, heterozygotes are transformed to homozygotes. Such homing and cleavage of recognition sites occurs with high efficiency through cellular DNA repair machinery using a homology-directed repair mechanism.

Recently, the CRISPR/Cas9 technology has been used to generate synthetic drive systems in a fashion similar to the homing nucleases wherein the original gene is disrupted or replaced with another DNA sequence by changing heterozygote drive carriers into homozygotes. In this system, specific RNA of about 17–20 nucleotide length guides the Cas nuclease to loci on the DNA, wherein the nuclease generates a double strand break in the DNA. Following this, homology-directed repair takes place, which copies the drive into the adjoining break site.

The power of the CRISPR/Cas system is the ease of gene editing, although caveats such as off-target effects by misaligning of the guide RNAs do exist. Additionally, these DNA breaks can be repaired via another mechanism, the nonhomologous end joining, which ligates the cleaved ends of the DNA together, which in most cases leads to loss of the drive element. So far, such synthetic CRISPR/Cas drive elements have been developed as devices to control many insect pests and vectors.[18,95,123,124] Table 6.2 shows some examples of CRISPR/Cas9-engineered insects for pest management.

## Use of synthetic gene drive constructs

Agricultural pest control measures that use suppression drives aim to cause lethality, induce flightlessness, or skew sex ratios such that there is a decline in the insect pest population.

## Genetic tools for management of *D. suzukii*

*Drosophila suzukii* is spread worldwide in recent times[129,130] wherein these insects oviposit inside ripening, undamaged fruits having soft skin and render the fruits unmarketable due to the developing larvae. Effectual pest control schemes are required to preclude serious crop and revenue losses in fruits such as vines and berries.[131,132] Nevertheless, the control of these insects poses a great challenge as the rapidly developing larvae inside the fruits are protected from the administration of insecticides. Also, the insects grow in several overlapping generations so that newly generated adults have to be eradicated multiple times.[133]

Of serious consideration to controlling *D. suzukii* is classic biological control by the use of larval parasitoids belonging to the families Figitidae and Braconidae.[134,135] An alternate strategy is to diminish the production of offspring in the wild females using SIT by mating with males that have been radiation-sterilized.[136] Further, Wolbachia can infect *D. suzukii*, causing the rapid dissemination of an incompatibility factor,[137] and this can be used to adopt IIT measures.

**TABLE 6.2** Examples of genetically engineered insects toward pest management using the CRISPR/Cas9 system.

| Name of the insects | Target gene | Editing | Transformation method | Reference |
|---|---|---|---|---|
| *Drosophila melanogaster* | Chitin synthase 1 | Substitution | Embryo injection | Douris et al.[125] |
| *Aedes aegypti* | Male-determining factor (M factor, Nix) | Knock-out | Embryo injection | Hall et al.[126] |
| *Drosophila suzukii* | White (w) and sex lethal (Sxl) genes | Knock-out | Embryo injection | Li and scott,[127] |
| *Spodoptera litura* | Abdominal-A (Slabd-A) gene | Knock-out | Embryo injection | Bi et al.[128] |

From Bisht et al. Improving plant-resistance to insect-pests and pathogens: the new opportunities through targeted genome editing. *Seminars in Cell and Developmental Biology* 2019;**96**:65–76.

Notwithstanding, such an approach would necessitate an effectual sexing methodology or sterilization of the *D. suzukii* females using low doses of radiation, since even a few females harboring the Wolbachia strain if accidently released could compromise the efficiency of the IIT.[105] But this seems to be a challenging factor for the control of D. *suzukii* and additionally, any horizontal gene transfer could be a detrimental issue wherein it could lead to integration of the whole genome of the Wolbachia.[137,138]

SIT using sterile insects that have been genetically modified is well established for several fruit flies (Tephritidae) living in habitats similar to that of *D. suzukii*. Studies concerned with efficacious mass-rearing of these flies and additionally, investigations to circumvent the decreased competitiveness of the released males could be applied to the control of *D. suzukii*.[139] Also, knowledge emerging from investigations of *D. melanogaster* to generate sterile males using CRISPR-based technologies could be adapted to *D. suzukii*.[140]

## Gene drives toward control of *D. suzukii*

Transgenes causing embryonic lethality have been introduced into *D. melanogaster*,[141] and this technique could be transferred to *D. suzukii*. Schetelig et al., 2018[142] have shown that Cas9 and various promoters developed for CRISPR/Cas9 editing in *D. melanogaster*[143] also work in *D. suzukii*, thus enabling genome editing in the latter.

For instance, in *D. suzukii*, site-specific mutations within a gene coding for eye color to mark individuals phenotypically were introduced using the CRISPR/Cas9 system. Mutations were also engineered in the master gene SxI encoding female development in order to hamper reproduction.[144,145] In an analogous study, it has been shown that females of *D. melanogaster* homozygous for this SxI mutation die early on during development.[146] *D. suzukii* embryos that were injected with a plasmid harboring this construct developed aberrant reproductive organs.[145] Earlier, this construct was adopted to engineer somatic and germline tissue in *D. melanogaster*.[143,147]

In yet another system, a lethality effector that is female-specific has been inserted into a transposon vector along with a fluorescent protein as marker, which was then introduced into *D. suzukii* so as to establish a conditional lethality scheme.[148] While such systems could be applied to eliminate females and to mark males for adopting SIT programs, they have not as yet been applied to construct a gene drive in *D. suzukii*. Recently, a Medea drive mechanism that was originally produced in *D. melanogaster* has been employed in D. suzukii and has been patented in the USA (WO2017132207-A1[149]). This system includes a DNA sequence that codes for a microRNA toxin under the control of a promoter that is specific to the maternal germline and another DNA sequence coding for an antidote under the control of a promoter that is specific for early embryos. This resulted in maternal-effect lethality. Further, crossing experiments were performed, which showed inheritance rates of up to as high as 100% depending on the strain. Long-term cage experiments on insect populations with a wild type strain harboring pre-existing resistance, and consequent mathematical modeling showed self-limiting dynamics of this scheme, which confers quite a large fitness cost.[120]

## 14. Risks incurred by gene drives

Despite the progress made in the generation and use of gene drives, several downsides exist, which preclude the use of these drives in an unbridled manner.[48,150] Underlying this technology, there exist several serious ethical constraints and unforeseen ecological outcomes concerned with the genetic manipulation of populations of wild plant species.[151] For instance, several invasive weeds play roles of key constituents of the flora and have important functions in the ecosystems in their native environments (e.g., Australian *Acacia spp.*).[152] Similarly, in agriculture, several weeds of crops are also essential pasture species and, in some instances, on the same farm as the crop and the weed species (e.g., *Lolium spp.*). Such contextual dependence poses ethical, management, ecological, and regulatory concerns as genetic suppression drives could probably lead to undesired declines and even species extinction if left to spread in an unbridled fashion.[153] Another probable compounding factor is the potential dissemination of the gene drive into nontarget species through hybridization.[154]

Depending on their design, different gene drives afford varying levels of risk of inadvertent spread into nontarget plant populations.[155] Many strategies have been put forth to limit temporal and spatial spread of these gene drives (e.g,[156,157]). While such technologies would be imperative for engineering drives that can be released into the wild, thorough socio-economic and ecological studies need to be executed to rigorously estimate risks in a comprehensive context.[158] To achieve this, research scientists, regulators, and managers should carefully assess the risks associated with the release of a given drive as against its potential benefits and clearly and indubitably articulate scientific findings to the broader community.[150]

Notwithstanding the efforts to engineer suppression-based gene drives having biological and geographical limits to their spread, these measures could lead to the extinction of species if they were allowed to spread in an unbridled manner.[15] Therefore, acknowledging the probable biological, regulatory, and ethical issues associated with uncontrolled population-suppression drives, several investigators have put forth strategies for preserving this technology to curtail the temporal and spatial dissemination of gene drive measures.[15,100,159] Among these, sensitizing drives focusing on augmenting the susceptibility of the target population to the consequent management and environmental mediation afford higher potential for regulatory approval and their ensuing use in weed control.

Before examining the application of gene drives for control of pests, several key questions should be addressed[160] such as: what would be the ecological effects of the extinction of the concerned pest species? How many pest species can we eradicate before encountering serious negative consequences? Who would determine which species deserves to be driven to become extinct? What are the ethical issues involved in deciding to do away with the existence of a given pest species? Therefore, risks regarding the use of gene drives globally would be too high, thus precluding their deployment.

## 15. Mitigating the risks posed by gene drives

Luckily, there are gene drives that are less risky: for instance, temporary gene drives or gene drives that switch on only after particular thresholds or environmental factors are reached.[161] Gene drives subject to the control of environmental conditions are called local gene drives which can be introduced into the existing integrated pest management schemes with lesser risk than that of the global gene drives.

The local gene drives must be initially tested and investigated on islands. Despite the possibility that species carrying the local gene drives could still escape from the islands, the species isolation afforded by the islands could add an additional level of containment.[162] Additionally, local gene drives could be controlled by acting only on specialized pests (for instance, western corn rootworm, *Diabrotica virgifera*, that feeds on European corn) or on generalist species of pests but only when they feed on a specific crop species (e.g., cotton bollworm, *Helicoverpa armigera*, that feeds on corn).

The execution of local gene drives within comparatively contained habitats will enable gathering of data on their efficacy while enhancing our capability to remedy any inadvertent consequences. If this proves successful, the knowledge acquired using these prototype control programs will ease communications with stakeholders as well as the public and lead to concerted decision-making regarding the administration of local gene drives. Succeeding generations of local gene drives could be employed to target invasive species that are less isolated. However, containment measures could fail, and therefore, it is essential to develop strategies to preclude the out of control spread of these local gene drives.

Control of gene drives necessitates precision gene drives or the generation of tested immunization or reverse gene drives prior to obtaining approval for use.[15,163] The impact of local gene drives on the diversity of target species over the long term needs to be examined.[163] To alleviate the effects of any possible losses in biodiversity, regulatory agencies should mandate procedures for the preservation of the genomes of pest populations that are under consideration for control using these gene drives. Representative genomes of the concerned pest populations should be preserved in genomic libraries run by the government.

Also, species categorized as pests by some people could be of value to others.[164] For instance, salt cedar is classified as a pest by landowners but is a treasured tree, as per wildlife protection personnel.[165] Likewise, the prickly pear is classified as a pest in Caribbean nations but is an invaluable plant in Mexico.[166,167] An additional interesting use of gene drives could be to disseminate pesticide resistance genes in wild pest populations. For example, herbicide-tolerant milkweeds could preclude the decline of monarch butterfly populations.

In another approach, gene drives can be used to modify the ability of pests to host important symbiotic microbes. In many arthropod pests, bacterial symbiosis is observed in both facultative and obligate modes.[168,169] For instance, aphids obligately require Buchnera aphidicola symbiotic bacteria for their survival.[170] Gene drives can be employed to disseminate genes that work to preclude the ability of aphids to host symbiotic bacteria. Likewise, vector competence of plant disease-transmitting arthropod species could be eliminated using gene drives that spread genes effecting such elimination.

## 16. Conclusions and future perspectives

The significant excitement caused by the rapid recent advancements in gene drive technology warrants the consideration of those involved in pest and agricultural weed management.[15,99,171] The compelling need for agroecological and technological innovations in crop protection schemes is considerable and gene drive affords extraordinary potential to precisely modify the genomes of pest species in order to introduce and disseminate fitness-diminishing traits into populations of wild pests toward achieving population suppression. Additionally, gene drives can operate to sensitize or re-sensitize pest populations to new and already practised control methodologies.

The societal, ethical, ecological, and biosecurity challenges concerned with the application and regulation of gene drive strategies are significant. Suppression gene drives serve as a mode of gene editing that can disperse potentially uncontrollably across wild populations with propensity to result in serious declines in pest populations (possibly even extinction) of these species in natural ecosystems. Stringent management of gene drives[48,163,172] is important to limit their unbridled spread or make them reversible at the laboratory level or in natural environments.[35,173,174] It should be acknowledged that the ethical and regulatory impediments to the application of gene drives are likely to be more determinant in their practice compared to that of any technical obstacles. Agricultural pests diminish crop yields through disease transmission, herbivory, pathogenicity, and competition for resources.

Broadly, most pest species including weeds are intrinsic constituents of agroecosystems and their ability to being inadmissible is a function of their plenitude and the accompanying potential for crop yield loss, rather than their implicit occurrence in those ecosystems. Certainly, many of the pest species can even play positive roles within agroecosystems pertaining to community structure and functioning of the ecosystem. Therefore, any gene drive that has potential to drive a given species to extinction would not be desired and rigorous regulations would help in prohibiting those objectives and their potential effects. Notwithstanding, local eradications would be feasible, and gene drives have the capability to facilitate more targeted control of pest species that significantly diminishes any inadvertent, off-target effects of current crop conservation strategies.

Sensitizing gene drives, in particular those that reverse the emergence of resistance may overcome intrinsic ecological issues and ethical challenges to gene drives. For instance, drives capable of reversing herbicide resistance have inherent "safeguarding" procedures. At the genetic scale, they address the reversal of individuals as well as populations that have generated resistance through human-driven selection of resistance-inducing alleles back to their wild-type, ancestral, sensitive form. These strategies do not need the introduced edited DNA that decreases the fitness of pests in "natural environments." Therefore, these drives would enable the restoration of populations to their respective genetic states they were in prior to human intervention. On a practical basis, the suppression of population conferred by these sensitizing gene drives is linked strongly to the exogenous administration of a chemical or some other management strategy, making it feasible to stringently limit the temporal and spatial control of populations. Hence, these sensitizing gene drives have the highest potential to confer novel, genetic control of weeds within the contemplated future regulatory schemes.

There is a compelling need for generation of cutting-edge tools to enable the management of wild plant populations in natural and agronomic systems. Gene drives show promise in solving a few of the plant-associated impediments in both of these ecosystems. Nevertheless, important issues such as efficiency of the engineered gene drives at the molecular genetic level, the accompanying potential for evolution of resistance, the influence of the genetic architecture, and life history of plants on the impact and spread of gene drives still remain to be addressed. Additionally, for several potential target species, there is a dearth of established molecular and genomic tools to execute feasible gene drives.

The availability of various natural gene drives in plant populations indicates that synthetic gene drives can be genetically engineered for some species of plants. Conversely, methodologies adapted from synthetic gene drives can enable researchers operating on natural gene drives to design constructs based on the gene drive of interest and precisely test hypotheses relevant to the fate and the evolutionary history of natural gene drives. Notwithstanding, important questions remain including the fitness of CRISPR as well as other endonuclease-enabled gene drives, given the probability for NHEJ and the particular types of cells, organs, and stages of development in which to target the expression of these gene drives.

Moreover, it is important to comprehend if toxin-antidote systems evolved in nature can be repurposed to engineer synthetic gene drives and the relative advantages of these alternate methodologies. In order for the synthetic gene drives to be efficiently accomplished, synergy between complex processes occurring at the cellular, molecular, individual, as well as population levels is required. In conclusion, transmission of gene drives is dependent on processes taking placed at the population level (for instance, mating, dispersal of the engineered species, and structure of the plant population). Plant species exhibit tremendous diversity in their life histories, which impact demographic and genetic processes occurring at the population level. Further, the diverse influence of these gene drives on the fitness of the target plants has to be factored in. These variations can impact the spread of the gene drives as well as the probability for the emergence of resistance, both of which are not currently well-deciphered. Therefore, the success of a given gene drive operation depends on population biology, autecology, and fitness impacts of the drive for any particular species.

Nevertheless, the regulatory scenario will finally determine the strategy for execution of this technology in various jurisdictions and the degree to which the public and the markets accept the employment of gene drives. The risk-cost-benefit studies will be critical in guiding the generation, regulation, and execution of gene drives in plant populations. Toward accomplishing this, models that address the evolutionary and ecological consequences of gene drives in different scenarios have to be developed. Such models, along with a lucid and candid communication of the scientific concepts involved, will be important in enabling public and market acceptance to the application of this significant scientific tool on plant populations. This will help in the achievement of purposeful agronomic and environmental outcomes.

## References

1. Ricroch AE, Hénard-Damave MC. Next biotech plants: new traits, crops, developers and technologies for addressing global challenges. *Critical Reviews in Biotechnology* 2016;**36**(4):675–90. https://doi.org/10.3109/07388551.2015.1004521.
2. James C. *ISAAA brief 46: executive summary global status of commercialised biotech/GM crops. The international Service for the Acquisition of agri-biotech applications.* 2013.
3. Kumar S, Chandra A, Pandey KC. Bacillus thuringiensis (Bt) transgenic crop: an environment friendly insect-pest management strategy. *Journal of Environmental Biology* 2008;**29**(5):641–53. http://www.jeb.co.in/index.php?page=journal_archives.
4. Hefferon KL, Herring RJ. The end of the GMO? Genome editing. *Gene Drives and New Frontiers of Plant Technology Review of Agrarian Studies* 2017;**7**.
5. Singh RP, Hodson DP, Jin Y, et al. Emergence and spread of new races of wheat stem rust fungus: continued threat to food security and prospects of genetic control. *Phytopathology* 2015;**105**(7):872–84. https://doi.org/10.1094/PHYTO-01-15-0030-FI.
6. Gonsalves D. Control of papaya ringspot virus in papaya: a case study. *Annual Review of Phytopathology* 1998;**36**:415–37. https://doi.org/10.1146/annurev.phyto.36.1.415.
7. Evanega S, Lynas M. *Dialectic of the pro-poor papaya.* The Oxford Handbook of Food, Politics, and Society; 2015.

8. Hamilton WD. Extraordinary sex ratios. *Science* 1967;**156**(3774):477–88. https://doi.org/10.1126/science.156.3774.477.
9. Burt A, Trivers R. *Genes in conflict.* Harvard University Press; 2006.
10. Curtis CF, Sinkins SP. Wolbachia as a possible means of driving genes into populations. *Parasitology* 1998;**116**(1):S111–5. https://doi.org/10.1017/s0031182000084997.
11. Lindholm AK, Dyer KA, Firman RC, et al. The ecology and evolutionary dynamics of meiotic drive. *Trends in Ecology and Evolution* 2016;**31**(4):315–26. https://doi.org/10.1016/j.tree.2016.02.001.
12. Romeis J, Collatz J, Glandorf DCM, Bonsall MB. The value of existing regulatory frameworks for the environmental risk assessment of agricultural pest control using gene drives. *Environmental Science & Policy* 2020;**108**:19–36. https://doi.org/10.1016/j.envsci.2020.02.016.
13. Sinkins SP, Gould F. Gene drive systems for insect disease vectors. *Nature Reviews Genetics* 2006;**7**(6):427–35. https://doi.org/10.1038/nrg1870.
14. Champer J, Buchman A, Akbari OS. Cheating evolution: engineering gene drives to manipulate the fate of wild populations. *Nature Reviews Genetics* 2016;**17**(3):146–59. https://doi.org/10.1038/nrg.2015.34.
15. Esvelt KM, Smidler AL, Catteruccia F, Church GM. Concerning RNA-guided gene drives for the alteration of wild populations. *eLife* 2014;**3**(2014):1–21. https://doi.org/10.7554/eLife.03401.
16. Gantz VM, Bier E. The mutagenic chain reaction: a method for converting heterozygous to homozygous mutations. *Science* 2015;**348**(6233):442–4. https://doi.org/10.1126/science.aaa5945.
17. Alphey L. Genetic control of mosquitoes. *Annual Review of Entomology* 2014;**59**:205–24. https://doi.org/10.1146/annurev-ento-011613-162002.
18. Hammond A, Galizi R, Kyrou K, et al. A CRISPR-Cas9 gene drive system targeting female reproduction in the malaria mosquito vector *Anopheles gambiae*. *Nature Biotechnology* 2016;**34**(1):78–83. https://doi.org/10.1038/nbt.3439.
19. Wieczorek A. *CRISPR-Cas9 and gene drives: the risks and benefits of game-changing technology.* Cooperative Extension Service; 2016.
20. Gurr GM, You M. Conservation biological control of pests in the molecular era: new opportunities to address old constraints. *Frontiers in Plant Science* 2016;**6**(2016). https://doi.org/10.3389/fpls.2015.01255.
21. Reid W, O'Brochta DA. Applications of genome editing in insects. *Current Opinion in Insect Science* 2016;**13**:43–54. https://doi.org/10.1016/j.cois.2015.11.001.
22. Singer M, Frischknecht F. Time for genome editing: next-generation attenuated malaria parasites. *Trends in Parasitology* 2017;**33**(3):202–13. https://doi.org/10.1016/j.pt..2016.09.012.
23. Languin K. *Genetic engineering to the rescue against invasive species?" National geographic*. 2014.
24. *Gene drives on the horizon: advancing science, navigating uncertainty, and aligning research with public values*. undefined: National Academies Press; 2016. p. 1–217. https://doi.org/10.17226/23405.
25. Callaway E. Gene drives thwarted by emergence of resistant organisms. *Nature* 2017;**542**(7639):15. https://doi.org/10.1038/542015a.
26. Lucas JA, Hawkins NJ, Fraaije BA. The evolution of fungicide resistance. *Advances in Applied Microbiology* 2015;**90**:29–92. https://doi.org/10.1016/bs.aambs.2014.09.001.

27. Powles SB, Yu Q. Evolution in action: plants resistant to herbicides. *Annual Review of Plant Biology* 2010;**61**:317–47. https://doi.org/10.1146/annurev-arplant-042809-112119.
28. Ranson H, N'Guessan R, Lines J, Moiroux N, Nkuni Z, Corbel V. Pyrethroid resistance in African anopheline mosquitoes: what are the implications for malaria control? *Trends in Parasitology* 2011;**27**(2):91–8. https://doi.org/10.1016/j.pt..2010.08.004.
29. Rudd JC, Horsley RD, McKendry AL, Elias EM. Host plant resistance genes for fusarium head blight: sources, mechanisms, and utility in conventional breeding systems. *Crop Science* 2001;**41**(3):620–7. https://doi.org/10.2135/cropsci2001.413620x.
30. Panda KG. *Host plant resistance to insects*. CAB International; 1995.
31. Lemerle D, Verbeek B, Cousens RD, Coombes NE. The potential for selecting wheat varieties strongly competitive against weeds. *Weed Research* 1996;**36**(6):505–13. https://doi.org/10.1111/j.1365-3180.1996.tb01679.x.
32. Belz RG. Allelopathy in crop/weed interactions - an update. *Pest Management Science* 2007;**63**(4):308–26. https://doi.org/10.1002/ps.1320.
33. James C. A global overview of biotech (GM) crops: adoption, impact and future prospects. *GM crops* 2010;**1**(1):8–12. https://doi.org/10.4161/gmcr.1.1.9756.
34. Duke SO. Perspectives on transgenic, herbicide-resistant crops in the United States almost 20 years after introduction. *Pest Management Science* 2015;**71**(5):652–7. https://doi.org/10.1002/ps.3863.
35. Neve P. Gene drive systems: do they have a place in agricultural weed management? *Pest Management Science* 2018;**74**(12):2671–9. https://doi.org/10.1002/ps.5137.
36. Thomas SG. Novel Rht-1 dwarfing genes: tools for wheat breeding and dissecting the function of della proteins. *Journal of Experimental Botany* 2017;**68**(3):354–8. https://doi.org/10.1093/jxb/erw509.
37. Nadir S, Xiong HB, Zhu Q, et al. Weedy rice in sustainable rice production. A review. *Agronomy for Sustainable Development* 2017;**37**(5). https://doi.org/10.1007/s13593-017-0456-4.
38. Ravet K, Patterson EL, Krähmer H, et al. The power and potential of genomics in weed biology and management. *Pest Management Science* 2018;**74**(10):2216–25. https://doi.org/10.1002/ps.5048.
39. Délye C, Jasieniuk M, Le Corre V. Deciphering the evolution of herbicide resistance in weeds. *Trends in Genetics* 2013;**29**(11):649–58. https://doi.org/10.1016/j.tig.2013.06.001.
40. Konishi S, Izawa T, Lin SY, et al. An SNP caused loss of seed shattering during rice domestication. *Science* 2006;**312**(5778):1392–6. https://doi.org/10.1126/science.1126410.
41. Unckless RL, Clark AG, Messer PW. Evolution of resistance against CRISPR/Cas9 gene drive. *Genetics* 2017;**205**(2):827–41. https://doi.org/10.1534/genetics.116.197285.
42. Barrett LG, Legros M, Kumaran N, Glassop D, Raghu S, Gardiner DM. Gene drives in plants: opportunities and challenges for weed control and engineered resilience. *Proc R Soc B* 2019:286.
43. Gould F. Broadening the application of evolutionarily based genetic pest management. *Evolution* 2008;**62**(2):500–10. https://doi.org/10.1111/j.1558-5646.2007.00298.x.
44. Hay BA, Chen CH, Ward CM, Huang H, Su JT, Guo M. Engineering the genomes of wild insect populations: challenges, and opportunities provided by synthetic Medea selfish genetic elements. *Journal of Insect Physiology* 2010;**56**(10):1402–13. https://doi.org/10.1016/j.jinsphys.2010.05.022.
45. Burt TR. *Genes in conflict: the biology of selfish genetic elements*. Harvard University Press; 2009.

46. Burt A. Site-specific selfish genes as tools for the control and genetic engineering of natural populations. *Proceedings of the Royal Society B* 2003;**270**:921–8.
47. Gould F. Broadening the application of evolutionarily based genetic pest management. *Evolution* 2008;**62**:500–10.
48. Webber BL, Raghu S, Edwards OR. Opinion: is CRISPR-based gene drive a biocontrol silver bullet or global conservation threat? *Proceedings of the National Academy of Sciences of the United States of America* 2015;**112**(34):10565–7. https://doi.org/10.1073/pnas.1514258112.
49. Gould F, Brown ZS, Kuzma J. Wicked evolution: can we address the sociobiological dilemma of pesticide resistance? *Science* 2018;**360**(6390):728–32. https://doi.org/10.1126/science.aar3780.
50. Epanchin-Niell RS, Hufford MB, Asian CE, Sexton JP, Port JD, Waring TM. Controlling invasive species in complex social landscapes. *Frontiers in Ecology and the Environment* 2010;**8**(4):210–6. https://doi.org/10.1890/090029.
51. Kreiner JM, Stinchcombe JR, Wright SI. Population genomics of herbicide resistance: adaptation via evolutionary rescue. *Annual Review of Plant Biology* 2018;**69**:611–35. https://doi.org/10.1146/annurev-arplant-042817-040038.
52. Tallmon DA, Luikart G, Waples RS. The alluring simplicity and complex reality of genetic rescue. *Trends in Ecology and Evolution* 2004;**19**(9):489–96. https://doi.org/10.1016/j.tree.2004.07.003.
53. Whiteley AR, Fitzpatrick SW, Funk WC, Tallmon DA. Genetic rescue to the rescue. *Trends in Ecology and Evolution* 2015;**30**(1):42–9. https://doi.org/10.1016/j.tree.2014.10.009.
54. Novak BJ, Maloney T, Phelan R. Advancing a new toolkit for conservation: from science to policy. *The CRISPR Journal* 2018;**1**(1):11–5. https://doi.org/10.1089/crispr.2017.0019.
55. Butler JB, Freeman JS, Vaillancourt RE, et al. Evidence for different QTL underlying the immune and hypersensitive responses of Eucalyptus globulus to the rust pathogen Puccinia psidii. *Tree Genetics and Genomes* 2016;**12**(3). https://doi.org/10.1007/s11295-016-0987-x.
56. Tobias PA, Guest DI, Külheim C, Hsieh JF, Park RF. A curious case of resistance to a new encounter pathogen: myrtle rust in Australia. *Molecular Plant Pathology* 2016;**17**(5):783–8. https://doi.org/10.1111/mpp.12331.
57. Andersson M, Turesson H, Nicolia A, Fält AS, Samuelsson M, Hofvander P. Efficient targeted multiallelic mutagenesis in tetraploid potato (Solanum tuberosum) by transient CRISPR-Cas9 expression in protoplasts. *Plant Cell Reports* 2017;**36**(1):117–28. https://doi.org/10.1007/s00299-016-2062-3.
58. Zhang Y, Liang Z, Zong Y, et al. Efficient and transgene-free genome editing in wheat through transient expression of CRISPR/Cas9 DNA or RNA. *Nature Communications* 2016;**7**. https://doi.org/10.1038/ncomms12617.
59. Wilson FM, Harrison K, Armitage AD, Simkin AJ, Harrison RJ. CRISPR/Cas9-mediated mutagenesis of phytoene desaturase in diploid and octoploid strawberry. *Plant Methods* 2019;**15**(1). https://doi.org/10.1186/s13007-019-0428-6.
60. Braatz J, Harloff HJ, Mascher M, Stein N, Himmelbach A, Jung C. CRISPR-Cas9 targeted mutagenesis leads to simultaneous modification of different homoeologous gene copies in polyploid oilseed rape (Brassica napus). *Plant Physiology* 2017;**174**(2):935–42. https://doi.org/10.1104/pp.17.00426.
61. Bravo Núñez MA, Nuckolls NL, Zanders SE. Genetic villains: killer meiotic drivers. *Trends in Genetics* 2018;**34**(6):424–33. https://doi.org/10.1016/j.tig.2018.02.003.

62. Yang WC, Shi DQ, Chen YH. Female gametophyte development in flowering plants. *Annual Review of Plant Biology* 2010;**61**:89–108. https://doi.org/10.1146/annurev-arplant-042809-112203.
63. Buckler IVES, Phelps-Durr TL, Buckler CSK, Dawe RK, Doebley JF, Holtsford TP. Meiotic drive of chromosomal knobs reshaped the maize genome. *Genetics* 1999;**153**(1):415–26.
64. Dawe RK, Lowry EG, Gent JI, et al. A kinesin-14 motor activates neocentromeres to promote meiotic drive in maize. *Cell* 2018;**173**(4):839–50. https://doi.org/10.1016/j.cell.2018.03.009. e18.
65. Higgins DM, Lowry EG, Kanizay LB, Becraft PW, Hall DW, Dawe RK. Fitness costs and variation in transmission distortion associated with the abnormal chromosome 10 meiotic drive system in maize. *Genetics* 2018;**208**(1):297–305. https://doi.org/10.1534/genetics.117.300060.
66. Fishman L, Saunders A. Centromere-associated female meiotic drive entails male fitness costs in monkeyflowers. *Science* 2008;**322**(5907):1559–62. https://doi.org/10.1126/science.1161406.
67. Fishman L, Willis JH. A novel meiotic drive locus almost completely distorts segregation in Mimulus (monkeyflower) hybrids. *Genetics* 2005;**169**(1):347–53. https://doi.org/10.1534/genetics.104.032789.
68. Fishman L, Kelly JK. Centromere-associated meiotic drive and female fitness variation in Mimulus. *Evolution* 2015;**69**(5):1208–18. https://doi.org/10.1111/evo.12661.
69. JONES RN. B chromosomes in plants. *New Phytologist* 1995;**131**(4):411–34. https://doi.org/10.1111/j.1469-8137.1995.tb03079.x.
70. Jones RN. Transmission and drive involving parasitic B chromosomes. *Genes* 2018;**9**(8):388. https://doi.org/10.3390/genes9080388.
71. Kanizay LB, Pyhäjärvi T, Lowry EG, et al. Diversity and abundance of the abnormal chromosome 10 meiotic drive complex in Zea mays. *Heredity* 2013;**110**(6):570–7. https://doi.org/10.1038/hdy.2013.2.
72. Werren JH. Selfish genetic elements, genetic conflict, and evolutionary innovation. *Proceedings of the National Academy of Sciences of the United States of America* 2011;**108**(2):10863–70. https://doi.org/10.1073/pnas.1102343108.
73. Lynch M, Marinov GK. The bioenergetic costs of a gene. *Proceedings of the National Academy of Sciences of the United States of America* 2015;**112**(51):15690–5. https://doi.org/10.1073/pnas.1514974112.
74. Endo TR. Gametocidal genes. In: *Alien introgression in wheat: cytogenetics, molecular biology, and genomics*. Czech Republic: Springer International Publishing; 2015. p. 121–31. https://doi.org/10.1007/978-3-319-23494-6_5.
75. Sweigart AL, Brandvain Y, Fishman L. Making a murderer: the evolutionary framing of hybrid gamete-killers. *Trends in Genetics* 2019;**35**(4):245–52. https://doi.org/10.1016/j.tig.2019.01.004.
76. Yu X, Zhao Z, Zheng X, et al. A selfish genetic element confers non-Mendelian inheritance in rice. *Science* 2018;**360**(6393):1130–2. https://doi.org/10.1126/science.aar4279.
77. Knight TM, Steets JA, Vamosi JC, et al. Pollen limitation of plant reproduction: pattern and process. *Annual Review of Ecology, Evolution, and Systematics* 2005;**36**:467–97. https://doi.org/10.1146/annurev.ecolsys.36.102403.115320.
78. Dance A. Core Concept: CRISPR gene editing. *Proceedings of the National Academy of Sciences* 2015;**112**(20):6245–6. https://doi.org/10.1073/pnas.1503840112.

79. Bax NJ, Thresher RE. Ecological, behavioral, and genetic factors influencing the recombinant control of invasive pests. *Ecological Applications* 2009;**19**(4):873–88. https://doi.org/10.1890/07-1588.1.
80. Burt A. Site-specific selfish genes as tools for the control and genetic engineering of natural populations. *Proceedings of the Royal Society B: Biological Sciences* 2003;**270**(1518):921–8. https://doi.org/10.1098/rspb.2002.2319.
81. Burt A, Koufopanou V. Homing endonuclease genes: the rise and fall and rise again of a selfish element. *Current Opinion in Genetics and Development* 2004;**14**(6):609–15. https://doi.org/10.1016/j.gde.2004.09.010.
82. Curtis CF. The case for malaria control by genetic manipulation of its vectors. *Parasitology Today* 1994;**10**(10):371–4. https://doi.org/10.1016/0169-4758(94)90222-4.
83. Davis S, Bax N, Grewe P. Engineered underdominance allows efficient and economical introgression of traits into pest populations. *Journal of Theoretical Biology* 2001;**212**(1):83–98. https://doi.org/10.1006/jtbi.2001.2357.
84. Davis SA, Catchpole EA, Fulford GR. Periodic triggering of an inducible gene for control of a wild population. *Theoretical Population Biology* 2000;**58**(2):95–106. https://doi.org/10.1006/tpbi.2000.1477.
85. Schliekelman P, Ellner S, Gould F. Pest control by genetic manipulation of sex ratio. *Journal of Economic Entomology* 2005;**98**(1):18–34. https://doi.org/10.1093/jee/98.1.18.
86. Spielman A. Why entomological antimalaria research should not focus on transgenic mosquitoes. *Parasitology Today* 1994;**10**(10):374–6. https://doi.org/10.1016/0169-4758(94)90223-2.
87. NASEM (National Academies of Sciences, Engineering, and Medicine). *Gene drives on the horizon: advancing science, navigating uncertainty, and aligning research with public values*. Washington, DC: The National Academies Press; 2016.
88. Eigenbrode SD, Bosque-Pérez NA, Davis TS. Insect-borne plant pathogens and their vectors: ecology, evolution, and complex interactions. *Annual Review of Entomology* 2018;**63**:169–91. https://doi.org/10.1146/annurev-ento-020117-043119.
89. Gilbertson RL, Batuman O, Webster CG, Adkins S. Role of the insect supervectors Bemisia tabaci and frankliniella occidentalis in the emergence and global spread of plant viruses. *Annual Review of Virology* 2015;**2**:67–93. https://doi.org/10.1146/annurev-virology-031413-085410.
90. Perilla-Henao LM, Casteel CL. Vector-borne bacterial plant pathogens: interactions with hemipteran insects and plants. *Frontiers in Plant Science* 2016;**7**(2016). https://doi.org/10.3389/fpls.2016.01163.
91. Pixley KV, Falck-Zepeda JB, Giller KE, et al. Genome editing, gene drives, and synthetic biology: will they contribute to disease-resistant crops, and who will benefit? *Annual Review of Phytopathology* 2019;**57**(1):165–88. https://doi.org/10.1146/annurev-phyto-080417-045954.
92. Lafforgue G, Martinez F, Sardanyes J, et al. Tempo and mode of plant RNA virus escape from RNA interference-mediated resistance. *Journal of Virology* 2011;**85**(19):9686–95. https://doi.org/10.1128/jvi.05326-11.
93. Zaidi SSeA, Tashkandi M, Mansoor S, Mahfouz MM. Engineering plant immunity: using CRISPR/Cas9 to generate virus resistance. *Frontiers in Plant Science* 2016;**7**(2016). https://doi.org/10.3389/fpls.2016.01673.
94. Godfray HCJ, North A, Burt A. How driving endonuclease genes can be used to combat pests and disease vectors. *BMC Biology* 2017;**15**(1). https://doi.org/10.1186/s12915-017-0420-4.

95. Kyrou K, Hammond AM, Galizi R, et al. A CRISPR–Cas9 gene drive targeting doublesex causes complete population suppression in caged *Anopheles gambiae* mosquitoes. *Nature Biotechnology* 2018;**36**(11):1062–6. https://doi.org/10.1038/nbt.4245.
96. Jacobson AL, Nault BA, Vargo EL, Kennedy GG. Restricted gene flow among lineages of thrips tabaci supports genetic divergence among cryptic species groups. *PLoS ONE* 2016;**11**(9). https://doi.org/10.1371/journal.pone.0163882.
97. Byrne FJ, Devonshire AL. Biochemical evidence of haplodiploidy in the whitefly Bemisia tabaci. *Biochemical Genetics* 1996;**34**(3–4):93–107. https://doi.org/10.1007/bf02396243.
98. Jairin J, Kobayashi T, Yamagata Y, et al. A simple sequence repeat- and single-nucleotide polymorphism-based genetic linkage map of the brown planthopper, Nilaparvata lugens. *DNA Research* 2013;**20**(1):17–30. https://doi.org/10.1093/dnares/dss030.
99. Baltzegar J, Cavin Barnes J, Elsensohn JE, et al. Anticipating complexity in the deployment of gene drive insects in agriculture. *Journal of Responsible Innovation* 2018;**5**(Suppl. 1):S81–97. https://doi.org/10.1080/23299460.2017.1407910.
100. Marshall JM, Akbari OS. Can CRISPR-based gene drive Be confined in the wild? A question for molecular and population biology. *ACS Chemical Biology* 2018;**13**(2):424–30. https://doi.org/10.1021/acschembio.7b00923.
101. Dyck VA, Hendrichs J. *Sterile insect technique*. Springer; 2005.
102. Krafsur ES. Sterile insect technique for suppressing and eradicating insect population: 55 years and counting. *Journal of Agricultural and Urban Entomology* 1998;**15**(4):303–17.
103. Bakri A, Mehta K, Lance DR. Sterilizing insects with ionizing radiation. In: *Sterile insect technique: principles and practice in area-wide integrated pest management*. Austria: Springer Netherlands; 2005. p. 233–68. https://doi.org/10.1007/1-4020-4051-2_9.
104. Phuc H, Andreasen MH, Burton RS, et al. Late-acting dominant lethal genetic systems and mosquito control. *BMC Biology* 2007;**5**. https://doi.org/10.1186/1741-7007-5-11.
105. Nikolouli K, Colinet H, Renault D, et al. Sterile insect technique and Wolbachia symbiosis as potential tools for the control of the invasive species Drosophila suzukii. *Journal of Pest Science* 2018;**91**(2):489–503. https://doi.org/10.1007/s10340-017-0944-y.
106. Robinson AS. Genetic basis of the sterile insect technique. In: *Sterile insect technique: principles and practice in area-wide integrated pest management*. Austria: Springer Netherlands; 2005. p. 95–114. https://doi.org/10.1007/1-4020-4051-2_4.
107. Werren JH, Baldo L, Clark ME. Wolbachia: master manipulators of invertebrate biology. *Nature Reviews Microbiology* 2008;**6**(10):741–51. https://doi.org/10.1038/nrmicro1969.
108. Hilgenboecker K, Hammerstein P, Schlattmann P, Telschow A, Werren JH. How many species are infected with Wolbachia? - a statistical analysis of current data. *FEMS Microbiology Letters* 2008;**281**(2):215–20. https://doi.org/10.1111/j.1574-6968.2008.01110.x.
109. Weinert LA, Araujo-Jnr EV, Ahmed MZ, Welch JJ. The incidence of bacterial endosymbionts in terrestrial arthropods. *Proceedings of the royal society B: biological sciences* 2015;**282**(1807). https://doi.org/10.1098/rspb.2015.0249.
110. Zug R, Hammerstein P, Cordaux R. Still a host of hosts for wolbachia: analysis of recent data suggests that 40% of terrestrial arthropod species are infected. *PLoS ONE* 2012;**7**(6):e38544. https://doi.org/10.1371/journal.pone.0038544.
111. Werren JH, Zhang W, Guo LR. Evolution and phylogeny of Wolbachia: reproductive parasites of arthropods. *Proceedings of the Royal Society B: Biological Sciences* 1995;**261**(1360):55–63. https://doi.org/10.1098/rspb.1995.0117.
112. O'Neill SL, Hoffmann AA, Werren JH. *Influential passengers*. Oxford University Press Inc; 1997.

113. Slade G, Morrison N. Developing GM insects for sustainable pest control in agriculture and human health. *BMC Proceedings* 2014;**8**(S4). https://doi.org/10.1186/1753-6561-8-s4-o43.
114. Beech CJ, Miller E. *Regulation of genetically modified insects.* 2013. ISBR News 20/08/2013, http://isbr.info/news?news_id=24. [Accessed 19 December 2019].
115. Beech CJ, Vasan SS, Quinlan MM, et al. Deployment of innovative genetic vector control strategies: progress on regulatory and biosafety aspects, capacity building and development of best-practice guidance. *Asia-Pacific Journal of Molecular Biology and Biotechnology* 2009;**17**(3):75–85. http://www.msmbb.org.my/apjmbb/html173/173c.pdf.
116. Bolton M, Collins HL, Chapman T, et al. Response to a synthetic pheromone source by OX4319L, a self-limiting diamondback moth (Lepidoptera: plutellidae) strain, and field dispersal characteristics of its progenitor strain. *Journal of Economic Entomology* 2019;**112**(4):1546–51. https://doi.org/10.1093/jee/toz056.
117. Simmons GS, Alphey LS, Vasquez T, et al. Potential use of a conditional lethal transgenic pink bollworm Pectinophora gossypiella in area-wide eradication or suppression programmes. In: *Area-wide control of insect pests: from research to field implementation.* United States: Springer Netherlands; 2007. p. 119–23. https://doi.org/10.1007/978-1-4020-6059-5_10.
118. Simmons GS, McKemey AR, Morrison NI, et al. *PLoS One*, vol. 6; 2011.
119. Beeman RW, Friesen KS, Denell RE. Maternal-effect selfish genes in flour beetles. *Science* 1992;**256**(5053):89–92. https://doi.org/10.1126/science.1566060.
120. Buchman A, Marshall JM, Ostrovski D, Yang T, Akbari OS. Synthetically engineered Medea gene drive system in the worldwide crop pest Drosophila suzukii. *Proceedings of the National Academy of Sciences of the United States of America* 2018;**115**(18):4725–30. https://doi.org/10.1073/pnas.1713139115.
121. Chen C-H, Huang H, Ward CM, et al. A synthetic maternal-effect selfish genetic element drives population replacement in Drosophila. *Science* 2007;**316**(5824):597–600. https://doi.org/10.1126/science.1138595.
122. Lorenzen MD, Gnirke A, Margolis J, et al. The maternal-effect, selfish genetic element Medea is associated with a composite Tc1 transposon. *Proceedings of the National Academy of Sciences of the United States of America* 2008;**105**(29):10085–9. https://doi.org/10.1073/pnas.0800444105.
123. Gantz VM, Jasinskiene N, Tatarenkova O, et al. Highly efficient Cas9-mediated gene drive for population modification of the malaria vector mosquito *Anopheles stephensi. Proceedings of the National Academy of Sciences of the United States of America* 2015;**112**(49):E6736–43. https://doi.org/10.1073/pnas.1521077112.
124. Huang Y, Chen Y, Zeng B, et al. CRISPR/Cas9 mediated knockout of the abdominal-A homeotic gene in the global pest, diamondback moth (Plutella xylostella). *Insect Biochemistry and Molecular Biology* 2016;**75**:98–106. https://doi.org/10.1016/j.ibmb.2016.06.004.
125. Douris V, Steinbach D, Panteleri R, Livadaras I, Pickett JA, Van Leeuwen T, Nauen R, Vontas J. Resistance mutation conserved between insects and mites unravels the benzoylurea insecticide mode of action on chitin biosynthesis. *Proceedings of the National Academy of Sciences of the United States of America* 2016;**113**:14692–7.
126. Hall AB, Basu S, Jiang X, Qi Y, Timoshevskiy VA, Biedler JK, Sharakhova MV, Elahi R, Anderson MAE, Chen X-G, Sharakhov IV, Adelman ZN, Tu Z. Sex determination. A male-determining factor in the mosquito *Aedes aegypti. Science* 2015;**348**:1268–70.

127. Li F, Scott MJ. CRISPR/Cas9-mediated mutagenesis of the white and sex lethal loci in the invasive pest, *Drosophila suzukii. Biochemical and Biophysical Research Communications* 2016;**469**:911–6.

128. Bi H-L, Xu J, Tan A-J, Huang Y-P. CRISPR/Cas9-mediated targeted gene mutagenesis in *Spodoptera litura. Journal of Insect Science* 2016;**23**:469–77.

129. Asplen MK, Anfora G, Biondi A, et al. Invasion biology of spotted wing Drosophila (Drosophila suzukii): a global perspective and future priorities. *Journal of Pest Science* 2015;**88**(3):469–94. https://doi.org/10.1007/s10340-015-0681-z.

130. Dos Santos LA, Mendes MF, Krüger AP, Blauth ML, Gottschalk MS, Garcia FRM. Global potential distribution of Drosophila suzukii (Diptera, Drosophilidae). *PLoS ONE* 2017;**12**(3). https://doi.org/10.1371/journal.pone.0174318.

131. Farnsworth D, Hamby KA, Bolda M, Goodhue RE, Williams JC, Zalom FG. Economic analysis of revenue losses and control costs associated with the spotted wing drosophila, Drosophila suzukii (Matsumura), in the California raspberry industry. *Pest Management Science* 2017;**73**(6):1083–90. https://doi.org/10.1002/ps.4497.

132. Mazzi D, Bravin E, Meraner M, Finger R, Kuske S. Economic impact of the introduction and establishment of Drosophila suzukii on sweet cherry production in Switzerland. *Insects* 2017;**8**(1):18. https://doi.org/10.3390/insects8010018.

133. Tochen S, Dalton DT, Wiman N, Hamm C, Shearer PW, Walton VM. Temperature-related development and population parameters for drosophila suzukii (Diptera: drosophilidae) on cherry and blueberry. *Environmental Entomology* 2014;**43**(2):501–10. https://doi.org/10.1603/EN13200.

134. Girod P, Lierhmann O, Urvois T, Turlings TCJ, Kenis M, Haye T. Host specificity of Asian parasitoids for potential classical biological control of Drosophila suzukii. *Journal of Pest Science* 2018;**91**(4):1241–50. https://doi.org/10.1007/s10340-018-1003-z.

135. Wang XG, Nance AH, Jones JML, Hoelmer KA, Daane KM. Aspects of the biology and reproductive strategy of two Asian larval parasitoids evaluated for classical biological control of Drosophila suzukii. *Biological Control* 2018;**121**:58–65. https://doi.org/10.1016/j.biocontrol.2018.02.010.

136. Lanouette G, Brodeur J, Fournier F, et al. The sterile insect technique for the management of the spotted wing drosophila, Drosophila suzukii: establishing the optimum irradiation dose. *PLOS ONE* 2017;**12**(9):e0180821. https://doi.org/10.1371/journal.pone.0180821.

137. Cattel J, Nikolouli K, Andrieux T, et al. Back and forth Wolbachia transfers reveal efficient strains to control spotted wing drosophila populations. *Journal of Applied Ecology* 2018;**55**(5):2408–18. https://doi.org/10.1111/1365-2664.13101.

138. Choi JY, Bubnell JE, Aquadro CF. Population genomics of infectious and integrated wolbachia pipientis genomes in *Drosophila ananassae. Genome Biology and Evolution* 2015;**7**(8):2362–82. https://doi.org/10.1093/gbe/evv158.

139. Gilchrist AS, Cameron EC, Sved JA, Meats AW. Genetic consequences of domestication and mass rearing of pest fruit fly Bactrocera tryoni (Diptera: tephritidae). *Journal of Economic Entomology* 2012;**105**(3):1051–6. https://doi.org/10.1603/EC11421.

140. Kandul NP, Liu J, Sanchez CHM, Wu SL, Marshall JM, Akbari OS. Transforming insect population control with precision guided sterile males with demonstration in flies. *Nature Communications* 2019;**10**(1). https://doi.org/10.1038/s41467-018-07964-7.

141. Horn C, Wimmer EA. A transgene-based, embryo-specific lethality system for insect pest management. *Nature Biotechnology* 2003;**21**(1):64–70. https://doi.org/10.1038/nbt769.

142. Schetelig MF, Lee K-Z, Otto S, et al. Environmentally sustainable pest control options for Drosophila suzukii. *Journal of Applied Entomology* 2018;**142**(1–2):3–17. https://doi.org/10.1111/jen.12469.
143. Port F, Chen HM, Lee T, Bullock SL. Optimized CRISPR/Cas tools for efficient germline and somatic genome engineering in Drosophila. *Proceedings of the National Academy of Sciences of the United States of America* 2014;**111**(29):E2967–76. https://doi.org/10.1073/pnas.1405500111.
144. Kalajdzic P, Schetelig MF. CRISPR/Cas-mediated gene editing using purified protein in Drosophila suzukii. *Entomologia Experimentalis et Applicata* 2017;**164**(3):350–62. https://doi.org/10.1111/eea.12599.
145. Li F, Scott MJ. CRISPR/Cas9-mediated mutagenesis of the white and Sex lethal loci in the invasive pest, Drosophila suzukii. *Biochemical and Biophysical Research Communications* 2016;**469**(4):911–6. https://doi.org/10.1016/j.bbrc.2015.12.081.
146. Cline TW. The Drosophila sex determination signal: how do flies count to two? *Trends in Genetics* 1993;**9**(11):385–90. https://doi.org/10.1016/0168-9525(93)90138-8.
147. Kondo S, Ueda R. Highly Improved gene targeting by germline-specific Cas9 expression in Drosophila. *Genetics* 2013;**195**(3):715–21. https://doi.org/10.1534/genetics.113.156737.
148. Schetelig MF, Handler AM. Germline transformation of the spotted wing drosophilid, Drosophila suzukii, with a piggyBac transposon vector. *Genetica* 2013;**141**(4–6):189–93. https://doi.org/10.1007/s10709-013-9717-6.
149. Akbari OS, Buchmann A. Use of medea elements for biocontrol of D. suzukii populations. In: *US patent number WO2017132207-A1*; 2017.
150. Collins JP. Gene drives in our future: challenges of and opportunities for using a self-sustaining technology in pest and vector management. *BMC Proceedings* 2018;**12**. https://doi.org/10.1186/s12919-018-0110-4.
151. Hochkirch A, Beninde J, Fischer M, et al. License to kill?-disease eradication programs may not be in line with the convention on biological diversity. *Conservation Letters* 2018;**11**(1):e12370. https://doi.org/10.1111/conl.12370.
152. Maitre DCL, et al. Impacts of invasive Australian acacias: implications for management and restoration. *Diversity and Distributions* 2011;**17**:1015–29.
153. Noble C, Adlam B, Church GM, Esvelt KM, Nowak MA. Current CRISPR gene drive systems are likely to be highly invasive in wild populations. *eLife* 2018;**7**. https://doi.org/10.7554/eLife.33423.
154. Whitney KD, Ahern JR, Campbell LG, Albert LP, King MS. Patterns of hybridization in plants. *Perspectives in Plant Ecology, Evolution and Systematics* 2010;**12**(3):175–82. https://doi.org/10.1016/j.ppees.2010.02.002.
155. Marshall JM, Hay BA. Confinement of gene drive systems to local populations: a comparative analysis. *Journal of Theoretical Biology* 2012;**294**:153–71. https://doi.org/10.1016/j.jtbi.2011.10.032.
156. Dhole S, Vella MR, Lloyd AL, Gould F. Invasion and migration of spatially self-limiting gene drives: a comparative analysis. *Evolutionary Applications* 2018;**11**(5):794–808. https://doi.org/10.1111/eva.12583.
157. Noble C, Min J, Olejarz J, et al. Daisy-chain gene drives for the alteration of local populations. *Proceedings of the National Academy of Sciences of the United States of America* 2019;**116**(17):8275–82. https://doi.org/10.1073/pnas.1716358116.
158. Hayes KR, Hosack GR, Dana GV, et al. Identifying and detecting potentially adverse ecological outcomes associated with the release of gene-drive modified organisms. *Journal of Responsible Innovation* 2018;**5**:S139–58. https://doi.org/10.1080/23299460.2017.1415585.

159. Noble C, Min J, Olejarz J, et al. Daisy-chain gene drives for the alteration of local populations. *BioRxiv* 2016. https://doi.org/10.1101/057307.
160. Medina RF. Gene drives and the management of agricultural pests. *Journal of Responsible Innovation* 2018;**5**(Suppl. 1):S255–62. https://doi.org/10.1080/23299460.2017.1407913.
161. Alphey L, Andreasen M. Dominant lethality and insect population control. *Molecular and Biochemical Parasitology* 2002;**121**(2):173–8. https://doi.org/10.1016/S0166-6851(02)00040-3.
162. O'Connor L, Plichart C, Sang AC, Brelsfoard CL, Bossin HC, Dobson SL. Open release of male mosquitoes infected with a wolbachia biopesticide: field performance and infection containment. *PLoS Neglected Tropical Diseases* 2012;**6**(11). https://doi.org/10.1371/journal.pntd.0001797.
163. Oye KA, Esvelt K, Appleton E, et al. Regulating gene drives. *Science* 2014;**345**(6197):626–8. https://doi.org/10.1126/science.1254287.
164. Turner CE. *Conflicting interests and biological control of weeds. International symposium of biological control of weeds*. 1985. p. 203–25.
165. Dudley TL, Bean DW. Tamarisk biocontrol, endangered species risk and resolution of conflict through riparian restoration. *BioControl* 2012;**57**(2):331–47. https://doi.org/10.1007/s10526-011-9436-9.
166. Zimmermann HG, Bloem S, Klein H. *Cactoblastis cactorum. The biology, history, threat, surveillance and control of the cactus moth." FAO/IAEA division of nuclear techniques in food and agriculture*. 2007.
167. Zimmermann HG, Klein H, Bloem S. The biology, history, threats, surveillance and control of the cactus moth, Cactoblastis cactorum. In: *Joint FAO/IAEA division of nuclear techniques in food and agriculture IAEA*; 2004.
168. Bourtzis K, Miller TA. *Insect symbiosis*. Greece: CRC Press; 2003. p. 1–347. https://doi.org/10.1201/9780203009918.
169. Oliver KM, Degnan PH, Burke GR, Moran NA. Facultative symbionts in aphids and the horizontal transfer of ecologically important traits. *Annual Review of Entomology* 2010;**55**:247–66. https://doi.org/10.1146/annurev-ento-112408-085305.
170. Douglas AE. Nutritional interactions in insect-microbial symbioses: aphids and their symbiotic bacteria Buchnera. *Annual Review of Entomology* 1998;**43**:17–37. https://doi.org/10.1146/annurev.ento.43.1.17.
171. Courtier-Orgogozo V, Morizot B, Boëte C. Agricultural pest control with CRISPR-based gene drive: time for public debate: should we use gene drive for pest control? *EMBO Reports* 2017;**18**(6):878–80. https://doi.org/10.15252/embr.201744205.
172. Adelman Z, Akbari O, Bauer J, et al. Rules of the road for insect gene drive research and testing. *Nature Biotechnology* 2017;**35**(8):716–8. https://doi.org/10.1038/nbt.3926.
173. Akbari OS, Bellen HJ, Bier E, et al. Safeguarding gene drive experiments in the laboratory. *Science* 2015;**349**(6251):927–9. https://doi.org/10.1126/science.aac7932.
174. DiCarlo JE, Chavez A, Dietz SL, Esvelt KM, Church GM. Safeguarding CRISPR-Cas9 gene drives in yeast. *Nature Biotechnology* 2015;**33**(12):1250–5. https://doi.org/10.1038/nbt.3412.

Chapter 7

# Phytoremediation strategies using biotechnology

## 1. Introduction

Heavy metal contamination of soil and ground water endangers the environmental conditions and is a great health risk for humans and livestock. Being nondegradable, heavy metals prevail in the environment causing food chain contamination which result in negative impacts on human health. Therefore, elimination of heavy metals is of prime concern to the human populace. Traditional practices for remediation of heavy metals are expensive, adverse and laborious. On the other hand, remediation technologies driven by plants, otherwise called phytoremediation, incur much lower costs as they only require solar energy and are compatible with the ecosystem which makes them worthy of public acceptance. Plants are intrinsically favorable for elimination of environmental pollutants as they remove the heavy metal contaminations by incorporating them within their own metabolic pathways while having symbiotic relationship with microbes. Therefore, there is an imperative requirement to employ plants to rehabilitate the ecosystem. This chapter discusses remediation of pollutants from air in addition to in situ phytoremediation to decontaminate groundwater, wastewater and soil. Hyperaccumulator plants, nanophytoremediaton and plants integrated with nanocomposites are addressed along with a discussion of the challenges and forthcoming prospects of phytoremediation technology.

## 2. Removal of air pollutants by phytoremediation: purification of indoor air pollutants by plants

It is well established that plants imbibe carbon dioxide and release oxygen via photosynthesis while enabling the degradation of volatile organic compounds (VOCs) in synergy with microorganisms in the rhizosphere. Also, plants absorb particulate matter in the atmosphere via leaves.[1–3] Earlier investigations have shown that plant stomata in the course of their normal gas

Agricultural Biotechnology. https://doi.org/10.1016/B978-0-323-88476-1.00007-2

exchange absorb the VOCs and subsequently convert them into amino acids by means of the Calvin cycle.[4,5] Additionally, leaves imbibe indoor air pollutants but nevertheless, physical impediments concerned with mesophyll and stomatal resistance circumscribe that.[6] Normally, plants control the porosity of their stomata through the production of abscisic acid in response to changes in the environment.[7] The elimination of air pollutants in the indoor atmosphere by plants is constrained by their total surface area, morphological, and anatomical attributes as well as their chemical properties and soil matrix characteristics.[8]

Specifically, the occurrence, density, and shape of leaf trichomes affect the leaf-mediated removal of atmospheric pollutants.[9] Microbes living in the rhizosphere and plant buds have been demonstrated to remove VOCs.[10] There exists an inverse relation between the area of the plant leaves and VOC removal, which substantiated that the removal of pollutants by plants is driven by microorganisms instead of the plants themselves.[11,12] Biodegradation of aromatic hydrocarbons by plants is mediated via a couple of microbial aerobic pathways in the toluene metabolism within the indoor plant rhizospheres.

The aromatic ring of toluene is recognized by the dioxygenase in the toluene dioxygenase (TOD) pathway, while its methyl substituent interacts with the monooxygenase in the toluene methyl monooxygenase (TOL) pathway.[13] The TOD pathway enables metabolization of benzene, while both the TOL and TOD pathways are responsible for metabolizing xylene and toluene.[14] Plants possess enzyme-based scavenging mechanisms, which enable pollutant degradation in the leaves while maintaining stable absorption at the time the stomata remain open.[6] Notwithstanding, thus far, only a few of the enzymes involved in metabolism of pollutants have been found, and the mechanism of removal of a majority of compounds in toxic gases is not yet clear.[15] Due to the complexity of the types of pollutants and the wide array of plants, the best scavenging effects of specific plants have not yet been discerned.

Hence, scientific criteria such as plant anatomy, in particular, the size and density of stomata, the composition of the epidermis and the mesophyll as well as plant morphological characteristics such as size, shape, and hairiness of the leaves have to be factored in when selecting plants for the indoor environments. Among air pollutants occurring in the indoor environment, a majority of them comprise of toxic gases such as carbon dioxide, carbon monoxide, nitrogen oxides, VOCs, benzene, formaldehyde, as well as aromatics.[16] Table 7.1 enlists the different rates of removal of air pollutants by plant species. As evident from the table, *Nephrolepis obliterata* shows the best formaldehyde removal efficacy, which can occur up to 90%–100% within 48 h.

**TABLE 7.1** Removal rate of air pollutants by plant species.

| Tree species | Types of pollutants | Pollutant concentration (ug/m3) | Removal rate | References |
|---|---|---|---|---|
| *Ficus elastica* | Benzene | 15,116 | 28%–91% in 12 h | Chun et al., 2010[17] |
| *Agave potatorum* | Formaldehyde | 15,000 | 2.21–4.60 mg/m3 in 7 days | Zhou et al., 2011[18] |
| *Ardisia crenata* | Toluene | 4896–13,181 | 2.5–40 μg/m3/h/m2 | Kim et al., 2012[19] |
| *Howea forsteriana* | n-hexane | 80,000–353,000 | 367–4032 mg/m3/day/m2 | Wood et al., 2002[20] |
| *Ardisia japonica* | Formaldehyde | 2488 | 0.14–0.88 μg/m3/cm2 in 5 h | Kim and kim, 2008[21] |
| *Azalea indica* | Toluene | 339,000 | 7–76 h to remove 95% of the initial dose | De kempeneer et al., 2004[22] |
| *Sansevieria trifasciata* | Toluene | 7,8,800–81,670 | 10.17 μmol in 72 h | Sriprapat et al., 2014[23] |
| *Nephrolepis obliterata* | Formaldehyde | 600–11000 | 90%–100% in 48 h | Teiri et al., 2018[24] |

From Wei et al. A review on phytoremediation of contaminants in air, water and soil. *Journal of Hazardous Materials* 2021; **403**:123,658.

The efficiency of pollutant removal by the different plant species varies with the species of the concerned plants and the concentration of the pollutants. The major means by which air pollutant removal is effected is through absorption of air-borne particles and their degradation and subsequent transformation within leaves and the stomata.[25] Plants enhance air quality by carbon dioxide absorption and oxygen release via photosynthesis and additionally augment air humidity through conductance by the stomata.[26] Nevertheless, it is still not clear by what mechanism plant stomata remove indoor air pollutants. Recent studies have shown that air pollutants are absorbed by the plants, and these chemicals are further metabolized as energy and carbon sources.[6] Green algae have been reported to perform the highest ethylbenzene removal rate.[26]

Important parameters such as light intensity, temperature, microbial organisms, and stomatal conductance determine the VOC removal rate in the indoor environment.[27–29] The rate of removal of particles including liquid or solid matter that is dispersed uniformly in the aerosol within the indoor atmosphere depends on the intensity of the ambient light and the area of the leaf.[30] Petit et al., 2019[31] found that green plants diminish the concentration of particulate materials in the indoor air. Pollutants in the air accumulate on the leaf surface and are absorbed within the plant tissues, and some are naturally metabolized into the plants.[32] The amount of particulate matter PM2.5 that is adsorbed onto to the leaf surface correlates with the degree of pollution[33] and is dependent on the leaf length, surface roughness, hairiness, and frequency of stomata.[34,35]

Trees accumulate PM mainly through their leaves. Xu et al., 2018,[36] showed that 65% of PM was deposited on the leaf surface and 35% on the wax. Out of the total PM absorbed, 28% was constituted by water-soluble ions. The concentration of trees in addition to tree coverage influences that rate of removal of air pollutants.[37] For instance, Conifers efficiently remove PM2.5, and Pinus trees are the most efficient in capturing PM. NO2 tolerance is exhibited by Magnolia, wherein the $NO_2$ is metabolized.[38,39] The efficiency of atmospheric remediation depends also on the introduction and composition of microbial organisms.[40]

Hence, to augment the purification of pollutants in the atmosphere, an array of biomass structure, as well as good management of greenery, plays important roles.[41] The removal of atmospheric PM is most efficient in trees containing expanded leaf areas.[42] The tree canopy structure and size determine the increase in airflow and deposition of PM on the leaves.[43] In comparison to leaves possessing smooth surfaces, leaf surfaces that are rough allow more PM deposition due to the presence of hairy structures.[44] The capability of PM deposition is dependent on the process of adhesion between the surface of the leaf blade and the PM, which is driven by the intermolecular force shared between the particles and the wax. The composition of the chemicals in the PM and those in the interface area to a large extent influence the absorption.[45] The efficacy of PM removal is augmented also by rainfall, which results in leaf expansion. The width of the tree crown and surface area of the leaves determine the differential purification effects of the different tree species.

Therefore, in order to enable more efficient PM removal from the atmosphere, tree species containing large, rough surfaces have to be considered along with increase in tree density so as to provide the crown with as high a space as possible. Along with these parameters, an appropriate environment has to be given for microbial organisms in order to sustain the trees.

## 3. Removal of water pollution by phytoremediation

Wetland areas lie in the zone of transition between aquatic and terrestrial ecosystems and are divided into natural and constructed wetlands. Within the wetland ecosystem, there occurs continuous recycling of nutrients, treatment of wastewater, facilitation of plant absorption, along with removal of chemicals, toxins and heavy metals from the water.[46] Fig. 7.1 shows a schematic for phytoremediation in air, soil, and water systems.

First, the heavy metals in the water combine with sediments and the soil to generate insoluble salts as precipitates following which they undergo absorption by plants.[47] The prevalent physical as well as chemical properties of

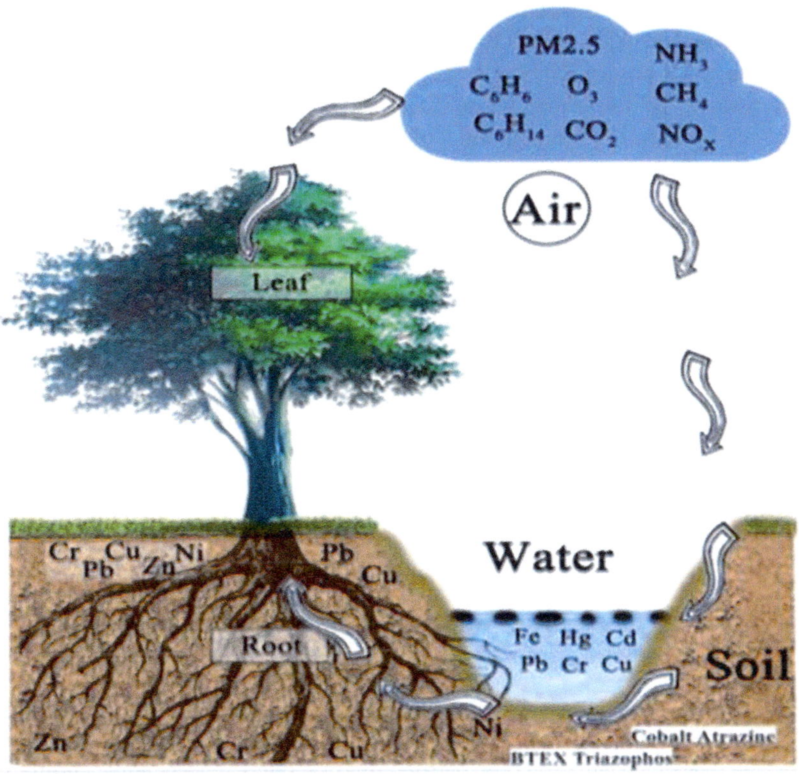


**FIGURE 7.1 A schematic drawing showing phytoremediation in air, soil, and water.** Phytoremediation in air (particulate matter and volatile organic compounds), soil (heavy metals), and water (heavy metals and organic matter). *From Wei et al. A review on phytoremediation of contaminants in air, water and soil.* Journal of Hazardous Materials *2021;**403**:123658.*

the wetland environment provide conducive conditions for removal of water pollution by phytoremediation and reinforce the redox reaction occurring between the rhizosphere microorganisms and the plants.[48] To diminish aquatic ecosystem eutrophication, plants are employed to remove phosphorus and nitrogen from the water.[49,50]

The Aspen and Litmus plants showed efficiencies of nitrate nitrogen and phosphorous removal rates to be 75%–99% and 40%–90%, respectively,[51] The depletion of nutrients enabled by plant absorption lead to considerable decrease in the natural growth of algae, thus augmenting water qualities such as pH and turbidity.[52] Additionally, plants possess enormous potential to remove water pollution due to heavy metals. For instance, *Eichhornia crassipes* as well as water slippery cotyledons could deplete organic matter, nutrients, and heavy metals contained in wastewater for as long as 10–17 days in heavy organic load feedlot effluents.[53] The plant *Eleocharis acicularis* been shown to efficiently deplete heavy metals such as zinc, copper, nickel, and lead from wastewater and urban river ecosystems.[54] Leaves and roots of decayed plants enabled strong decrease in heavy metal concentrations in aquatic systems.[55] It was found that the uptake of Cr occurred rapidly and was completed within the first 24 h. Nevertheless, Cr translocation from roots to leaves was very low. Therefore, the major reason for the enhancement of Cr content in the leaves was due to Cr absorption by direct contact. Jayaweera et al., 2008[56] found that the removal of iron from wastewater by the water hyacinth was primarily caused by chemical precipitation of $Fe_2O_3$ and Fe $(OH)_3$ as well as root filtration. Various plant species show differential effects of removal of heavy metals. For instance, among *Pistia stratiotes, Elodea densa, and Eichhornia crassipes, Pistia stratiotes* shows the highest Hg removal effect, while the *Egeria densa* has the best Cd removal effect. Importantly, construction of artificial wetland systems should be considered along with phytoremediation in order to achieve high efficiency of removal of water pollution.

The plant species *Mentha aquatic*, *Ludwigia palustris,* and *Myriophyllum aquaticum* showed an average removal efficiency for Hg to be 99.8% when they were grown in contaminated water with up to 0.5 mg Hg per liter.[57] Plants such as *Pistia stratiotes*, *Colocasia esculenta*, *Eichornia crassipes,* and *Scirpus tabernaemontani* diminished water Hg concentration of 0.5 or 2 mg Hg L1 to nontoxic levels after 30 days.[58] *Pistia stratiotes* accumulated up to 83.2 mg Hg kg1 dw.

## 4. Removal of organic pollutants from the soil by phytoremediation

With the ongoing progress of contemporary industrialization and social advancements, there is an ever-increasing requirement of petroleum-driven fuel energy. The widespread use of petroleum compounds causes undesirable effects such as pollution of valuable water and land resources.[59] Conventional strategies to confront oil pollution at the soil level include treatment of the soils by physical, chemical, heat, and biological methods.[60,61] However, these methods incur high costs and cause several side effects impacting the ecosystems, which calls for the use of phytoremediation strategies as they are more efficient, economical, and environment-friendly.[62,63]

Plant-based mechanisms for depletion of organic pollutants in the soil depend on the concerned species of the plants but nevertheless occur within the rhizomes wherein there exists symbiosis between microorganisms and the plants which enable soil recovery.[64] In this symbiotic association, plants confer a conducive environment for the growth of microorganisms and enhance soil structure, while, in turn, the microbes diminish soil toxicity and augment the growth and metabolism of the plants by depleting organic pollutants. Therefore, soil stability and fertility in mining areas can be efficiently improved by selecting pertinent species of plants and microbes. This improves ecosystem sustainability as well as the quality of the surrounding habitat through plant microbial remediation.[65]

Crude oil spills lead to the generation of monocyclic aromatic hydrocarbons of low molecular weight, which are volatilized into the surrounding atmosphere and at the same time cause the permeation of polycyclic aromatic hydrocarbons containing greater than 20 carbon atoms into the immediate soil environment, which disrupts the soil structure.[66] Plants possess inherent abilities to breakdown soil organic hydrocarbons by means of secretion and relationships with microbial organisms.[67] Plants such as *Zea mays* can deplete 97% of the pesticide atrazine from the soil.

Hence, the adoption of phytoremediation to deplete soil organic pollutants conserves resources and precludes environmental pollution at the secondary level. Additionally, plants add desirable aesthetic values. Plants promote the decomposition of organic pollutants through their own metabolism, which generates nutrients and triggers the action of rhizosphere microbial organisms. Notwithstanding, a majority of phytoremediation research has been conducted at the experimental level, and this calls for more long-term research in the field environment. Since there exist differences in the nature of organic pollutants in different pollution sites, this necessitates appropriate selection of plant species for phytoremediation. Furthermore, because plants convert organic pollutants into nontoxic compounds, this precludes the requirement to manage potential waste substances.

## 5. In situ phytoremediation

The uptake of heavy metals by plants via roots, leaves, and shoots is termed "in situ phytoremediation." A giant reed of the Poaceae plant family, *Arundo donax* (L.) as well as the mycorrhized *A. donax* are both favorable heavy metal bioaccumulators. *A. donax* in symbiosis with the fungus *Trichoderma harzianum* has been shown to perform good bioaccumulation of metals such as Hg, Pb, and Ni.[68] Phytoremediation is inherently eco-friendly by virtue of being driven by the plants and the sun and is a cost-effective green technology for recovery of contaminated sites while at the same time being aesthetic and feasible for long-term applications. Pollutants including a wide array of inorganic and organic elements can be removed from soil, water, and air through phytoremediation. High concentrations of metals can be accumulated within the biomass of these plants growing in metal-contaminated soils without any effects of toxicity.[69] Moreover, in situ phytoremediation incurs lesser infrastructure costs and labor compared to conventional methodologies.[70,71]

## Decontamination of soil contaminated with heavy metals by in situ phytoremediation

With the advent of intensive industrialization, the plenitude of mining operations for coal as well as other minerals has resulted in several contaminated lands the world over. Plants can be used to preclude soil erosion and fix toxic compounds in order to enable the removal of hazardous materials and facilitate the restoration of vegetation and soil fertility.[72] Heavy metals a major hazard in mining areas and cause detrimental effects on human health and surrounding ecosystems. Such environments can be restored by planting fast-growing trees that are long-rooted and adapted to drought and metal contaminations.[73,74] Nevertheless, due to the limitation of the contaminated soil several plants may not be appropriate for survival in these soils. Therefore, in addition to the problem of heavy metal accumulation, localized climatic conditions have to be factored in while using phytoremediation.[75]

Importantly, in situ phytoremediation can be applied through phytoextraction as well as phytostabilization. Phytoextraction is a useful remediation methodology for uptake of toxic heavy metals and metalloids from soils, sediments, and water.[76] The efficiency of phytoextraction is dependent on several factors such as toxic metal bioavailability in the soil, various properties of the soil including heavy metal speciation, and plant species particularly hyperaccumulators. The potential for phytoextraction is mainly driven by the shoot biomass and the shoot total metal concentration.[77]

Phytostabilization is another efficient phytoremediation technique to lower the mobility of metals such as As, Cu, Cr, Cd, Pb, and Zn in contaminated soils. However, the major drawback with this methodology is that these toxic heavy metals remain persistently in the soil while only their mobility is circumscribed. Hence, phytostabilization is not a permanent fix but only a management technique for stabilization or immobilization of toxic contaminating elements.[78] This strategy is useful for mitigating the ecological threats of contaminated sites, which works in the short-term especially in areas devoid of vegetation due to high concentrations of metals in the soil. Nevertheless, phytostabilization is not applicable in extremely contaminated sites. Generally, phytostabilization is administered in conjunction with chemical stabilization in order to rehabilitate abandoned mining areas.

Some chemical modifications such as the application of lime, phosphates, and compost are necessary for phytostabilization in order to diminish toxicity and increase heavy metal bioavailability.[79] Some plants such as hippo grass (*Vossia cuspidata* (Roxb.) Griff.), wiregrass (*Gentiana pennelliana* Fernald),

Syrian bean-caper (*Zygophyllum fabago* L.), and red fescue (*Festuca rubra* L.) show great potential for phytostabilization of soil areas contaminated with Cu, Zn, Cr, and Pb.[80,81]

Plants capable of highly efficient growth and possessing metal sequestration properties are typically selected for purposes of heavy metal phytoremediation. Upon investigation of mining areas, it was found that high level accumulation of Cd and Mn occurred in the above-ground regions of plants such as *Camellia oleifera, Celosia argentea, Polygonum pubescens and Solanum nigrum.* These species significantly impact soil phytoremediation in heavy metal contaminated lands in China.[82] Also, phytostabilization may be augmented through soil amendments that cause immobilization of metal(loid)s accumulated within plants that are highly tolerant to contaminants and poor fertility soils or tailings.[83] There is steady buildup of metal soil contamination because of continued emission activities due to industrial production operations.[84] Such heavy metal contaminations also impact the food chain that lead to detrimental health effects in livestock and humans.[85,86]

Various weeds accumulate metals from contaminated soils to different extents.[87] Weed plant mobility (>1) mirrors their capability for metal restoration. Hence, weed species are employed as economically viable and eco-friendly strategies for phytoremediation of land regions contaminated with toxic metals. Plants such as *Brassica napus* were found to augment the diversity and plenitude of soil microbes that reduced detrimental effects of Cd on the soil.[88] The cultivation of metal-tolerant plants in metal polluted soils supports microbial organisms to augment fertility of the soil and the growth of plants.[89,90] Therefore, targeted and selective application of rhizobia tolerant to heavy metals and capable of promoting the growth of plants including microbes such as nitrogen fixing symbiotic organisms could help enhance the efficacy of plant restoration (Table 7.2).[91]

**TABLE 7.2 Plant species accumulating heavy metals in the sewage.**
*Some examples of plant species accumulating heavy metals in the sewage.*

| Plant | Heavy metal | Accumulation (d. w.) | Reference |
|---|---|---|---|
| *Eichhornia crassipes* | Hg | 119 ng g-1 | Molisani et al., 2006[92] |
| *Pistia stratiotes* | Hg | 215 ng g-1 | Molisani et al., 2006[92] |
| *Egeria densa* | Cd | 70.25 mg g-1 | Pietrobelli et al., 2009[93] |
| *Salvinia natans* | Cr | 7.4 mg g-1 | Dhir et al., 2009[94] |
| *Eleocharis acicularis* | Fe | 59,500 μg g-1 | Ha et al., 2009[95] |
| *Vallisneria spiralis* | Cr | 2.85 mg g-1 | Verma et al., 2008[96] |

From Wei et al. A review on phytoremediation of contaminants in air, water and soil. *Journal of Hazardous Materials* 2021;**403**:123658.

Table 7.2 enlists plants showing differential capabilities to remove heavy metals. Genetic engineering approaches and the use of nanoparticles to function as catalysts can improve phytoremediation of heavy metal contaminated soils. There exists a unique ratio between nanomaterial volume and single molecular size, which influences the plant chlorophyll content.[97]

Nanomaterials enable augmentation of protein levels in plants in addition to enhancing interactions with microbial organisms in the plant rhizosphere to boost heavy metal absorption and increase soil fertility.[98] Nanoparticles greatly increase the Pb and Cd residual concentration in heavy metal contaminated soils augmenting the fixation of these metals in the improved soil.[99] Under conditions of cadmium stress,[100] found that carbon nanotubes diminish the production of organic solutes, re-establish plant growth, augment, and decrease Cd accumulation in the shoot and the root.

In this context, compost is often employed as a soil enhancer to reinforce the ability of plants to retrieve soils due to its beneficial ecology.[101] Compost contributes organic matter, elemental nutrients, microelement nutrients to the soil in addition to favorably impacting its physical and chemical properties. This in turn furthers soil fertility and bolsters plant growth.[102] Furthermore,[103] showed that composting could enable the aggregation of plants and heavy metal purification by altering the quantity of soil functional groups, thereby recirculating heavy metals. Organic modifications can augment heavy metal removal from the soil by influencing the adsorption, reduction, complexation, and volatilization of the metals.[104] Similarly, phytoremediation can be improved through the addition of sawdust and domestic compost to the forest soil.[105] Nevertheless, despite phytoremediation of soil heavy metals being a green and habitat friendly technology, there still are several challenges to be overcome in practicing it including the slow growth of the plants, poor efficacy of metal utilization, and hence phytoremediation in the long run. In this context, it would be beneficial to identify appropriate catalysts for plants and expedite the environmental phytorestoration efficacy.

## Decontamination of groundwater contaminated with heavy metals by in situ phytoremediation

Rhizofiltration is a highly efficient strategy for remediation of groundwater contaminated with heavy metals. This technique is useful to filter water contaminated with effluents as well as groundwater contaminated with heavy metals by means of plants such as macroalgae and aquatic macrophytes.[106] Among these plants, *Lemna gibba* (L.) is an efficient as an accumulator.[107] *Hydrilla verticillata* (L.f.) Royle plant is a bioindicator of as pollution present in aquatic ecosystems. About 28 macrophytes have been identified from the As-contaminated river from New Zealand, which showed high as uptake concentrations.[108] The aquatic macrophytes *L. gibba* (L.), *Ceratophyllum demersum* (L.), and *H. verticillata* were shown to remove high amounts of As. A medley of at least a couple of macrophytes from among *Lemna minor* (L.), *H. verticillata, C. demersum, Eichhornia crassipes,* (Mart.) Solms, and *Wolffia globosa* (Roxb.) Hartog, were found to effectively remove as contaminations at levels better than individual plants.

## 6. Improved integrated phytoremediation measures

### Sustainable phytoremediation

Economic, social, and environmental factors influence sustainable phytoremediation and can be attained by using natural colonizer plants to enable spontaneous vegetation growth. It is a well-known finding that heavy metals can neither be destroyed or degraded and undergo bioaccumulation into the food chain causing detrimental health effects such as cancers as well as dysfunctions and organ damage in humans.[109] Out of the plenitude of technologies currently available, phytoremediation is favored due to its low-cost compatibility with natural ecosystems. Several plant species that are natural colonizers could be cultivated easily on soils contaminated with heavy metals.

These natural colonizer plants would necessarily have to be perennial, economically important and unpalatable, and should be amenable to ecological succession. Sustainable phytoremediation is beneficial in promoting carbon sequestration, conservation of biodiversity, improvement of substrate quality, and soil reclamation.[110] Therefore, phytoremediation is a dynamic technology, and while remediating contaminated sites, it yields economically valuable phytoproducts such as biochar, biomass, aromatic essential oils, paper and pulp, ornamental plants, biodiesel, various derived woods, and improved crops by augmenting biofortification.

Aromatic, inedible plants help in mediating a viable approach to phytoremediation as they bioaccumulate heavy metals precluding their entry into the food chain. Also, they are unpalatable for animals due to their essence. Therefore, they offer a safe, economically viable, and ecologically feasible strategy for phytoremediation. *Chrysopogon zizanioides* (L.) Roberty), an aromatic plant vetiver is a perennial grass of the Poaceae family and functions as a heavy metal bioaccumulator. It is highly effectual and can be used both in irrigation measures and wetland systems. Due to its high rate of proliferation and high yielding biomass, it can remove high amounts of heavy metal contamination from toxic sites. The deployment of a vetiver system costs only a fraction of that of traditional technologies and requires little infrastructure as well as minimal maintenance. The sweet basil (*Ocimum basilicum* L.) is a popular aromatic herb used as spice, medicine and essential oil and has been found to be a hyperaccumulator of Cr, Pb and Cd.[111] Citronella (*Cymbopogon winterianus* Jowitt) is a perennial evergreen grass that yields essential oils of use on cosmetic industries, and this can be repurposed to preclude soil erosion. Palmarosa (*Cymbopogon martini* Roxb. Wats.) yields insect repellants[112] and lemon grass (*Cymbopogon flexuosus Nees* ex Steud. W.Watson) is another aromatic plant used in the edible product industry while acting as a barrier to preclude weed growth and functioning as a heavy metal accumulator.

## Microbially-enabled phytoremediation

Microbe-driven phytoremediation is a new and favorable technology for removal of toxic metals from the soil. A wide array of algae, bacteria, and fungi are capable of toxic metal remediation at different concentrations. In specific symbiotic plant–microbe systems, each partner in the symbiotic relationship collaborates in various ways.[113] Significant phytoremediation is achieved by the microbially enabled endurance capability of the plants in soils containing high levels of toxic metal contaminations.[114] Table 7.3 enlists a few microbial species effecting remediation of heavy metals.

**TABLE 7.3 Heavy metal removal by microbes.**
*Removal of heavy metals by microbes.*

| Microbial species | Heavy metals removed | Percentage removal efficiency |
|---|---|---|
| **Bacteria** | | |
| Iron oxidizing bacteria | $Cr^{6+}$, $Cu^{2+}$, $Zn^{2+}$, $Ni^{2+}$, $Pb^{2+}$ | 16.2–91.5 |
| *Pseudomonas aeruginosa* | $Cd^{2+}$ | 99 |
| *Enterobacter cloacae* | $Cd^{2+}$, $Cu^{2+}$, $Co^{2+}$ | 1.43–65.95 |
| **Algae** | | |
| *Microspora* | $Pb^{2+}$ | 97 |
| *Spirogyra* sp. | $Pb^{2+}$ | 80 |
| **Fungi** | | |
| *Aspergillus niger* | $Pb^{2+}$, $Cu^{2+}$, $Ni^{2+}$, $Zn^{2+}$ | High removal efficiency |
| *Mucor rouxii* | $Pb^{2+}$, $Ni^{2+}$, $Cd^{2+}$, $Zn^{2+}$ | 60 |

From Srivastava S, Agrawal SB, Mondal MK. A review on progress of heavy metal removal using adsorbents of microbial and plant origin. *Environmental Science and Pollution Research* 2015;**22**(20):15386–15415. https://doi.org/10.1007/s11356,-015-52,78-9.

## Remediation of heavy metals by plant-associated microbial organisms

Even as heavy metals are useful and are incorporated in several biochemical pathways of microbes and plants,[115] these metals at high concentrations can lead to the generation of nonspecific compounds that can have negative cytotoxic effects.[116] Metals such as molybdenum, iron, and manganese are essential trace elements that have low toxicity levels. Nevertheless, metals like Ce, Ga, Sn, Th, and Zr are ineffectual for any biological function. Besides, certain metals including Co, Cu, Zn, Ni, W, and V are useful while also being capable of causing toxic effects. Metals such as As, Cd, Hg, Pb, and U play limited biological roles and are very toxic.[117,118]

Toxic heavy metals in soils, at elevated concentrations influence the growth of microbes and plants by having direct adverse implications on population size and species diversity.[119] The rhizosphere is formed by the assembly of plant species with microbial organisms, wherein the latter are mutually integrated with the roots of the plants and are termed rhizospheric microbes. These can include algae, bacteria, protozoa, nematodes, and fungi all of which collaborate with plants and the soil toward a mutually beneficial association. Additionally, the rhizosphere structure has a dense configuration wherein the root exudate composition and that of the various root zones as well as the plant type and quantity determine their co-existence in a mutually integrated manner.[120,121] The rhizosphere can detoxify metals by metal immobilization[122] and stimulate plant growth by improving the nutritional value of the plants.[123]

The plenitude of rhizosphere microflora and fauna differs from that of their equivalents in the bulk soil and varies with the soil as well as the species of the plants. Certain bacterial microbes are effectual for both the growth rate and the uptake of phosphorous and toxic metals into plants via plant hyperaccumulators such as *Sedum alfredii* Hance.[124] The effectuality of metallophyte plants is primarily controlled by means of the total metal bioavailability in the environment. Certain microbes have the capability to reduce metals such as Cr, Fe, Mn, U, and Te and enable their bioavailability for absorption by the plants. They can also influence the levels of metal solubility, mobility, and phytotoxicity[125] whereby they act as bio-surfactants. *Pseudomonas aeruginosa* (Gram-negative bacteria) generates bio-surfactant that is useful for remediation of soils contaminated with Cd and Pb.[126] Rhizosphere bacteria have been implicated in reducing Zn toxicity[127] and Cr toxicity, the latter by the *Enterobacter cloacae*, a Gram-negative facultative bacterium.[128]

## Bacterial siderophores

Plants and microorganism secrete small iron-chelating compounds called siderophores that enhance Fe solubility in the soil and uptake by the plants. Siderophores have also been shown to augment the tolerance/resistance of the plants to heavy metal contamination. *Thlaspi goesingense*, a microbial organism that has been shown to have naturally occurring Ni resistance in Ni-contaminated sites, was able to synthesize siderophores.[129] The plant *Solanum nigrum,* which is a Cd hyperaccumulator benefits from interaction with its endophyte, the Pseudomonas sp. Lk9, which produces siderophores, organic acids, and biosurfactants that increase Cu, Zn, and Cd availability, thus promoting phytoextraction.[130] Bacterial siderophores have been suggested as agents of chelation in the remediation of heavy metals. The synthesis of siderophores was induced in metal-resistant Streptomyces spp. in the presence of metals such as Cd2+, Al3+, Cu2+, and Ni2+ and the interaction of Ni and Cd to these siderophores was ascertained.[131] Cd uptake was shown to stimulate siderophore synthesis in *Pseudomonas aeruginosa* strain PAO1.[132] Extracellular Hg2+ chelation was observed in the *Pseudomonas aeruginosa* siderophore that formed a complex that was internalized through a transporter molecule located at the cell surface.[133] It has been demonstrated that Hg does not stimulate the production of siderophores, although occurrence of Hg siderophore hinders the uptake of Fe-siderophore complex.[133]

## 7. Nanophytoremediation: an inimitable strategy

Several studies have shown that remediation of heavy metal contaminated soils can be augmented when phytoremediation is supplemented with nanotechnology. Nanomaterials such as nano-Ni, nano-Zn, nano-Fe3O4, nano-Ag, nano-SnO2, nZVI (nanozero valent iron), nanocarbon black, and biochar-supported nano-hydroxyapatite can be used for remediation of Cd, Cr, Cu, Pb, and Zn contaminants from soil.[134–136] Different nanomaterials cause different nanophytoremediation effects. Also, by varying the particle size of these nanomaterials, toxic heavy metal immobilization is effected from contaminated soils.[137]

When mustard plants are supplemented with nanozero valent iron (nZVI), they can remove Cr in contaminated soils while at the same time, enhancing plant growth rate.[138] nZVI has been shown to improve phytoremediation of Pb-contaminated sediments.[139] At lower concentrations, nZVI promoted higher uptake of Pb along with enhancement of the concentration of plant biomass. Pb phytoremediation effected by *Lolium perenne* (L.) is enabled by low nZVI (100–500 mg/kg) concentration. Significantly, nZVI at higher concentrations inhibited plant growth, negatively impacted the leaf physiological structure, and induced oxidative stress in *L. perenne*. Hence, caution should be exercised when applying nanomaterials for augmenting phytoremediation.

### Nano-$TiO_2$ integrated plants

$TiO_2$ nanoparticles can augment plant protein and promote 133Cs uptake in roots and in soil surrounding the plants.[140] $TiO_2$ has also been shown to enable increase in average root length of cucumber and onion plants.[141] Nano-$TiO_2$ at

0.01% concentration in the soil significantly augments leaf area, root dry weight, and stem height in legumes.[142] Cd toxicity has been shown to be remediated by nano-$TiO_2$ while improving plant growth rate and the rate of photosynthesis.[143] This study also showed Cd accumulation in flower buds and roots of *Glycine max* L. Merr., and increase in concentration of the nano-$TiO_2$ concomitantly increased Cd concentration, which predominantly accrued in roots than buds.

### Plants integrated with nanocomposites

Nanocomposite materials such as nanocarbon black (NCB) and nano-hydroxyapatite (NHAP) when added to soil can promote Pb absorption by plants. NCB when used along with nZVI, reduced Pb toxicity, and enhanced L. perenne biomass.[144] Biochar-supported nanohydroxyapatite caused reduction in Pb availability in the soil and positively impacts soil remediation.[145,146] Bentonite at higher amounts diminishes the Pb levels in brown rice and straw.[147] The mobility of Pb in the soil can be effectually decreased by using zeolites and nano-$SiO_2$.[134] Absorption of heavy metals by plants was increased by administration of biochar and nanohydroxyapatite.[148]

Rehabilitation of plants and ecological role of southern salt marshes were favored by Ag particles.[149] Cu–Zn double metal nanomaterials were shown to improve precipitation of heavy metals in the soil.[150] Nano-$SiO_2$ improved the solubility of Pb and Cd in the rhizosphere and enabled easy accumulation of these heavy metals by the plants thus enhancing phytoremediation.[134]

## 8. Genetic engineering measures used for phytoremediation

In recent times, remediation of heavy metals is increasingly being achieved by the overexpression of genes encoding metal chelators and metal transporters for augmenting the metal metabolism of plants.[151,152] Besides, the ATP binding cassette gene termed the ABC gene is overexpressed in order to enhance the mobile ion translocation or to bind the nonmobile ions with parts of the plants. The hyperaccumulator plant, *Noccaea caerulescens* encodes a metal transporter, which enables metal remediation from toxic contaminated soils in an efficient manner.[153]

In a reported study by Lin et al., 2016,[154] it has been shown that overexpression of the transgene NcZNT1 that encodes a Zn transporter resulted in increased Cd and Zn accumulation in modified plants of *Arabidopsis thaliana* (L.) Heynh. The heavy metal tolerance capacity can be enhanced by the overexpression of phytochelatins synthase without augmenting accumulation in *Nicotiana glauca* and *Nicotiana tabacum* L. wherein both of these plants were shown to be Cd tolerant.[155] Genetic engineering–based phytoremediation has been found to be most feasible in Nicotiana species and

*Arabidopsis thaliana*, although it has also been demonstrated in Oryza, Populus, and *Brassica juncea* species. However, heavy metal tolerance and accumulation undergo regulation by two independent mechanisms. This inherently incurs some risks such as decrease in nutrient absorption by the plants and increase in heavy metal sensitivity of the plants. Therefore, comprehensive risk management measures have to be adopted while using genetic engineering techniques toward phytoremediation.

## 9. In situ phytoremediation mechanisms and examples of hyperaccumulator plants

Phytoremediation is primarily driven by technologies called phytoextraction, phytofiltration, or rhizofiltration, phytostabilization and phytovolatilization each of which possesses unique mechanisms for remediation of metal-polluted sediments, soil or water. Fig. 7.2 shows a schematic of these processes. In the phytoextraction technique, plants absorb the metals from the soil and transfer them to the plant biomass (mainly in the shoots and roots) where they remain. In phytostabilization, plants efficiently stabilize the contaminants through metal retention in the roots instead of removing the metals. The rhizofiltration

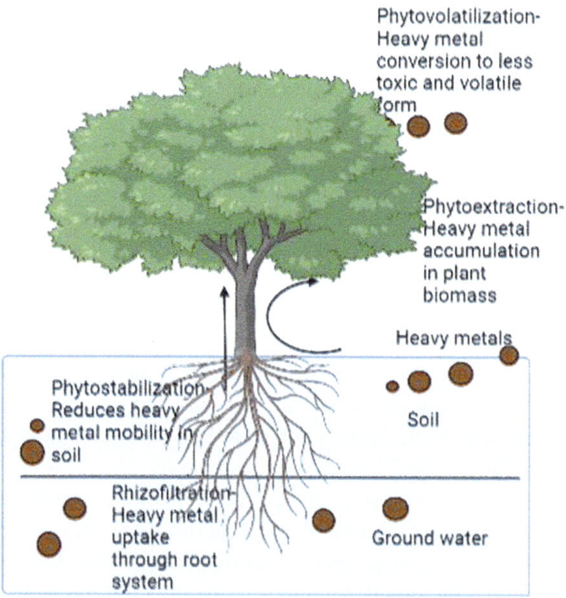


**FIGURE 7.2 Mechanisms of in situ phytoremediation.** Schematic of mechanisms of in situ phytoremediation. *From Shikha D, Singh PK. In situ phytoremediation of heavy metal–contaminated soil and groundwater: a green inventive approach.* Environmental Science and Pollution Research *2021;**28**(4):4104–4124. https://doi.org/10.1007/s11356-020-11600-7.*

or phytofiltration is accomplished through the roots of aquatic plants to remove contaminants from water. In phytovolatilization, plants extract metals from the soil via the leaves, which release the contaminants into the ambient atmosphere through the process of volatilization.

## Phytoextraction

In phytoextraction, toxic pollutants from soil or water are accumulated in plant roots and plant biomass above the ground such as shoots.[156,157] Phytoextraction is a highly favorable technique for remediation of radionuclide wastes and hazardous metals.[57] The concentration of metals in the soil is an important determinant of the phytoextraction of toxic heavy metals.[77] Several hydrophilic or aquatic plant species like *Ludwigia stolonifera*,[158] *Myriophyllum aquaticum*,[159] and *Mentha aquatica* (L.)[160] accumulate toxic metals by phytoextraction.

Accumulation of copper from contaminated water is performed by macrophyte species such as *E. crassipes* and *Centella asiatica* (L.)[161]) and reportedly their efficiency of copper accumulation are 99.6% and 97.3%, respectively. Certain species of *Brassica juncea* were found to accumulate Cu, Ni, and Pb from contaminated areas into their biomass above the ground.[162] Cr and Cd accumulation was found in the harvestable parts of the plant *Cyperus rotundus* (L.), although another study found that the Cr remediation was more than the Cd removal.[163]

## Phytostabilization

Phytostabilization is a preferred technique due to its benefits over other remediation measures. In phytostabilization, the densely rooted plants stabilize or immobilize the metal contaminants by absorption enabled through dense root hairs, and these metals are precipitated within the rhizosphere and are adsorbed onto the surface of the roots.[164] Therefore, this circumscribes the movement of the contaminants as they are trapped within the root region of the plants thereby impeding their entry into the shoots and leaves, which precludes their entry into the food chain and hence diminishes contaminant bioavailability. This technique can help in vegetation restabilization in contaminated sites.[165]

Phytostabilization is suitable for remediation of contaminated soils, which are rich in organic matter and are fine textured. Phytostabilization precludes the leaching of heavy metals into the groundwater by limiting their accumulation in the plant biomass. Grasses such as *Agrostis tenuis* cv. Goginan are able to remediate Cu, Pb, and Zn from toxic mine fields, while *Festuca rubra* cv. Merlin removes Zn and Pb mine wastes and *A. tenuis* cv. Parys removes Cu mine wastes. Nevertheless, phytostabilization is only a feasible management technology for stabilization of toxic metals.

## Phytofiltration

Phytofiltration is a technique that removes contaminants from contaminated surface water and sewage water through the plant–root system.[166] This includes blastofiltration (remediation of contaminants through seedlings,[167] caulofiltration (remediation of pollutants through shoots), and rhizofiltration (remediation through plant roots). The plant roots absorb the contaminants, predominantly metals from groundwater and wastewater by rhizofiltration.[168] Partial treatment of agricultural runoff, effluent discharge, and drainage from acid mines can be achieved using rhizofiltration. Also, rhizofiltration can be used in the remediation of metals such as Cr, Cu, Cd, and Ni, which are initially absorbed into the root system.[169]

Several species of macrophytes possess the capability to remove toxic metals from polluted water. These include pennywort (*C. asiatica*), a perennial herbaceous plant[161]; the common water hyacinth (*E. crassipes*), which is an invasive species of macrophyte,[170] and water lentils or duckweeds.[171] However, plants that perform phytofiltration have certain drawbacks such as complexities in the process of composting, drying, and incineration. Despite this, the water hyacinth has been shown to remediate trace elements in wastewater streams.[172] Terrestrial plants such as the sunflower (*Helianthus annuus* L.) and the Indian mustard (*B. juncea*) have also been demonstrated to remediate contaminated water, the former for remediating toxic metals such as Pb[69] and the latter for absorbing Ni, Cd, Cr, Cu, Pb, and Zn.[173,174]

## Phytovolatilization

In phytovolatilization, the plants absorb the toxic contaminants and convert them into their volatile form after which these contaminants undergo evaporation via the leaves of the plants into the surrounding atmosphere. Toxic metals such as arsenic, mercury, and selenium, which occur in the environment as volatile forms namely, methyl arsines, mercuric oxide, and dimethyl selenide, respectively, are volatilized into the atmosphere. Nevertheless, phytovolatilization happens to be a controversial technique as the contaminants are not sufficiently removed using this method.[69]

*B. juncea* (Indian mustard) is able to accumulate Se and Hg,[175] which are further converted into their volatilized forms in the plant leaves and released into the atmosphere. LeDuc and Terry 2005[176] report that Se phytovolatilization from contaminated soils can be achieved by the macrophyte, *Typa latifolia* (L.). Additionally, phytovolatilization of Se from contaminated soils is very advantageous particularly during crop rotation, wherein many crop species can be harvested. Also, the generated Se-contaminated plant biomass serves as good food supplements for livestock. Macrophytes are often used to remove Se from polluted sites.[177] In this context, elemental Se in soil is more toxic compared to its volatile form.[178] Treatment of irrigation drainage streams is feasible through phytotvolatilization.[179] Table 7.4 enlists the major advantages and disadvantages associated with phytoremediation.

**TABLE 7.4 Phytoremediation: advantages and disadvantages.**
*Advantages and disadvantages of phytoremediation.*

| Phytoremediation mechanism | Advantages | Disadvantages |
|---|---|---|
| Phytoextraction | Promotes high growth plant biomass | Slow mechanism |
| | Incurs low cost | Low metal absorption rates from roots |
| | Appropriate in laboratory and field studies | Low metal bioavailability in rhizosphere |
| | Functions to conserve the ecosystem | |
| Phytostabilization | No requirement for contaminant disposal | Contaminant remain in soil until hyperaccumulator plants eliminated |
| | Reduces contaminant mobility | |
| | Diminishes soil erosion | |
| | Metal-tolerant hyperaccumulator plants | |
| Phytofiltration | Contaminant remediation from surface water | To a great extent, species-specific |
| | Contaminant remediation from ground water | Adjustment in pH impedes plant growth |
| | Remediation of radioactive elements | Rapid plant growth hampers subsequent disposal |
| | Aquatic and terrestrial plants used | |
| | Remediation of toxic heavy metals | |
| Phytovolatilization | Contaminants converted into less toxic forms | Contaminants resettle back into surrounding environment |
| | Suitable for sediments, sludges, and soils | |

From Shikha D, Singh PK. In situ phytoremediation of heavy metal—contaminated soil and groundwater: a green inventive approach. *Environmental Science and Pollution Research* 2021;**28**(4):4104–4124. https://doi.org/10.1007/s11356-020-11600-7.

## 10. Challenges faced by phytoremediation procedures

Phytoremediation, despite being an eco-friendly, reliable, plant-based technology for treatment of toxic soils contaminated with heavy metals, has several drawbacks which have to be overcome: all phytoremediation procedures require protracted lengths of time; most metal hyperaccumulator plants are constrained by their low biomass and delayed growth rate; the bioavailability of toxic contaminants in the soil is limited due to the slow mobility of metal ions through the soil; also, the accumulation capability of plants used for metal decontamination is circumscribed due to prevalence of pests and plant diseases; the growth and development of these plants are not readily sustained in highly contaminated soils containing toxic metals; the establishment of invasive hyperaccumulator plants can disrupt the native floral diversity; several agronomic practices and soil amelioration measures can impact the mobility of contaminants in an adverse manner, and there is the problem of disposing plant biomass containing high concentrations of toxic metals; most of the phytoremediation procedures are influenced by weather and climatic conditions; lack of diligence in the handling of the plant biomass obtained from these hyperaccumulators can result in undesirable contamination of the food chain; introduction of new invasive plant bioaccumulators may negatively impact the surrounding biodiversity; the reach of the phytoremediation technique is limited to a few meters from the water surface for the groundwater and 1 m from the soil surface; and choosing and identifying plant species that are locally available for the purpose of phytoremediation is a major challenge.

## 11. Conclusions and future directions

This chapter provides a broad overview of phytoremediation. Phytoremediation is a rapidly emerging field and has various applications. It is low-cost and eco-friendly; however, studies at the field level are required for optimizing phytoremediation measures. Different chemicals in the environment may be metabolized by different enzymatic transformation mechanisms, and more research in this area is warranted. The use of plant growth-promoting bacteria toward in situ phytoremediation also needs to be further explored.[180] The generation of transgenic plants with desirable high metal accumulation capability, increased biomass production, and adaptability toward different climatic conditions is required by means of genetic engineering procedures. Heavy metal bioavailability can be improved to increase the solubility rate of these metals and their absorption by the plant–root system. The fate of the toxic metals within the plant tissues needs to be explored in order to augment the mechanisms involved in hyperaccumulation of metals by the plants.

The contamination of soil and groundwater with toxic heavy metals is a life-endangering environmental constraint. This necessitates the practice of efficient remediation procedures. Considering that conventional methods for physical and chemical treatment of wastewaters and soils incur high costs and are complicated by climatic conditions, phytoremediation is favorable due to its low-cost, biodegradability and environmental friendliness in addition to its solar-driven approach. The human population is incurring several health ailments due to the intake of food and water contaminated with heavy metals, which calls for improved discernment of phytoremediation as a technology. Sustainable phytoremediation procedures are required using nanomaterials and rhizospheric microbial organisms. Traditional phytoremediation would not be economically viable especially for large-scale applications, and therefore, this calls for rapid genetic engineering methodologies to improve heavy metal accumulation capacity of these plants and for production of high plant biomass.

Plants are by and large utilized to retrieve the ecological balance of environments such as soil, water, and air because of their capabilities of absorption and metabolism mediated by the relationship between plant roots and microbial organisms. Plants remediate contaminants without impacting the topsoil, thereby conserving and even augmenting soil fertility via root exudation.[181] Considering the extensive diversity of plant biology, a cautious selection of plant species is called for to efficiently rectify and restore individual ecosystems. Therefore, despite compost and nanomaterials being highly effective in plant restoration to the surrounding environment, an elevated dose of these could be detrimental to the plants. Additionally, while using nanomaterials in combination with compost and other biotechnological methods to deplete pollutants, the effect of these nanomaterials as well as compost on plant-inhabiting microbes should be factored in. Toward improving the applicability of phytoremediation for restoring the environment, future research is necessary to investigate the impacts of various kinds of catalysts on phytoremediation efficacy.

## References


[1] Gawrońska H, Bakera B. Phytoremediation of particulate matter from indoor air by Chlorophytum comosum L. plants. *Air Quality, Atmosphere and Health* 2015;**8**(3):265–72. https://doi.org/10.1007/s11869-014-0285-4.

[2] Orwell RL, Wood RA, Burchett MD, Tarran J, Torpy F. The potted-plant microcosm substantially reduces indoor air VOC pollution: II. Laboratory study. *Water, Air, and Soil Pollution* 2006;**177**(1–4):59–80. https://doi.org/10.1007/s11270-006-9092-3.

[3] Torpy F, Clements N, Pollinger M, et al. Testing the single-pass VOC removal efficiency of an active green wall using methyl ethyl ketone (MEK). *Air Quality, Atmosphere and Health* 2018;**11**(2):163–70. https://doi.org/10.1007/s11869-017-0518-4.

[4] Peterson G, Jones T, Rispoli D, Stitt D, Haddadi S, Niri V. *Monitoring volatile organic compound removal by common indoor plants using solid phase microextraction and gas chromatography-mass spectrometry*. 2016.

[5] Yoo MH, Kwon YJ, Son KC, Kays SJ. Efficacy of indoor plants for the removal of single and mixed volatile organic pollutants and physiological effects of the volatiles on the plants. *Journal of the American Society for Horticultural Science* 2006;**131**(4):452–8. https://doi.org/10.21273/jashs.131.4.452.

[6] Brilli F, Fares S, Ghirardo A, et al. Plants for sustainable improvement of indoor air quality. *Trends in Plant Science* 2018;**23**(6):507–12. https://doi.org/10.1016/j.tplants.2018.03.004.
[7] Tardieu F, Parent B, Simonneau T. Control of leaf growth by abscisic acid: hydraulic or non-hydraulic processes? *Plant, Cell and Environment* 2010;**33**(4):636–47. https://doi.org/10.1111/j.1365-3040.2009.02091.x.
[8] Irga PJ, Torpy FR, Burchett MD. Can hydroculture be used to enhance the performance of indoor plants for the removal of air pollutants? *Atmospheric Environment* 2013;**77**:267–71. https://doi.org/10.1016/j.atmosenv.2013.04.078.
[9] Li S, Tosens T, Harley PC, et al. Glandular trichomes as a barrier against atmospheric oxidative stress: relationships with ozone uptake, leaf damage, and emission of LOX products across a diverse set of species. *Plant Cell and Environment* 2018;**41**(6):1263–77. https://doi.org/10.1111/pce.13128.
[10] Khaksar G, Treesubsuntorn C, Thiravetyan P. Effect of endophytic Bacillus cereus ERBP inoculation into non-native host: potentials and challenges for airborne formaldehyde removal. *Plant Physiology and Biochemistry* 2016;**107**:326–36. https://doi.org/10.1016/j.plaphy.2016.06.020.
[11] Kim KJ, Kil MJ, Song JS, Yoo EH, Son K-C, Kays SJ. Efficiency of volatile formaldehyde removal by indoor plants: contribution of aerial plant parts versus the root zone. *Journal of the American Society for Horticultural Science* 2008;**133**(4):521–6. https://doi.org/10.21273/jashs.133.4.521.
[12] Orwell RL, Wood RL, Tarran J, Torpy F, Burchett MD. Removal of benzene by the indoor plant/substrate microcosm and implications for air quality. *Water, Air, and Soil Pollution* 2004;**157**(1–4):193–207. https://doi.org/10.1023/B:WATE.0000038896.55713.5b.
[13] Mikesell MD, Kukor JJ, Olsen RH. Metabolic diversity of aromatic hydrocarbon-degrading bacteria from a petroleum-contaminated aquifer. *Biodegradation* 1993;**4**:249–59.
[14] Tsao CW, Song HG, Bartha R. Metabolism of benzene, toluene, and xylene hydrocarbons in soil. *Applied and Environmental Microbiology* 1998;**64**(12):4924–9. https://doi.org/10.1128/aem.64.12.4924-4929.1998.
[15] Korte F, Kvesitadze G, Ugrekhelidze D, et al. Organic toxicants and plants. *Ecotoxicology and Environmental Safety* 2000;**47**(1):1–26. https://doi.org/10.1006/eesa.2000.1929.
[16] Davardoost F, Kahforoushan D. Modeling the dispersion of volatile organic compounds in indoor environment. *Chemical Engineering and Technology* 2019;**42**(3):549–59. https://doi.org/10.1002/ceat.201800244.
[17] Chun SC, Yoo MH, Moon YS, Shin MH, Son KC, Chung IM, et al. Effect of Bacterial Population from Rhizosphere of Various Foliage plants on removal of indoor volatile Organic Compounds. *Korean Horticultural Science and Technology* 2010;**28**(3):476–83.
[18] Zhou JH, Qin FF, Su J, Liao JW, Xu HL. Purification of formaldehyde-polluted air by indoor plants of Araceae, Agavaceae and Liliaceae. *Journal of Food, Agriculture and Environment* 2011;**9**(3–4):1012–8.
[19] Kim KJ, Yoo EH, Kays SJ. Decay Kinetics of toluene Phytoremediation Stimulation. *Hortscience* 2012;**47**(8):1195–8.
[20] Wood RA, Orwell RL, Tarran J, Torpy F, Burchett M. Potted-plant/growth media interactions and capacities for removal of volatiles from indoor air. *Journal of Horticultural Science and Biotechnology* 2002;**77**(1):120–9.
[21] Kim KJ, Kim HD. Development of Model and Calculating Equation for rate of volatile formaldehyde removal of IndoorPlants. *Horticulture Environment and Biotechnology* 2008;**49**(3):155–61.

[22] De Kempeneer L, Sercu B, Vanbrabant W, Van Langenhove H, Verstraete W. Bioaugmentation of the phyllosphere for the removal of toluene from indoor air. *Applied Microbiology and Biotechnology* 2004;**64**(2):284–8.

[23] Sriprapat W, Suksabye P, Areephak S, et al. Uptake of toluene and ethylbenzene by plants: removal of volatile indoor air contaminants. *Ecotoxicology and Environmental Safety* 2014;**102**(1):147–51. https://doi.org/10.1016/j.ecoenv.2014.01.032.

[24] Teiri H, Pourzamzni H, Hajizadeh Y. Phytoremediation of formaldehyde from indoor environment by ornamental plants: An approach to promote occupants health. *International Journal of Preventive Medicine* 2018;**14**(9):70.

[25] Zhang WK, Wang B, Niu X. Adsorption capacity of the air particulate matter in urban landscape plants in different polluted regions of Beijing. *Huanjing Kexue/Environmental Science*. 2015;**36**(7):2381–8. https://doi.org/10.13227/j.hjkx.2015.07.008.

[26] Rodriguez-Hernandez MC, Flores-Chaparro CE, Rangel-Mendez JR. Influence of dissolved organic matter and oil on the biosorption of BTEX by macroalgae in single and multi-solute systems. *Environmental Science and Pollution Research* 2017;**24**(26):20922–33.

[27] Dela Cruz M, Christensen JH, Thomsen JD, Müller R. Can ornamental potted plants remove volatile organic compounds from indoor air?—a review. *Environmental Science and Pollution Research* 2014;**21**(24):13909–28. https://doi.org/10.1007/s11356-014-3240-x.

[28] Park HM, Lee AK. Efficiency of removal of indoor pollutants by pistia stratiotes, eichhornia crassipes and hydrocotyle umbellata. *Journal of People, Plants, and Environment* 2020;**23**(1):15–21. https://doi.org/10.11628/ksppe.2020.23.1.15.

[29] Zhao S, Zhao Y, Liang H, Su Y. Formaldehyde removal in the air by six plant systems with or without rhizosphere microorganisms. *International Journal of Phytoremediation* 2019;**21**(13):1296–304. https://doi.org/10.1080/15226514.2019.1586036.

[30] Yoon JW, Son KC, Yang DS, Kays SJ. Removal of indoor tobacco smoke under light and dark conditions as affected by foliage plants. *Korean Journal of Horticultural Science and Technology* 2009;**27**(2):312–8.

[31] Pettit T, Irga PJ, Torpy FR. The in situ pilot-scale phytoremediation of airborne VOCs and particulate matter with an active green wall. *Air Quality, Atmosphere and Health* 2019;**12**(1):33–44. https://doi.org/10.1007/s11869-018-0628-7.

[32] Gawronski SW, Gawronska H, Lomnicki S, Sæbo A, Vangronsveld J. Plants in air phytoremediation. *Advances in Botanical Research* 2017;**83**:319–46. https://doi.org/10.1016/bs.abr.2016.12.008.

[33] Lu S, Yang X, Li S, et al. Effects of plant leaf surface and different pollution levels on PM2.5 adsorption capacity. *Urban Forestry and Urban Greening* 2018;**34**:64–70. https://doi.org/10.1016/j.ufug.2018.05.006.

[34] Ram SS, Majumder S, Chaudhuri P, et al. A review on air pollution monitoring and management using plants with special reference to foliar dust adsorption and physiological stress responses. *Critical Reviews in Environmental Science and Technology* 2015;**45**(23):2489–522. https://doi.org/10.1080/10643389.2015.1046775.

[35] Ram SS, Majumder S, Chaudhuri P, et al. Plant canopies: bio-monitor and trap for resuspended dust particulates contaminated with heavy metals. *Mitigation and Adaptation Strategies for Global Change* 2014;**19**(5):499–508. https://doi.org/10.1007/s11027-012-9445-8.

[36] Xu Y, Xu W, Mo L, Heal MR, Xu X, Yu X. Quantifying particulate matter accumulated on leaves by 17 species of urban trees in Beijing, China. *Environmental Science and Pollution Research* 2018;**25**(13):12545–56. https://doi.org/10.1007/s11356-018-1478-4.

[37] Selmi W, Weber C, Rivière E, Blond N, Mehdi L, Nowak D. Air pollution removal by trees in public green spaces in Strasbourg city, France. *Urban Forestry and Urban Greening* 2016;**17**:192–201. https://doi.org/10.1016/j.ufug.2016.04.010.

[38] Gourdji S. Review of plants to mitigate particulate matter, ozone as well as nitrogen dioxide air pollutants and applicable recommendations for green roofs in Montreal, Quebec. *Environmental Pollution* 2018;**241**:378–87. https://doi.org/10.1016/j.envpol.2018.05.053.

[39] Yang J, Chang Y, Yan P. Ranking the suitability of common urban tree species for controlling PM2.5pollution. *Atmospheric Pollution Research* 2015;**6**(2):267–77. https://doi.org/10.5094/APR.2015.031.

[40] Wang Z, Pei J, Zhang JS. Experimental investigation of the formaldehyde removal mechanisms in a dynamic botanical filtration system for indoor air purification. *Journal of Hazardous Materials* 2014;**280**:235–43. https://doi.org/10.1016/j.jhazmat.2014.07.059.

[41] Jim CY, Chen WY. Assessing the ecosystem service of air pollutant removal by urban trees in Guangzhou (China). *Journal of Environmental Management* 2008;**88**(4):665–76. https://doi.org/10.1016/j.jenvman.2007.03.035.

[42] McDonald AG, Bealey WJ, Fowler D, et al. Quantifying the effect of urban tree planting on concentrations and depositions of PM10 in two UK conurbations. *Atmospheric Environment* 2007;**41**(38):8455–67. https://doi.org/10.1016/j.atmosenv.2007.07.025.

[43] Fowler D, Cape JN, Unsworth MH. Deposition of atmospheric pollutants on forests. *Philosophical Transactions—Royal Society of London, B.* 1989;**324**(1223):247–65.

[44] Burkhardt J. Hygroscopic particles on leaves: nutrients or desiccants? *Ecological Monographs* 2010;**80**(3):369–99. https://doi.org/10.1890/09-1988.1.

[45] Wang L, Gong H, Liao W, Wang Z. Accumulation of particles on the surface of leaves during leaf expansion. *Science of the Total Environment* 2015;**532**:420–34. https://doi.org/10.1016/j.scitotenv.2015.06.014.

[46] Williams JB. Phytoremediation in wetland ecosystems: progress, problems, and potential. *Critical Reviews in Plant Sciences* 2002;**21**(6):607–35. https://doi.org/10.1080/0735-260291044386.

[47] Weis JS, Weis P. Metal uptake, transport and release by wetland plants: implications for phytoremediation and restoration. *Environment International* 2004;**30**(5):685–700. https://doi.org/10.1016/j.envint.2003.11.002.

[48] Macek T, Macková M, Káš J. Exploitation of plants for the removal of organics in environmental remediation. *Biotechnology Advances* 2000;**18**(1):23–34. https://doi.org/10.1016/S0734-9750(99)00034-8.

[49] Liu JL, Liu JK, Anderson JT, Zhang R, Zhang ZM. Potential of aquatic macrophytes and artificial floating island for removing contaminants. *Plant Biosystems* 2016;**150**(4):702–9. https://doi.org/10.1080/11263504.2014.990535.

[50] Yu S, Miao C, Song H, Huang Y, Chen W, He X. Efficiency of nitrogen and phosphorus removal by six macrophytes from eutrophic water. *International Journal of Phytoremediation* 2019;**21**(7):643–51. https://doi.org/10.1080/15226514.2018.1556582.

[51] Sundaralingam T, Gnanavelrajah N. Phytoremediation potential of selected plants for nitrate and phosphorus from ground water. *International Journal of Phytoremediation* 2014;**16**(3):275–84. https://doi.org/10.1080/15226514.2013.773279.

[52] Han P, Vijayaraghavan K, Reuben S, Estrada ES, Joshi UM. Reduction of nutrient contaminants into shallow eutrophic waters through vegetated treatment beds. *Water Science and Technology* 2013;**68**(6):1280–7. https://doi.org/10.2166/wst.2013.361.

[53] Rizzo PF, Bres PA, Arreghini S, Crespo DE, Serafini RJM, Fabrizio de Iorio AR. Remediation of feedlot effluents using aquatic plants. *Rev. FCA UNCUYO.* 2012;**44**(2):47–64.

[54] Ha NTH, Sakakibara M, Sano S, Hori RS, Sera K. The potential of eleocharis acicularis for phytoremediation: case study at an abandoned mine site. *Clean—Soil, Air, Water* 2009;**37**(3):203–8. https://doi.org/10.1002/clen.200900009.

[55] Burezq H, Aliewi A. Using phytoremediation by decaying leaves and roots of reed (Phragmites austrates) plant uptake to treat polluted shallow groundwater in Kuwait. *Environmental Science and Pollution Research* 2018;**25**(34):34570–82. https://doi.org/10.1007/s11356-018-3385-0.

[56] Jayaweera MW, Kasturiarachchi JC, Kularatne RKA, Wijeyekoon SLJ. Contribution of water hyacinth (*Eichhornia crassipes* (Mart.) Solms) grown under different nutrient conditions to Fe-removal mechanisms in constructed wetlands. *Journal of Environmental Management* 2008;**87**(3):450–60. https://doi.org/10.1016/j.jenvman.2007.01.013.

[57] Kamal M, Ghaly AE, Mahmoud N, CoteCôté R. Phytoaccumulation of heavy metals by aquatic plants. *Environment International* 2004;**29**(8):1029–39. https://doi.org/10.1016/S0160-4120(03)00091-6.

[58] Skinner K, Wright N, Porter-Goff E. Mercury uptake and accumulation by four species of aquatic plants. *Environmental Pollution* 2007;**145**(1):234–7. https://doi.org/10.1016/j.envpol.2006.03.017.

[59] Brassington KJ, Hough RL, Paton GI, et al. Weathered hydrocarbon wastes: a risk management primer. *Critical Reviews in Environmental Science and Technology* 2007;**37**(3):199–232. https://doi.org/10.1080/10643380600819625.

[60] Dave D, Ghaly AE. Remediation technologies for marine oil spills: a critical review and comparative analysis. *American Journal of Environmental Sciences* 2011;**7**(5):424–40. https://doi.org/10.3844/ajessp.2011.424.440.

[61] Walls WD. Petroleum refining industry in China. *Energy Policy* 2010;**38**(5):2110–5. https://doi.org/10.1016/j.enpol.2009.06.002.

[62] Ndimele PE. A review on the phytoremediation of petroleum hydrocarbon. *Pakistan Journal of Biological Sciences* 2010;**13**(15):715–722. https://doi.org/10.3923/pjbs.2010.715.722.

[63] YANG SZ, JIN HJ, WEI Z, et al. Bioremediation of oil spills in cold environments: a review. *Pedosphere* 2009;**19**(3):371–81. https://doi.org/10.1016/S1002-0160(09)60128-4.

[64] Meagher RB. Phytoremediation of toxic elemental and organic pollutants. *Current Opinion in Plant Biology* 2000;**3**(2):153–62. https://doi.org/10.1016/S1369-5266(99)00054-0.

[65] Juwarkar AA, Jambhulkar HP. Phytoremediation of coal mine spoil dump through integrated biotechnological approach. *Bioresource Technology* 2008;**99**(11):4732–41. https://doi.org/10.1016/j.biortech.2007.09.060.

[66] Yavari S, Malakahmad A, Sapari NB. A review on phytoremediation of crude oil spills. *Water, Air, and Soil Pollution* 2015;**226**(8). https://doi.org/10.1007/s11270-015-2550-z.

[67] Feng NX, Yu J, Zhao HM, et al. Efficient phytoremediation of organic contaminants in soils using plant–endophyte partnerships. *Science of the Total Environment* 2017;**583**:352–68. https://doi.org/10.1016/j.scitotenv.2017.01.075.

[68] Cristaldi A, Oliveri Conti G, Cosentino SL, et al. Phytoremediation potential of Arundo donax (Giant Reed) in contaminated soil by heavy metals. *Environmental Research* 2020;**185**:109427. https://doi.org/10.1016/j.envres.2020.109427.

[69] Padmavathiamma PK, Li LY. Phytoremediation technology: hyper-accumulation metals in plants. *Water, Air, and Soil Pollution* 2007;**184**(1–4):105–26. https://doi.org/10.1007/s11270-007-9401-5.

[70] Burges A, Alkorta I, Epelde L, Garbisu C. From phytoremediation of soil contaminants to phytomanagement of ecosystem services in metal contaminated sites. *International Journal of Phytoremediation* 2018;**20**(4):384–97. https://doi.org/10.1080/15226514.2017.1365340.

[71] Jaskulak M, Grobelak A, Grosser A, Vandenbulcke F. Gene expression, DNA damage and other stress markers in Sinapis alba L. exposed to heavy metals with special reference to sewage sludge application on contaminated sites. *Ecotoxicology and Environmental Safety* 2019;**181**:508–17. https://doi.org/10.1016/j.ecoenv.2019.06.025.

[72] Rout GR, Samantaray S, Das P. Effects of chromium and nickel on germination and growth in tolerant and non-tolerant populations of Echinochloa colona (L.) Link. *Chemosphere* 2000;**40**(8):855–9.

[73] Tőzsér D, Magura T, Simon E. Heavy metal uptake by plant parts of willow species: a meta-analysis. *Journal of Hazardous Materials* 2017;**336**:101–9. https://doi.org/10.1016/j.jhazmat.2017.03.068.

[74] Wong MH. Ecological restoration of mine degraded soils, with emphasis on metal contaminated soils. *Chemosphere* 2003;**50**(6):775–80. https://doi.org/10.1016/S0045-6535(02)00232-1.

[75] Arreghini S, de Cabo L, Serafini R, de Iorio AF. Effect of the combined addition of Zn and Pb on partitioning in sediments and their accumulation by the emergent macrophyte Schoenoplectus californicus. *Environmental Science and Pollution Research* 2017;**24**(9):8098–107. https://doi.org/10.1007/s11356-017-8478-7.

[76] Milić D, Luković J, Ninkov J, et al. Heavy metal content in halophytic plants from inland and maritime saline areas. *Central European Journal of Biology* 2012;**7**(2):307–17. https://doi.org/10.2478/s11535-012-0015-6.

[77] Li JT, Liao B, Lan CY, Ye ZH, Baker AJM, Shu WS. Cadmium tolerance and accumulation in cultivars of a high-biomass tropical tree (Averrhoa carambola) and its potential for phytoextraction. *Journal of Environmental Quality* 2010;**39**(4):1262–8. https://doi.org/10.2134/jeq2009.0195.

[78] Vangronsveld J, Herzig R, Weyens N, et al. Phytoremediation of contaminated soils and groundwater: lessons from the field. *Environmental Science and Pollution Research* 2009;**16**(7):765–94. https://doi.org/10.1007/s11356-009-0213-6.

[79] Chaney RL, Baklanov IA. Phytoremediation and phytomining: status and promise. *Advances in Botanical Research* 2017;**83**:189–221. https://doi.org/10.1016/bs.abr.2016.12.006.

[80] Galal TM, Gharib FA, Ghazi SM, Mansour KH. Phytostabilization of heavy metals by the emergent macrophyte Vossia cuspidata (Roxb.) Griff.: a phytoremediation approach. *International Journal of Phytoremediation* 2017;**19**(11):992–9. https://doi.org/10.1080/15226514.2017.1303816.

[81] Radziemska M, Vaverková MD, Baryła A. Phytostabilization-management strategy for stabilizing trace elements in contaminated soils. *International Journal of Environmental Research and Public Health* 2017;**14**(9). https://doi.org/10.3390/ijerph14090958.

[82] Liu J, Zhang XH, Li TY, Wu QX, Jin ZJ. Soil characteristics and heavy metal accumulation by native plants in a Mn mining area of Guangxi, South China. *Environmental Monitoring and Assessment* 2014;**186**(4):2269–79. https://doi.org/10.1007/s10661-013-3535-2.

[83] Bolan NS, Park JH, Robinson B, Naidu R, Huh KY. Phytostabilization. A green approach to contaminant containment. *Australia* 2011;**112**:145–204. https://doi.org/10.1016/B978-0-12-385538-1.00004-4.

[84] Weissmannova HD, Mihocova S, Chovanec P, Pavlovsky J. Potential ecological risk and human health risk assessment of heavy metal pollution in industrial affected soils by coal mining and metallurgy in ostrava. *International Journal of Environmental Research and Public Health* 2019;**16**(22).

[85] O'Connor D, Hou D, Ok YS, Lanphear BP. The effects of iniquitous lead exposure on health. *Nature Sustainability* 2020;**3**(2):77–9. https://doi.org/10.1038/s41893-020-0475-z.

[86] Pizarro I, Gómez-Gómez M, León J, Román D, Palacios MA. Bioaccessibility and arsenic speciation in carrots, beets and quinoa from a contaminated area of Chile. *Science of the Total Environment* 2016;**565**:557–63. https://doi.org/10.1016/j.scitotenv.2016.04.199.

[87] Singh A, Prasad SM, Singh S, Singh M. Phytoremediation potential of weed plants' oxidative biomarker and antioxidant responses. *Chemistry and Ecology* 2016;**32**(7):684–706. https://doi.org/10.1080/02757540.2016.1182994.

[88] Wang X, Bai J, Wei T, et al. Oilseed rape cultivation increases the microbial richness and diversity in soils contaminated with cadmium. *Journal of Soils and Sediments* 2018;**18**(7):2451–62. https://doi.org/10.1007/s11368-018-1938-y.

[89] Kumar A, Chaturvedi AK, Yadav K, et al. *Fungal phytoremediation of heavy metal-contaminated resources: current scenario and future prospects*. Springer Science and Business Media LLC; 2019. p. 437–61. https://doi.org/10.1007/978-3-030-25506-0_18.

[90] Kumar A, Patel JS, Meena VS, Srivastava R. Recent advances of PGPR based approaches for stress tolerance in plants for sustainable agriculture. *Biocatalysis and Agricultural Biotechnology* 2019;**20**. https://doi.org/10.1016/j.bcab.2019.101271.

[91] Khan MS, Zaidi A, Wani PA, Oves M. Role of plant growth promoting rhizobacteria in the remediation of metal contaminated soils. *Environmental Chemistry Letters* 2009;**7**(1):1–19. https://doi.org/10.1007/s10311-008-0155-0.

[92] Molisani, M.M., Rocha, R., Machado, W., Barreto, R.C., Lacerda, L.D., 2006. Mercury contents in aquatic macrophytes from two reservoirs in the Paraiba do Sul: Guandu riversystem, SE Brazil. Brazilian journal of biology = Revista brasleira de biologia 66 (1a), 101–107.

[93] Pietrobelli JMTD, Modenes AN, Fagundes-Klen MR, Espinoza-Quinones FR. *Cadmium, Copper and Zinc Biosorption study by Non-Living Egeria densa biomass water,* **(1–4)**,. Air and Soil Pollution 202; 2009. p. 385–92.

[94] Dhir B, Sharmila P, Saradhi PP, Nasim SA. Physiological and antioxidant responses of Salvinia natans exposed to chromium-rich wastewater. *Ecotoxicology and environmental safety* 2009;**72**(6):1790–7.

[95] Ha NTH, Sakakibara M, Sano S. Phytoremediation of Sb, As, Cu, and Zn from contaminated water by the aquatic macrophyte Eleocharis acicularis. *Clean–Soil, Air, Water* 2009;**37**(9):720–5.

[96] Verma VK, Tewari S, Rai JPN. Ion exchange during heavy metal bio-sorption from aqueous solution by dried biomass ofmacrophytes. *Bioresource Technology* 2008;**99**(6):1932–8.

[97] Xing J, Hu T, Cang L, Zhou D. Remediation of copper contaminated soil by using different particle sizes of apatite: a field experiment. *SpringerPlus* 2016;**5**(1). https://doi.org/10.1186/s40064-016-2492-y.

[98] Zhu Y, Xu F, Liu Q, et al. Nanomaterials and plants: positive effects, toxicity and the remediation of metal and metalloid pollution in soil. *Science of the Total Environment* 2019;**662**:414–21. https://doi.org/10.1016/j.scitotenv.2019.01.234.

[99] Elkhatib EA, Sherif F, Kandil M, Mahdy A, Moharem M, Al-Basri AA. Using nanoparticles from water treatment residuals to reduce the mobility and phytoavailability of Cd and Pb in biosolid-amended soils. *Environmental Geochemistry and Health* 2017;**40**(4):1573–84. https://doi.org/10.1007/s10653-018-0072-5.

[100] Chai M, Shi F, Li R, Liu L, Liu Y, Liu F. Interactive effects of cadmium and carbon nanotubes on the growth and metal accumulation in a halophyte Spartina alterniflora (Poaceae). *Plant Growth Regulation* 2013;**71**(2):171–9. https://doi.org/10.1007/s10725-013-9817-4.

[101] Zhang X, Zhong Z, Bian F, Yang C. Effects of composted bamboo residue amendments on soil microbial communities in an intensively managed bamboo (Phyllostachys praecox) plantation. *Applied Soil Ecology* 2019;**136**:178–83. https://doi.org/10.1016/j.apsoil.2018.12.025.

[102] Gondek K, Mierzwa-Hersztek M, Kopeć M. Mobility of heavy metals in sandy soil after application of composts produced from maize straw, sewage sludge and biochar. *Journal of Environmental Management* 2018;**210**:87–95. https://doi.org/10.1016/j.jenvman.2018.01.023.

[103] Huang H, Luo L, Huang L, Zhang J, Gikas P, Zhou Y. Effect of manure compost on distribution of Cu and Zn in rhizosphere soil and heavy metal accumulation by Brassica juncea. *Water, Air, and Soil Pollution* 2020;**231**(5). https://doi.org/10.1007/s11270-020-04572-4.

[104] Park JH, Lamb D, Paneerselvam P, Choppala G, Bolan N, Chung JW. Role of organic amendments on enhanced bioremediation of heavy metal(loid) contaminated soils. *Journal of Hazardous Materials* 2011;**185**(2–3):549–74. https://doi.org/10.1016/j.jhazmat.2010.09.082.

[105] Helmisaari HS, Salemaa M, Derome J, Kiikkilä O, Uhlig C, Nieminen TM. Remediation of heavy metal-contaminated forest soil using recycled organic matter and native woody plants. *Journal of Environmental Quality* 2007;**36**(4):1145–53. https://doi.org/10.2134/jeq2006.0319.

[106] Mazej Z, Germ M. Trace element accumulation and distribution in four aquatic macrophytes. *Chemosphere* 2009;**74**(5):642–7. https://doi.org/10.1016/j.chemosphere.2008.10.019.

[107] Mkandawire M, Dudel EG. Accumulation of arsenic in Lemna gibba L. (duckweed) in tailing waters of two abandoned uranium mining sites in Saxony, Germany. *Science of the Total Environment* 2005;**336**(1–3):81–9. https://doi.org/10.1016/j.scitotenv.2004.06.002.

[108] Robinson B, Kim N, Marchetti M, et al. Arsenic hyperaccumulation by aquatic macrophytes in the taupo volcanic zone, New Zealand. *Environmental and Experimental Botany* 2006;**58**(1–3):206–15. https://doi.org/10.1016/j.envexpbot.2005.08.004.

[109] Khan A, Khan S, Khan MA, Qamar Z, Waqas M. The uptake and bioaccumulation of heavy metals by food plants, their effects on plants nutrients, and associated health risk: a review. *Environmental Science and Pollution Research* 2015;**22**(18):13772–99. https://doi.org/10.1007/s11356-015-4881-0.

[110] Pandey VC. Suitability of Ricinus communis L. cultivation for phytoremediation of fly ash disposal sites. *Ecological Engineering* 2013;**57**:336–41. https://doi.org/10.1016/j.ecoleng.2013.04.054.

[111] Chand S, Singh S, Singh VK, Patra DD. Utilization of heavy metal-rich tannery sludge for sweet basil (Ocimum basilicum L.) cultivation. *Environmental Science and Pollution Research* 2015;**22**(10):7470–5. https://doi.org/10.1007/s11356-015-4446-2.

[112] Kumar R, Srivastava M, Dubey NK. Evaluation of Cymbopogon martinii oil extract for control of postharvest insect deterioration in cereals and legumes. *Journal of Food Protection* 2007;**70**(1):172–8. https://doi.org/10.4315/0362-028X-70.1.172.

[113] Smith D, Gravel V, Yergeau E. Signaling in the phytomicrobiome. *Front Plant Sci* 2017;**8**.

[114] Bell TH, Cloutier-Hurteau B, Al-Otaibi F, et al. Early rhizosphere microbiome composition is related to the growth and Zn uptake of willows introduced to a former landfill. *Environmental Microbiology* 2015;**17**(8):3025–38. https://doi.org/10.1111/1462-2920.12900.

[115] Chatterjee C, Gopal R, Dube BK. Physiological and biochemical responses of French bean to excess cobalt. *Journal of Plant Nutrition* 2006;**29**(1):127–36. https://doi.org/10.1080/01904160500416513.

[116] Lin Q, Wang Z, Ma S, Chen Y. Evaluation of dissipation mechanisms by Lolium perenne L, and Raphanus sativus for pentachlorophenol (PCP) in copper co-contaminated soil. *Science of the Total Environment* 2006;**368**(2–3):814–22. https://doi.org/10.1016/j.scitotenv.2006.03.024.

[117] Gall JE, Boyd RS, Rajakaruna N. Transfer of heavy metals through terrestrial food webs: a review. *Environmental Monitoring and Assessment* 2015;**187**(4). https://doi.org/10.1007/s10661-015-4436-3.

[118] Yılmaz AB, Sangün MK, Yağlıoğlu D, Turan C. Metals (major, essential to non-essential) composition of the different tissues of three demersal fish species from İskenderun Bay, Turkey. *Food Chemistry* 2010;**123**(2):410–5. https://doi.org/10.1016/j.foodchem.2010.04.057.

[119] Pacwa-Płociniczak M, Płociniczak T, Yu D, et al. Effect of silene vulgaris and heavy metal pollution on soil microbial diversity in long-term contaminated soil. *Water, Air, and Soil Pollution* 2018;**229**(1). https://doi.org/10.1007/s11270-017-3655-3.

[120] Marschner P, Crowley D, Yang CH. Development of specific rhizosphere bacterial communities in relation to plant species, nutrition and soil type. *Plant and Soil* 2004;**261**(1–2):199–208. https://doi.org/10.1023/B:PLSO.0000035569.80747.c5.

[121] Yang CH, Crowley DE. Rhizosphere microbial community structure in relation to root location and plant iron nutritional status. *Applied and Environmental Microbiology* 2000;**66**(1):345–51. https://doi.org/10.1128/AEM.66.1.345-351.2000.

[122] Wu SC, Cheung KC, Luo YM, Wong MH. Effects of inoculation of plant growth-promoting rhizobacteria on metal uptake by Brassica juncea. *Environmental Pollution* 2006;**140**(1):124–35. https://doi.org/10.1016/j.envpol.2005.06.023.

[123] Vogel-Mikuš K, Pongrac P, Kump P, Nečemer M, Regvar M. Colonisation of a Zn, Cd and Pb hyperaccumulator Thlaspi praecox Wulfen with indigenous arbuscular mycorrhizal fungal mixture induces changes in heavy metal and nutrient uptake. *Environmental Pollution* 2006;**139**(2):362–71. https://doi.org/10.1016/j.envpol.2005.05.005.

[124] Li WC, Ye ZH, Wong MH. Effects of bacteria on enhanced metal uptake of the Cd/Zn-hyperaccumulating plant, Sedum alfredii. *Journal of Experimental Botany* 2007;**58**(15–16):4173–82. https://doi.org/10.1093/jxb/erm274.

[125] Lasat M. Phytoextraction of metals from contaminated soil: a review of plant/soil/metal interaction and assessment of pertinent agronomic issues. *Journal of Hazardous Substance Research* 2000;**2**(1).

[126] Juwarkar AA, Nair A, Dubey KV, Singh SK, Devotta S. Biosurfactant technology for remediation of cadmium and lead contaminated soils. *Chemosphere* 2007;**68**(10):1996–2002. https://doi.org/10.1016/j.chemosphere.2007.02.027.

[127] Bennisse R, Labat M, ElAsli A, et al. Rhizosphere bacterial populations of metallophyte plants in heavy metal-contaminated soils from mining areas in semiarid climate. *World Journal of Microbiology and Biotechnology* 2004;**20**(7):759–66. https://doi.org/10.1007/s11274-004-5812-2.

[128] Pattnaik S, Dash D, Mohapatra S, et al. Improvement of rice plant productivity by native Cr(VI) reducing and plant growth promoting soil bacteria *Enterobacter cloacae*. *Chemosphere* 2020;**240**. https://doi.org/10.1016/j.chemosphere.2019.124895.

[129] Idris R, Trifonova R, Puschenreiter M, Wenzel WW, Sessitsch A. Bacterial communities associated with flowering plants of the Ni hyperaccumulator Thlaspi goesingense. *Applied and Environmental Microbiology* 2004;**70**(5):2667–77. https://doi.org/10.1128/AEM.70.5.2667-2677.2004.

[130] Chen L, Luo S, Li X, Wan Y, Chen J, Liu C. Interaction of Cd-hyperaccumulator Solanum nigrum L. and functional endophyte Pseudomonas sp. Lk9 on soil heavy metals uptake. *Soil Biology and Biochemistry* 2014;**68**:300–8. https://doi.org/10.1016/j.soilbio.2013.10.021.

[131] Dimkpa CO, Svatoš A, Dabrowska P, Schmidt A, Boland W, Kothe E. Involvement of siderophores in the reduction of metal-induced inhibition of auxin synthesis in Streptomyces spp. *Chemosphere* 2008;**74**(1):19–25. https://doi.org/10.1016/j.chemosphere.2008.09.079.

[132] Dao KHT, Hamer KE, Clark CL, Harshman LG. Pyoverdine production by *Pseudomonas aeruginosa* exposed to metals or an oxidative stress agent. *Ecological Applications* 1999;**9**(2):441–8. https://doi.org/10.1890/1051-0761(1999)009[0441:PPBPAE]2.0.CO;2.

[133] Braud A, Hoegy F, Jezequel K, Lebeau T, Schalk IJ. New insights into the metal specificity of the *Pseudomonas aeruginosa* pyoverdine-iron uptake pathway. *Environmental Microbiology* 2009;**11**(5):1079–91. https://doi.org/10.1111/j.1462-2920.2008.01838.x.

[134] Mousavi SM, Motesharezadeh B, Hosseini HM, Alikhani H, Zolfaghari AA. Root-induced changes of Zn and Pb dynamics in the rhizosphere of sunflower with different plant growth promoting treatments in a heavily contaminated soil. *Ecotoxicology and Environmental Safety* 2018;**147**:206–16. https://doi.org/10.1016/j.ecoenv.2017.08.045.

[135] Yadav KK, Singh JK, Gupta N, Kumar V. A review of nanobioremediation technologies for environmental cleanup: a novel biological approach. *Journal of Materials and Environmental Science* 2017;**8**(2):740–57. http://www.jmaterenvironsci.com/Document/vol8/vol8_N2/78-JMES-2831-Yadav.pdf.

[136] Zhang R, Zhang N, Fang Z. In situ remediation of hexavalent chromium contaminated soil by CMC-stabilized nanoscale zero-valent iron composited with biochar. *Water Science and Technology* 2018;**77**(6):1622–31. https://doi.org/10.2166/wst.2018.039.

[137] Srivastav A, Yadav KK, Yadav S, et al. Nano-phytoremediation of pollutants from contaminated soil environment: current scenario and future prospects. In: *Phytoremediation: management of environmental contaminants*. vol. 6. India: Springer International Publishing; 2019. p. 383–401. https://doi.org/10.1007/978-3-319-99651-6_16.

[138] Su H, Fang Z, Tsang PE, Fang J, Zhao D. Stabilisation of nanoscale zero-valent iron with biochar for enhanced transport and in-situ remediation of hexavalent chromium in soil. *Environmental Pollution* 2016;**214**:94–100. https://doi.org/10.1016/j.envpol.2016.03.072.

[139] Huang D, Qin X, Peng Z, et al. Nanoscale zero-valent iron assisted phytoremediation of Pb in sediment: impacts on metal accumulation and antioxidative system of Lolium perenne. *Ecotoxicology and Environmental Safety* 2018;**153**:229–37. https://doi.org/10.1016/j.ecoenv.2018.01.060.

[140] Singh J, Lee BK. Effects of Nano-TiO2 particles on bioaccumulation of 133Cs from the contaminated soil by Soybean (Glycine max). *Process Safety and Environmental Protection* 2018;**116**:301–11. https://doi.org/10.1016/j.psep.2018.02.016.

[141] Andersen CP, King G, Plocher M, et al. Germination and early plant development of ten plant species exposed to titanium dioxide and cerium oxide nanoparticles. *Environmental Toxicology and Chemistry* 2016;**35**(9):2223–9. https://doi.org/10.1002/etc.3374.

[142] Abdel Latef AAH, Srivastava AK, El-sadek MSA, Kordrostami M, Tran LSP. Titanium dioxide nanoparticles improve growth and enhance tolerance of broad bean plants under saline soil conditions. *Land Degradation and Development* 2018;**29**(4):1065–73. https://doi.org/10.1002/ldr.2780.

[143] Singh J, Lee BK. Influence of nano-TiO2 particles on the bioaccumulation of Cd in soybean plants (Glycine max): a possible mechanism for the removal of Cd from the contaminated soil. *Journal of Environmental Management* 2016;**170**:88–96. https://doi.org/10.1016/j.jenvman.2016.01.015.

[144] Liang Sx, Jin Y, Liu W, Li X, Shen Sg, Ding L. Feasibility of Pb phytoextraction using nano-materials assisted ryegrass: results of a one-year field-scale experiment. *Journal of Environmental Management* 2017;**190**:170–5. https://doi.org/10.1016/j.jenvman.2016.12.064.

[145] Yang SX, Liao B, Yang ZH, Chai LY, Li JT. Revegetation of extremely acid mine soils based on aided phytostabilization: a case study from southern China. *Science of the Total Environment* 2016;**562**:427–34. https://doi.org/10.1016/j.scitotenv.2016.03.208.

[146] Yang Z, Fang Z, Zheng L, et al. Remediation of lead contaminated soil by biochar-supported nano-hydroxyapatite. *Ecotoxicology and Environmental Safety* 2016;**132**:224–30. https://doi.org/10.1016/j.ecoenv.2016.06.008.

[147] Sun Y, Li Y, Xu Y, Liang X, Wang L. In situ stabilization remediation of cadmium (Cd) and lead (Pb) co-contaminated paddy soil using bentonite. *Applied Clay Science* 2015;**105–106**:200–6. https://doi.org/10.1016/j.clay.2014.12.031.

[148] Jin Y, Liu W, Li Xl, et al. Nano-hydroxyapatite immobilized lead and enhanced plant growth of ryegrass in a contaminated soil. *Ecological Engineering* 2016;**95**:25–9. https://doi.org/10.1016/j.ecoleng.2016.06.071.

[149] Fernandes JP, Mucha AP, Francisco T, Gomes CR, Almeida CMR. Silver nanoparticles uptake by salt marsh plants—implications for phytoremediation processes and effects in microbial community dynamics. *Marine Pollution Bulletin* 2017;**119**(1):176–83. https://doi.org/10.1016/j.marpolbul.2017.03.052.

[150] Antonoglou O, Moustaka J, Adamakis IDS, et al. Nanobrass CuZn nanoparticles as foliar spray nonphytotoxic fungicides. *ACS Applied Materials and Interfaces* 2018;**10**(5):4450–61. https://doi.org/10.1021/acsami.7b17017.

[151] Das N, Bhattacharya S, Maiti MK. Enhanced cadmium accumulation and tolerance in transgenic tobacco overexpressing rice metal tolerance protein gene OsMTP1 is promising for phytoremediation. *Plant Physiology and Biochemistry* 2016;**105**:297–309. https://doi.org/10.1016/j.plaphy.2016.04.049.

[152] Mani D, Kumar C. Biotechnological advances in bioremediation of heavy metals contaminated ecosystems: an overview with special reference to phytoremediation. *International Journal of Environmental Science and Technology* 2014;**11**(3):843–72. https://doi.org/10.1007/s13762-013-0299-8.

[153] Koźmińska A, Wiszniewska A, Hanus-Fajerska E, Muszyńska E. Recent strategies of increasing metal tolerance and phytoremediation potential using genetic transformation of plants. *Plant Biotechnology Reports* 2018;**12**(1):1–14. https://doi.org/10.1007/s11816-017-0467-2.

[154] Lin Y-F, Hassan Z, Talukdar S, Schat H, Aarts MGM, Baxter I. Expression of the ZNT1 zinc transporter from the metal hyperaccumulator Noccaea caerulescens confers enhanced zinc and cadmium tolerance and accumulation to *Arabidopsis thaliana*. *PLOS ONE* 2016;**11**(3):e0149750. https://doi.org/10.1371/journal.pone.0149750.

[155] Chen Y, Liu Y, Ding Y, Wang X, Xu J. Overexpression of PtPCS enhances cadmium tolerance and cadmium accumulation in tobacco. *Plant Cell, Tissue and Organ Culture (PCTOC)* 2015;**121**(2):389–96. https://doi.org/10.1007/s11240-015-0710-x.

[156] Rafati M, Khorasani N, Moattar F, Shirvany A, Moraghebi F, Hosseinzadeh S. Phytoremediation potential of populus alba and morus alba for cadmium, chromuim and nickel absorption from polluted soil. *International Journal of Environmental Research* 2011;**5**(4):961–70. http://ijer.ut.ac.ir/images/Issues/Vol5.No4/16.rar.

[157] Yoon J, Cao X, Zhou Q, Ma LQ. Accumulation of Pb, Cu, and Zn in native plants growing on a contaminated Florida site. *Science of the Total Environment* 2006;**368**(2–3):456–64. https://doi.org/10.1016/j.scitotenv.2006.01.016.

[158] Hosam MS, Refaat FA, Hazem HM. *Ludwigia stolonifera* for remediation of toxic metals from simulated wastewater. *Chemistry and Ecology* 2019;**35**(2):164–78. https://doi.org/10.1080/02757540.2018.1546296.

[159] Harguinteguy CA, Schreiber R, Pignata ML. Myriophyllum aquaticum as a biomonitor of water heavy metal input related to agricultural activities in the Xanaes River (Córdoba, Argentina). *Ecological Indicators* 2013;**27**:8–16. https://doi.org/10.1016/j.ecolind.2012.11.018.

[160] Zurayk R, Sukkariyah B, Baalbaki R, Ghanem DA. Ni phytoaccumulation in Mentha aquatica L. And Mentha sylvestris L. *Water, Air, and Soil Pollution* 2002;**139**(1–4):355–64. https://doi.org/10.1023/A:1015840601761.

[161] Mokhtar H, Morad N, Fizri F. Hyperaccumulation of copper by two species of aquatic plants. *International Conference of Environmental Science and Engineering* 2011;vol. 8:115–8.

[162] Singh S, Sinha S. Accumulation of metals and its effects in Brassica juncea (L.) Czern. (cv. Rohini) grown on various amendments of tannery waste. *Ecotoxicology and Environmental Safety* 2005;**62**(1):118–27. https://doi.org/10.1016/j.ecoenv.2004.12.026.

[163] Subhashini V, Swamy A. Phytoremediation of cadmium and chromium contaminated soils by Cyperus rotundus L. *American International Journal of Research in Science and Technology Engineering and Mathematics* 2014;**6**:97–101.

[164] Mendez MO, Maier RM. Phytostabilization of mine tailings in arid and semiarid environments—an emerging remediation technology. *Environmental Health Perspectives* 2008;**116**(3):278–83. https://doi.org/10.1289/ehp.10608.

[165] Regvar M, Vogel-Mikuš K, Kugonič N, Turk B, Batič F. Vegetational and mycorrhizal successions at a metal polluted site: indications for the direction of phytostabilisation? *Environmental Pollution* 2006;**144**(3):976–84. https://doi.org/10.1016/j.envpol.2006.01.036.

[166] Mukhopadhyay S, Maiti S. Phytoremediation of metal enriched mine waste: a review. *Global Journal of Environmental Research* 2010;**4**:135–50.

[167] Mesjasz-Przybyłowics J, Nakonieczny M, Migula P, et al. Uptake of cadmium, lead, nickel and zinc from soil and water solutions by the nickel hyperaccumulator berkheya coddii. *Acta Biologica Cracoviensia Series Botanica* 2004;**46**:75–85.

[168] Prasad MNV, De Oliveira Freitas HM. Metal hyperaccumulation in plants—biodiversity prospecting forphytoremediation technology. *Electronic Journal of Biotechnology* 2003;**6**(3):110–46. https://doi.org/10.2225/vol6-issue3-fulltext-6.

[169] USEPA (United State Environmental Protection Agency) (2000). Introduction to phytoremediation.Cincinnati, Ohio EPA 600/R-99/107.

[170] Newete SW, Byrne MJ. The capacity of aquatic macrophytes for phytoremediation and their disposal with specific reference to water hyacinth. *Environmental Science and Pollution Research* 2016;**23**(11):10630–43. https://doi.org/10.1007/s11356-016-6329-6.

[171] Rai PK. Heavy metals/metalloids remediation from wastewater using free floating macrophytes of a natural wetland. *Environmental Technology & Innovation.* 2019;**15**:100393. https://doi.org/10.1016/j.eti.2019.100393.

[172] Mishra S, Maiti A. The efficiency of *Eichhornia crassipes* in the removal of organic and inorganic pollutants from wastewater: a review. *Environmental Science and Pollution Research* 2017;**24**(9):7921–37. https://doi.org/10.1007/s11356-016-8357-7.

[173] Tian LY, Yang JY, Huang JH. Uptake and speciation of vanadium in the rhizosphere soils of rape (Brassica juncea L.). *Environmental Science and Pollution Research* 2015;**22**(12):9215–23. https://doi.org/10.1007/s11356-014-4031-0.

[174] Wang L, Lin H, Dong Y, Li B, He Y. Effects of endophytes inoculation on rhizosphere and endosphere microecology of Indian mustard (Brassica juncea) grown in vanadium-contaminated soil and its enhancement on phytoremediation. *Chemosphere* 2020;**240**:124891. https://doi.org/10.1016/j.chemosphere.2019.124891.

[175] Ghosh M, Singh SP. A review on phytoremediation of heavy metals and utilization of its byproducts. *Applied Ecology and Environmental Research* 2005;**3**(1):1–18. https://doi.org/10.15666/aeer/0301_001018.

[176] LeDuc DL, Terry N. Phytoremediation of toxic trace elements in soil and water. *Journal of Industrial Microbiology and Biotechnology* 2005;**32**(11–12):514–20. https://doi.org/10.1007/s10295-005-0227-0.

[177] Pilon-Smits EAH, De Souza MP, Hong G, et al. Selenium volatilization and accumulation by twenty aquatic plant species. *Journal of Environmental Quality* 1999;**28**(3):1011–8. https://doi.org/10.2134/jeq1999.00472425002800030035x.

[178] Terry N, Zayed AM, De Souza MP, Tarun AS. Selenium in higher plants. *Annual Review of Plant Biology* 2000;**51**:401–32. https://doi.org/10.1146/annurev.arplant.51.1.401.

[179] Houda Z, Bejaoui Z, Albouchi A, Gupta DK, Corpas FJ. Comparative study of plant growth of two poplar tree species irrigated with treated wastewater, with particular reference to accumulation of heavy metals (Cd, Pb, As, and Ni). *Environmental Monitoring and Assessment* 2016;**188**(2):1–10. https://doi.org/10.1007/s10661-016-5102-0.

[180] Arshad M, Saleem M, Hussain S. Perspectives of bacterial ACC deaminase in phytoremediation. *Trends in Biotechnology* 2007;**25**(8):356–62. https://doi.org/10.1016/j.tibtech.2007.05.005.

[181] Mench M, Schwitzguébel JP, Schroeder P, Bert V, Gawronski S, Gupta S. Assessment of successful experiments and limitations of phytotechnologies: contaminant uptake, detoxification and sequestration, and consequences for food safety. *Environmental Science and Pollution Research* 2009;**16**(7):876–900. https://doi.org/10.1007/s11356-009-0252-z.

Chapter 8

# Engineering crop resistance to biotic stresses

## 1. Introduction

Plants being sessile organisms, undergo several biotic and abiotic stresses, leading to great losses in yield, growth, and development, which could threaten their very survival.[1] These stresses include drought, extremes of temperature, light conditions, high salt concentrations, presence of heavy metals, flooding or physical wounding.[2–5] Additionally, plants are subjected to biotic stresses such as viral, bacterial, and fungal pathogens, pests including insects, nematodes, and rodents which circumscribe plant productivity.[6–10] As a result, there are several negative outcomes such as loss of photosynthetic pigments, scorching and sunburn of leaves, stunted growth,[11,12] diminished synthesis of photo-assimilates, and dwindling of carbohydrate reserves leading to starvation.[13–16]

These stresses induce the production of reactive oxygen species (ROS) in the form of superoxide radicals ($O_2$-), singlet oxygen ($^1O_2$), hydroxyl radical (OH), and hydrogen peroxide ($H_2O_2$), which are generated in cellular organelles like chloroplasts, peroxisomes, and mitochondria, and these ROS react with different macromolecules such as DNA, lipids, proteins, and pigments.[17,18] At the onset of stress conditions, plants change the levels of hormones such as ethylene, abscisic acid, jasmonic acid, and salicylic acid that act as signaling molecules toward systemic acquired resistance and stress tolerance.[19] Additionally, secondary metabolites like flavonoids, carotenoids, isopropanoids, isoprenoids, lignin, and anthocyanin are generated and accumulated.

Plants have also developed sophisticated molecular and cellular mechanisms wherein during the emergence of stress situations, plants diminish the production of normal proteins as well as drive the transcription and translation of heat shock proteins (HSPs). Plants have evolved sophisticated post-transcriptional adaptations such as microRNA (miRNA) and alternative splicing, which enable them to deal with these stresses. The miRNAs bind to mRNAs and result in repression of translation or cause direct cleavage of the

Agricultural Biotechnology. https://doi.org/10.1016/B978-0-323-88476-1.00005-9

mRNA species complementary to the miRNA. Some of the miRNAs also are involved with abiotic stress tolerance.[20]

About 35% yield losses are estimated to occur due to biotic stresses.[21] It is important for plants to evolve multiple resistance genes to combat the emergence of new virulence in organisms causing biotic stresses. Plant breeders look for the development of new resistance genes to cope with the evolving organisms that threaten susceptible crops. Plant geneticists have discovered resistance genes in the various crop species, and molecular geneticists have identified molecular markers for many of those genes.[22]

Biotic stresses negatively impact crop productivity, global agriculture, and food security. Development of cultivars resistant to diseases and insects is the most environmentally friendly and viable method for crop protection. Nevertheless, lack of source of resistance within the available gene pool circumscribes the scope of developing resistant cultivars.[23,24] When genetic resistance is not available, crop protection primarily relies on the use of toxic agrochemicals that are cost-intensive and hazardous to the environment.

Remarkable advances in modern molecular biology are enabling the translation of biological concepts into new avenues for evolving genetic resistance against pathogens and insect pests. Introduction of RNAi technology has provided new channels for trait improvement by developing "loss of function" in a specific, targeted fashion beyond the narrow scope of breeding based on gene introgression. Nevertheless, a major drawback of RNAi technology is the absence of complete silencing of the target genes. In recent times, widening arrays of genome editing techniques as tools for gene editing in a site-specific manner have opened enormous avenues for engineering favorable agronomic traits such as biotic stress resistance.[25,26]

## 2. Cellular machinery of plant defense against pathogens and insect pests

Plants have developed complex immune response networks to overcome most of the pathogen and insect attacks. In the initial stages, attack by pathogens is sensed by the pathogen recognition receptors (PRRs) on the cell surface. These PRRs identify and bind to pathogen associated molecular patterns (PAMPs), which are conserved microbial molecules some of which are the well characterized fungal chitins and bacterial flagellins.[27] These molecules are called herbivore-associated molecular patterns (HAMPs) in instances of insect invasion.[28] The PRRs are primarily lysine motif kinases and leucine rich repeat kinases.[29]

The interaction of PAMP/HAMP-PRR elicits the initial layer of plant defense known as pathogen triggered immunity (PTI) that curtails microbial colonization. Nevertheless, a successful pathogen overcomes PTI to transport specific effectors into plant cells and appropriate the plant cellular machinery for their own sustenance. To confront this, plants have elicited another layer of defense known as effector triggered immunity (ETI).[29] ETI is facilitated by the intracellular resistance receptor (R) protein possessing the NBS-LRR domain that is stimulated against specific effectors.[30,31] The effector stimulated activation of the R gene frequently results in localized death of cells or hypersensitive response (HR) and therefore precludes further pathogen colonization.[32] Fig. 8.1 presents a schematic of the different levels of plant defense responses elicited upon pathogen attack.

Plants are highly impacted by several biotic stress factors, which lead to great losses in crop productivity. Nevertheless, plants have evolved response

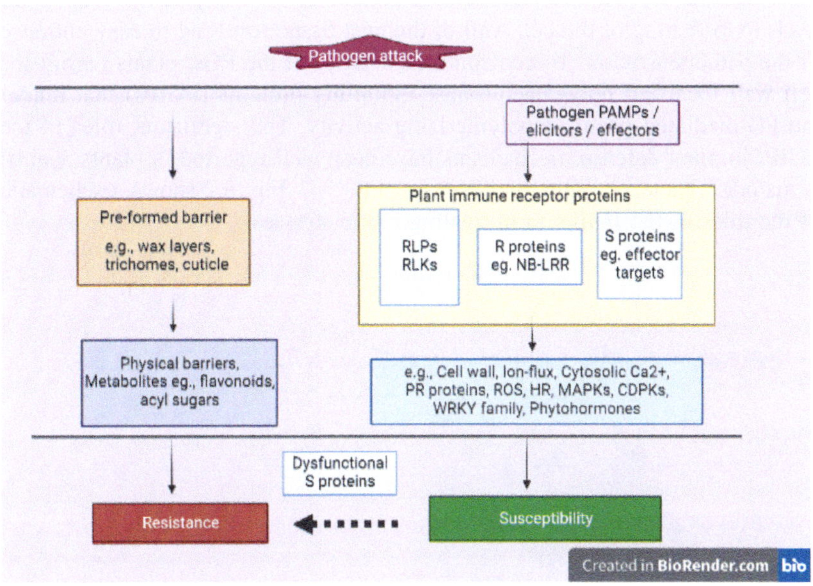


**FIGURE 8.1 Various layers of defense responses in plants following pathogen attack.** Various layers of defense responses in plants following pathogen attack. These consist of preformed physical barriers as well as two pathogen-elicited layers wherein recognition between receptors in the plant cells and pathogen molecules such as elicitors, pathogen-associated molecular patterns (PAMPs) and effectors takes place. The plant receptors include receptor-like kinases (RLKs), receptor-like proteins (RLPs), proteins coded for by plant susceptibility and resistance genes. Following receptor recognition, various signaling cascades are induced to advance local and systemic plant defense responses. These include strength of the cell wall, plasma membrane ion flux alterations, increase in cytosolic $Ca^{2+}$ levels, generation of extracellular ROS, $Ca^{2+}$-dependent protein kinase (CDPK) activation, pathogen-related protein induction, mitogen stimulated protein kinase (MAPK) cascades, hypersensitive defense responses (HRs) and induction of phytohormones such as ethylene, JA and SA. Some of the plant proteins are used by the pathogen to complete its life cycle wherein these proteins are expressed by the plant susceptibility (S) genes. Occurrence of S gene loss-of-function mutations results in recessive resistance. *Adapted from Bai Y, Kissoudis C, Yan Z, Visser RGF, van der Linden G. Plant behavior under combined stress: tomato responses to combined salinity and pathogen stress.* Plant Journal *2018;**93**(4):781–793. https://doi.org/10.1111/tpj.13800.*

mechanisms to these stresses including structural and morphological barriers, proteins, enzymes, and chemical compounds. Timely elicitation of these immune responses is essential in determining whether plants are capable of resisting or succumbing to the various challenges caused by biotic stress agents.[33,34] The plant cell wall is the initial and most obvious obstacle met with by the biotic stress factors and pathogens have evolved several mechanisms to overcome this barrier in order to enter into the plant cell.[33]

Foremost among these are the cell wall degrading enzymes synthesized by the biotic stress factors, which are vital in cell entry and one important example of these enzymes is the endo-polygalacturonases (PGs), which cleave the alpha-1,4 linkages between D-galacturonic acid moieties of homogalacturonan, the major component of pectin. The activity of these enzymes leads to loosening of the cell wall of the host tissue resulting in easy entrance of the pathogens/pests. To confront this activity of the PGs, plants recruit the cell wall localized polygalacturonase-inhibiting proteins (PGIPs) that hinder the PG-mediated pectin depolymerizing activity. The significant roles of the PGIPs in plant defense mechanisms have been well reported in plants against nematodes, bacteria, fungi, and even insects.[35–40] Fig. 8.2 shows a schematic of the roles of the PGIPs in mitigating biotic stresses.

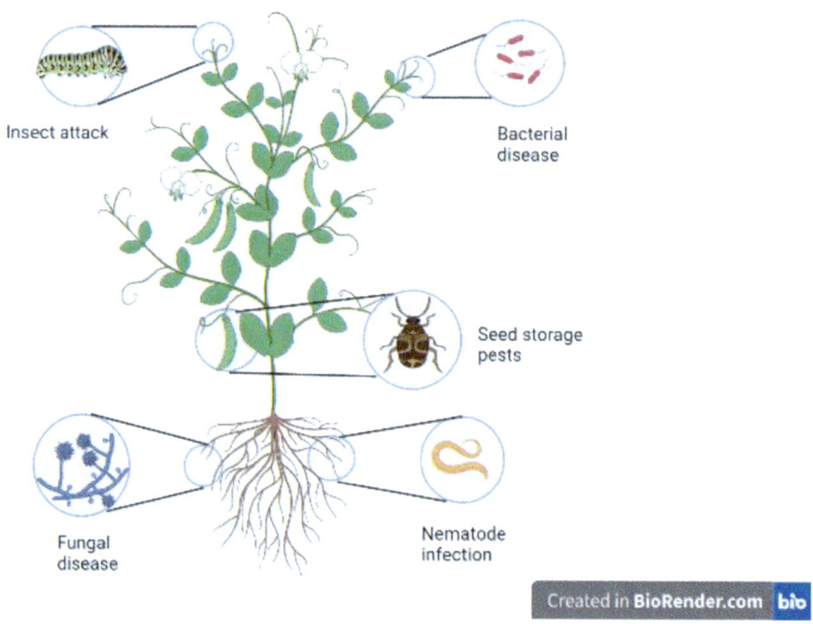


**FIGURE 8.2** Schematic representation of the function of plant polygalacturonase inhibitors (PGIPs) in ameliorating biotic stresses. *Adapted from Rathinam M, Rao U, Sreevathsa R. Novel biotechnological strategies to combat biotic stresses: polygalacturonase inhibitor (PGIP) proteins as a promising comprehensive option.* Applied Microbiology and Biotechnology *2020;**104**(6): 2333–2342. https://doi.org/10.1007/s00253-020-10396-3.*

## 3. Translational application of PGIP toward mitigation of biotic stresses

Fungal resistance mediated by PGIP has been shown in several commercially important crops such as papaya, alfalfa, grapevine, potato, cotton, tomato, common bean, pea, and wheat[35,37,41–43] as well as in model plants such as Arabidopsis and tobacco.[44–46] Moreover, PGIP expression from a spatio-temporal context and their association with fungal resistance have been demonstrated in various crops.[44,46–50] Further, the significance of PGIP-mediated fungal resistance was observed in raspberry and pear plants wherein young fruits expressed PGIPs in greater amounts and showed augmented resistance to infections caused by fungi when compared to mature fruits, which showed lower expression. Fig. 8.2 depicts a schematic of the role of plant PGIPs in alleviating biotic stresses.

The bean PvPGIP2 has been expressed in several heterologous systems.[51–54] This PGIP was the first one to be investigated in detail in various crops that exhibited wide-ranging fungal disease resistance.[51–53] Recently, it was shown that PGIP sourced from sugar beet when expressed in *Nicotiana benthamiana* transgenic plants afforded resistance in the latter to infections by the fungi *Fusarium solani*, *Botrytis cinerea*, and *Rhizoctonia solani*.[40] Likewise, PGIP1 sourced from cotton was overexpressed in transgenic Arabidopsis, which provided augmented resistance to fungal diseases caused by *Fusarium* and *Verticillium*. Silencing of this PGIP1 gene increased susceptibility of cotton toward both of these fungi.[43] Therefore, there has been an ongoing identification of more PGIPs effectual against fungal pathogens impacting crop productivity.

Aside from fungi, PGIPs have been implicated in conferring oomycete resistance. Expression of PvPGIP2 in transgenic tobacco showed robust resistance against oomycetes such as *Peronospora hyoscyami sp. Tabacina* and *Phytophthora parasitica var. nicotianae*, as well as *Rhizoctonia solani*, a fungal pathogen.[55] When gene expression was analyzed between susceptible and resistant varieties of pearl millet, this showed the role of PGIP in developing resistance against *Sclerospora graminicola* (Sacc.) *Schroet* that causes downy mildew disease.[56] This finding corroborates earlier work showing the repression of PG from *S. graminicola (Sacc.) Schroet*. by PGIP sourced from pearl millet.[57]

Bacterial disease mitigation is yet another function of PGIPs that has been shown in various crops.[58–62] Transgenic grapevine plants expressing pgip gene sourced from the pear *Pyrus communis* showed resistance to Pierce's disease caused by the bacterium *Xylella fastidiosa*.[63] Moreover, transgenic tobacco and Chinese cabbage overexpressing the PGIP2 sourced from Chinese cabbage demonstrated enhanced resistance against *Pectobacterium carotovorum* ssp. Carotovorum.[59] PGIP activity assay was performed using extracts from tomato stem, which showed inhibition of *Ralstonia solanacearum* PGs.[60]

A recent investigation in rice plants authenticated that the pgip4 chromosome locus was responsible for generating resistance to bacterial leaf streak disease, while RNAi-enabled silencing of this gene increased susceptibility to the bacterium *Xanthomonas oryzae pv. oryzicola*.[58] Also, several other phytopathogens such as parasitic nematodes produce PGs, which recognize plant PGIPs,[39] leading to resistance in some instances. Genotypes of *Pisum sativum* plants expressing PGIPs showed resistance or susceptibility to infection by *Heterodera goettingiana*, a nematode pathogen.[39] Among these, the plants showing resistant phenotypes exhibited high level of PGIP expression when compared to those showing susceptible phenotypes. In situ PGIP localization in roots of the plants showing resistant phenotype revealed augmented PGIP expression in the area infected by nematodes[39] Transgenic *Arabidopsis* that overexpressed PGIP demonstrated improved resistance to *H. schachtii*, a cyst nematode.[64] Nevertheless despite all these above findings, in-depth and elaborate discernment of the role of PGIPs in plant-nematode relationships is still required.

PGIP expression in plants in defense responses to attack and feeding by insects suggested the significance of the PG-PGIP interaction.[65,66] It is well established that PGs are obligately required in the insect digestive system.[67] In mirid bugs, enhanced PG activity was reported in the salivary glands when compared to the insect gut, establishing the function of PGs in insect feeding.[68] Following the demonstration of the importance of PGIPs from *Phaseolus vulgaris*, two other proteins showing efficacy against insects were described.[68] The role of PGIPs in generating insect resistance was further demonstrated in the mirid insects *Adelphocoris lineolatus* and *Lygus rugulipennis* whose PGs were actively inhibited by the PvPGIP3 and PvPGIP4.[68,69] Two of the *P. vulgaris* PGIPs namely, PvPGIP3 and PvPGIP4, showed inhibition of the PGs in all of the mirid bugs investigated.

Confirmed evidence of the involvement of PGIPs in insect resistance was acquired from reports regarding PGIP functionality localized in the chromosome five of the mung bean plants, which afforded resistance to bruchids.[70] The wild relative of the plant *Vigna radiata* showed PGIP-driven resistance to insects, which revealed the significance of the crop wild relatives in engendering many agronomic traits toward crop improvement.

The V2709 accession of *Vigna radiata* demonstrated the role of PGIP1 and PGIP2 in conferring bruchid resistance.[36] Furthermore, PGIP2 obtained from the accession V2709 when sequentially aligned with PvPGIP2 showed variation in the amino acid sequence within the sites involved in PG recognition.[36,71] These novel PGIP genes sourced from both V2709 and V2802 accessions can be used in mung bean as well as other crop enhancement programs. Considering the role of the PGIPs in containing bruchids and mirid bugs, it would be interesting to identify the importance of PGIPs in inhibiting herbivorous insects bearing in mind their various feeding patterns. It would also be intriguing to identify PGIPs to contain chewing insects for successful application of PGIPs in protecting crops against several other insects.

## 4. Regulatory activities of plant transcription factors in response to biotic stress

Plants have developed rapid response mechanisms towards unfavorable stresses, and these responses involve interconnected molecular networks that are under the control of signal cascades. The various factors concerned with stress responses include perception of stress signals, signal transduction, and induction of genes involved in stress response.[72] Upon perceiving a stress signal, sensors or receptors in the plant cell wall or cell membrane recognize the stimulus and elicit rapid response transducing the external stress signal to intracellular signaling responses. In this process, activation of signaling cascades and kinase cascades involving intracellular ions and molecules take place characteristically in the cytoplasm. The main cascades involve calcium ions ($Ca^{2+}$) and reactive oxygen species (ROS). Phytohormones such as abscisic acid (ABA), salicylic acid (SA), jasmonic acid (JA), and ethylene (ET) act as robust second messengers coordinating signal transduction pathways at the time of stress responses. These molecular signals stimulate many simultaneous transduction pathways involving protein kinases and phosphatases.[72]

Subsequent to the first stage of signal perception, plants elicit two main signal cascades, namely, the calcium-dependent protein kinase (CDPK) and the mitogen-activated protein kinase (MAPK) pathways.[73,74] Following this, the expression of specific TFs is increased or decreased by means of phosphatases or protein kinases wherein these TFs recognize and interact with cis elements present in the stress-associated genes to augment or inhibit their transcription.

One of the major plant TF families is the NAC (NAM (for no apical meristem), ATAF1 and −2, and CUC2 (for cup-shaped cotyledon)) TF family.[75] NAC genes are implicated in regulating immune responses to invasion by pathogens, feeding by insects and wounding processes. NAC TFs are also shown to be elicited as part of biotic stress responses.[76] Defense responses to infestation by greenbugs in sorghum involve the upregulation of NAC TFs.[77] RNA sequencing revealed that several NAC genes were induced in defense responses to greenbugs. This study found that the SbNAC (*Sorghum bicolor* NAC) genes serve diverse functions in the defense response to infestation by greenbugs and hence may enable genetic resistance to the greenbugs. NACs in wheat have been shown to play roles in defense responses to the powdery mildew disease. For instance, overexpression of TaNAC6 (*Triticum aestivum* NAC6) decreased the haustorium content of the pathogen and improved pathogen resistance in transgenic plants.[78] Transgenic barley plants were generated to identify the functions of HvNAC6 (*Hordeum vulgare* NAC6) in immune response to *Blumeria graminis f. sp. hordei (Bgh)* wherein the phytohormone ABA was involved. Silencing of HvNAC6 by RNAi diminished the HvNAC6 transcript levels, and these barley plants showed increased susceptibility to the Bgh pathogen when compared to wild-type plants.[79]

In a study by Liu et al., 2018[80] it was found that in rice, the transcription factor ONAC66 was highly induced due to blast disease infection. Overexpression of ONAC66 augmented resistance to blast disease and bacterial blight, and this increase was accompanied by diminished expression of the genes related to ABA. This showed that ONAC66 regulates the components of the ABA signaling pathway to advance disease resistance. MYB TFs (myeloblastosis family of transcription factors) have been demonstrated to be involved in responses to biotic stresses in sorghum. Sorghum combats attack due to *Colletotrichum sublineolum*, a fungal pathogen by expressing 3 deoxyanthocyanidin phytoalexins whose synthesis needs the MYB TF of the sorghum yellow seed 1 (Y1).[81]

On the other hand, 3-deoxyanthocyanidin was induced in transgenic maize plants overexpressing the MYB TF gene, wherein increased resistance was observed against the ingress of the leaf blight pathogen.[82] Also, enhanced resistance to infection by *Rhizoctonia cerealis* was demonstrated in wheat overexpressing the R2R3-MYB TF gene TaRIM1.[83] TaRIM1 regulated expression of defense genes in transgenic wheat by interacting with the MYB binding site. In yet another study, the involvement of the MYB TF was observed in response to infestation by insects in chrysanthemum. Aphid infestation induced the expression of the CmMYB19 gene and overexpressing this gene contained aphid proliferation through the accumulation of lignin in transgenic chrysanthemum.[84]

WRKY TFs (containing the highly conserved 60 amino acid long WRKY domains of TFs, and having a conserved heptapeptide motif WRKYGQK at the N-terminus and a novel zinc-finger-like motif at the C-terminus) also play important roles in defense regulation in response to pathogen attack. Improved resistance to rice sheath blight disease was observed in rice plants expressing the WRKY4 and WRKY80 genes. The WRKY80 factor interacts with the W-box located within the WRKY4 promoter region, which functions as a positive regulator for inducing *Rhizoctonia solani* resistance in rice. Overexpression of genes WRKY58, WRKY7, WRKY64, WRKY62, and WRKY76 was observed in rice blast disease.[85] Also, WRKY67 overexpression improved resistance of transgenic rice to two important rice diseases, bacterial blight, and blast, thus making WRKY67 an excellent target for rice enhancement.[86]

Silencing of wheat TaWRKY62[87] augmented susceptibility to the stripe rust disease while silencing of the TaWRKY49 improved resistance to this disease compared to non-silenced plants. This latter effect by TaWRKY49 silencing was associated with augmented expression of SA- and JA-responsive genes such as TaPR1.1, TaAOS, as well as the ROS-related genes TaPOD and TaCAT while suppressed expression of ET-responsive gene, TaPIE1 was observed. On the other hand, diminished resistance subsequent to TaWRKY62 silencing was associated with overexpression of the TaPR1.1 and TaPOD genes as well as inhibition of the TaAOS and TaPIE1 genes.[87] The WRKY TFs were also implicated in defense responses against plant viruses: WRKY8 is involved in defense responses to the TMV-cg (crucifer-infecting Tobacco

Mosaic Virus) in *Arabidopsis* by coordinating both the ET and ABA signaling pathways to induce resistance against the TMV infection.[88] Moreover, the WRKY TFs could enable cross-talk between pathways involved in biotic and abiotic stress responses. In rice plants, the OsWRKY11 enhanced drought tolerance and pathogen defense by regulating both the biotic and abiotic stress-associated genes.[89]

Also, factors ERF/DREB (ethylene-responsive factor/dehydration-responsive element-binding protein) could have conserved functions essential for the regulation of pathways involved in disease resistance. The grape VvERF (*Vitis vinifera* ERF) was shown to be overexpressed in defense response to infection by *Botrytis cinerea.*[90] Overexpression of VaERF20 (*Vitis amurensis* ERF20) gene sourced from Amur grape in transgenic *Arabidopsis* increased resistance to *Botrytis cinerea* and *Pseudomonas syringae.*[91]

The bZIP TFs have been shown to positively or negatively influence disease resistance in response to pathogen infection.[92] In soybean, bZIP genes GmbZIPE1, GmbZIPE2, GmbZIP105, and GmbZIP62 (*Glycine max* basic leucine zipper transcription factors) were shown to mediate defense response to Asian soybean rust disease (ASR) by regulating the ASR-associated genes. In cassava (*Manihot esculenta*), gene expression of MebZIP5 and MebZIP3 was stimulated by hydrogen peroxide, salicylic acid, and *Xanthomonas axonopodis Pv. manihotis.*[93] Also, transgenic tobacco plants overexpressing both the MebZIP5 and MebZIP3 genes showed augmented resistance to the cassava bacterial blight disease. On the other hand, silencing of these two genes in the transgenic plants diminished the levels of transcripts encoded by the defense response genes, which resulted in disease sensitivity. Arabidopsis plants transgenic for the CabZIP gene sourced from pepper were shown to overexpress the CabZIP protein, which improved plant tolerance to *Pseudomonas syringae* Pv. tomato DC3000.[94] NAC21/22 transcription factor gene regulation by mi164 in wheat was shown to affect stripe rust disease resistance.[95] miRNAs are favorable targets to study the crosstalk between signaling pathways and gene expression networks as several miRNAs target the TFs, which act as chief regulators of gene expression.[96]

Many long noncoding RNAs (lncRNAs) have been shown to be involved in gene regulation in biotic stress responses. In rice lines that were resistant to *Magnaporthe oryzae*, the rice blast pathogen, thousands of lncRNAs were characterized.[97] These lncRNAs upon functional classification revealed that the amounts of long intergenic ncRNAs were more abundant in the rice lines that were resistant and LncRNA might be associated in JA-mediated resistance of rice to blast fungus. Identification of lncRNAs associated with resistance to *Verticillium dahlia*, a fungus, was demonstrated in cotton.[98]

Comparisons were made with respect to the lncRNA expression profiles of *Gossypium barbadense*, the disease-resistant cotton and *Gossypium hirsutum*, the susceptible cotton variety, which revealed apparent disease response pathways. Intriguingly, some of the lncRNAs were shown to play roles in multiple stress responses. In transgenic rice, OsNAC6 (*Oryza sativa* NAC6) expression augmented tolerance to blast disease.[99] Plant TF-based responses to biotic stresses are highly complex, and many of the TF families are apparently involved in combating single or several stresses in addition to manifold crosstalk between various signal transduction pathways.

## 5. Role of plant hormones in plant defense response

SA, JA, and ET are important hormones in mediating plant defense responses against several pests and pathogens.[100] SA activates defense responses against hemi-biotrophic and biotrophic pathogens,[101] while JA and ET are involved in defense response against herbivorous insects and necrotrophic pathogens.[102,103] Synthesis of SA is activated in response to the detection of plant pathogens. Upon initiation of the SA pathway at the infection site, a defense response is frequently elicited in distal parts of the plants that protect undamaged tissues. This broad-spectrum long-lasting induced resistance is called as systemic acquired resistance (SAR).

SA accumulation-defective or insensitive mutants show augmented pathogen susceptibility. Additionally, enhanced SA levels in tissues that are exposed to pathogens leads to the elicitation of PATHOGENESIS RELATED (PR) genes. These PR genes encode many proteins containing antimicrobial activity and therefore provide increased resistance against a wide array of pathogens.[104] A crucial regulatory element in SA-driven PR gene activation is the NON-EXPRESSOR OF PR GENE 1 (NPR1). SA coordinates the NPR1 deoligomerization into its monomeric, active forms, which undergo nuclear localization whereupon they interact with the TGA class of the transcription factor bZIP that in turn enable PR gene expression and consequent defense response.[101] Nevertheless, recent reports have demonstrated that NPR1 serves as a receptor for SA and SA directly controls the NPR1 conformation by deoligomerizing NPR1 into its dimeric form.[105] Many WRKY transcription factors play significant roles in coordinating plant defense responses downstream of NPR1.[106,107]

Augmentation of JA levels during pathogen infection highlights its role in plant defense response. JA also plays significant roles in protecting plants against a gamut of herbivores such as thrips, caterpillars, beetles, spider mites, and mirid bugs.[103] Many representatives of the APETALA2/ETHYLENE-RESPONSIVE FACTOR (AP2/ERF) family have also been shown to be involved in stress responses regulated by JA.[108,109] Factors such as ERF1, ERF2, ERF5, and ERF6 regulate the expression levels of PLANT DEFENSIN 1.2 (PDF1.2), a JA-responsive marker gene and confer resistance to necrotrophic pathogens.[110,111] Moreover, JASMONATE-JIM-DOMIN (JAZ) repressor protein plays a critical role in JA response to stress conditions.[112]

In the absence of the bioactive JA, namely, JA-Ile,[113] JAZ proteins bind with JIN1/MYC2 and repress transcriptional regulation of genes responsive to

JA. During JA stimulation, JA-Ile interacts with its receptor, CORONATINE INSENSITIVE1 (COI1), and results in 26S proteasome-driven degradation of JAZ enabling MYC2 to increase the expression level of genes targeted by JA.[112] MYC2 is modified post-translationally by Thr328 phosphorylation to induce its transcription activity.[114] Nevertheless, modified MYC2 is inherently unstable and is degraded by the Plant U-box protein (PUB10) that acts as an E3 ligase.[115] This enables MYC2 turnover thereby promoting fine-tuning of MYC2-mediated JA responses.

ET is involved in diverse functions in plant defense response.[100,102] In the ET signaling pathway involved in stress responses, the ERFs are the main downstream regulatory factors. The ETHYLENE INSENSITIVE3 (EIN3) transcription factor was implicated in the induction of ERF1 gene expression in response to ET, activating defense responses.[116] EIN2 is another positive determinant of ET signaling. When ET is absent, EIN2 is phosphorylated and repressed by CONSTITUTIVE TRIPLE RESPONSE (CTR1) Raf-like protein kinase. Upon recognition of ET by its receptor ETHYLENE RESPONSE 1 (ETR1), the CTR1 is inhibited and EIN2 repression is relieved by cleavage and nuclear localization of its C-terminus, resulting in the activation of ET signaling.[117] Moreover, ET is involved in cross-talking with SA and JA pathways antagonistically or by enhancing them to accomplish tailored plant defense responses.

## 6. Hormonal crosstalk involved in plant defense response

The SA and JA signaling pathways intersect at multiple points as they both antagonistically regulate biotic stress responses,[100] and NPR1 is the key molecule involved in this antagonistic crosstalk. The SA-mediated inhibition of JA-responsive genes such as VEGETATIVE STORAGE PROTEIN (VSP), LIPOXYGENASE 2 (LOX2), and PDF1.2 is abrogated in npr1 mutant plants impaired in transduction of SA signals.[118] This showed that the SA-mediated antagonistic effect on JA signaling requires the NPR1 regulatory protein. The transcription factor WRKY70 is also a key determinant in facilitating the antagonistic molecular interaction between these two hormones. WRKY70 overexpression has been reported to drive constitutive expression of the PR (pathogen-related) genes responsive to SA, thus resulting in resistance to virulent pathogens while causing inhibition of the JA-responsive PDF1.2 gene.[119] Similarly, in *Arabidopsis*, the mpk4 (MAP kinase 4) knock-out mutants exhibiting constitutive SAR had enhanced SA-responsive PR gene expression levels, whereas the JA-responsive gene (PDF1.2 and THI2.1) expression levels were repressed.[120]

Despite many studies demonstrating SA and JA antagonistic interaction, synergistic relationships have also been reported at low concentrations of SA and JA in addition to simultaneous stimulation of both defenses.[119,120] Whereas SA and JA are involved in antagonistic interactions, the JA and ET interact synergistically in controlling defense-related genes following pathogen infection. Both the JA and ET hormonal pathways stimulate/stabilize EIN3 and thereby show synergy in the development of root hairs and necrotroph resistance.[121] A positive interaction between JA and ET induces genes encoding proteinase inhibitors in wounding responses in tomato.[122] Similarly, both JA and ET are necessary to induce simultaneous expression of ERF1 and result in activation of PR genes.[108] In *Arabidopsis*, studies have shown that both JA and ET molecular signaling pathways could also act antagonistically against herbivore and insect attacks. The JA-induced MYC2 was found to recognize ET-stabilized EIN3 and inhibit its downstream activities. Contrarily, EIN3 inhibits MYC2 and therefore represses JA-mediated herbivore defense response.[123]

Most auxin responsive genes were also inhibited after SAR induction leading to the suggestion that auxin augments disease susceptibility while improved disease resistance would require the repression of auxin-mediated signaling. Therefore, auxin functions as a key element in the hormonal signaling network mediating defense regulation. Cytokinins (CK) have also been shown to be involved in biotic stress responses.[124–126] Transgenic plants of *Arabidopsis* containing stabilized CK levels showed improved resistance against infection by *Verticillium longisporum*, a hemi-biotrophic pathogen.[126] Cytokinins have also been reported to be involved in interaction with the SA signaling cascade to coordinate plant defense responses. Cytokinin-induced transcription factor ARABIDOPSIS RESPONSE REGULATOR 2 (ARR2), recognizes the bZIP-type transcription factor TGA3, and induces SA-mediated defense responses based on NPR1 dependency.[124] Likewise in rice plants, blast fungus *Magnaporthe oryzae* resistance is augmented by the synergistic relationship between SA and CK in a WRKY45- and OsNPR1-dependent manner.[127]

## 7. Generation of plants resistant to biotic stresses

In the development of plant breeding resistance, R genes have been widely used and are mostly introgressed from closely linked species or wild relatives that are sexually compatible. Notwithstanding, because of the emergence of novel virulence determinants in the pathogens, the endurance of the resistance provided by the R gene is frequently compromised.[128] Alternately, transgenic strategies that access genetic elements from distant organisms have opened a potential avenue for surmounting several bottlenecks inherent to breeding resistance.[129]

Many strategies exist for engineering disease resistance in crops such as expression of pathogenesis-related (PR)/antimicrobial genes and R genes, pathogen virulence factor detoxification, enhancing structural barriers, etc. For engineering insect resistance, expression of insecticidal bacterial genes, RNAi delivered by the host, alteration of pathways involved in defense signaling, expression of antinutritional proteins such as protease inhibitors, etc. have been developed.[128,129]

Nevertheless, in spite of the expanded cultivation of virus- and insect-resistant transgenic plant varieties, transgenics possessing fungal and bacterial resistance have met with limited success.[130] This failure is due to diminished levels of resistance conveyed by the transgenes or due to the resistance being mostly confined to a single strain.[130,131] Also, despite circumventing technical glitches, field level testing and introduction of transgenics engineered for fungal, bacterial, and insect resistance significantly suffer from unfavorable public perception.

Such negative perception is mostly because of opposition to genetically engineered crops and ignorance of the public regarding scientific principles. Use of genome editing for augmenting biotic stress resistance could change on a case-by-case basis.[132] Such strategies could include changing the interaction between the effector and the target, susceptibility gene knockout, regulating the expression of genes concerned with the propagation of pathogen infection progression or insect feeding, and the engineering of R gene synthetic variants specific for recognition of wide ranges of pathogens or insects. Conventional mutagenesis strategy of knock-down after loss of function mutation is an unequivocal approach to improve crop disease resistance but nevertheless highly time consuming and laborious.

A well-known example is mutation generated in barley *MLO* recessive locus (*mlo*) that confers wide spectrum resistance to the disease Powdery mildew.[133] In this context, a more advantageous and quick method of generating Powdery mildew resistance was shown in wheat in which *Ta-MLO* homeoalleles were knocked down efficiently by the use of genome editing tools.[134]

## Genome editing strategies for augmenting resistance of plants to insect pests

Insect pests are responsible for greatly damaging growth and development of plants in a direct or indirect manner by transmission of pathogenic viruses ultimately causing major yield losses in crops. The existing control measures based on agrochemicals are costly and hazardous to the eco-system.[132] Additionally, they negatively influence nontarget insects like pollinators and bio-control agents. They also meet with gradual inefficiency due to the emergence of insecticide resistance. In several cases, the breeding of crops resistant to insect pests is limited in scope mainly due to the lack of well-defined source of resistance within a given crossable gene pool. In alleviating this obstacle, many measures aim to access genes from related wild plant species and uncharacterized crop plant accessions.

Nevertheless, much progress could not be made due to poorly characterized genetics of resistance traits within undefined accessions.[135] Alternately, transgenic strategies have been employed for introducing insect resistant genes sourced from distant species into the crops, for instance, the bacterially derived

Bt gene. However, these transgenic plant lines face serious ethical, political, and societal resistance on account of dearth in scientific knowledge.

In this circumstance, the main challenge for contemporary agriculture is to generate crop breeding tools that are environmentally friendly with the ability to develop de novo resistance where applicable R-genes are not accessible and to regulate pest dynamics by disrupting insecticide resistance, destroying, or introducing sterility in the affecting insects. In this context, genome editing techniques have the potential to produce designer plants specifically in situations wherein a targeted gene knockout is likely to evolve superior and elite traits or to initiate gene drives for selective propagation of mutations that lead to lethality of females in a given insect population.

## Targets for genome editing toward augmenting resistance of plants to insect pests

Genome editing has been profitably used in a wide range of species including insects over the past several years.[132] Nevertheless, recent progress in the precise engineering ability of this technology affords replacement of the conventional radiation and chemical mutagenesis methodologies with genome editing measures. Genome editing techniques for control of insects operate by disseminating a gene drive to necessitate knockout of insect genome target sites that could lead to lethality, impairment of fecundity or sex ratio of the destructive insects. Alternately, editing key plant immunity genes, i.e., resistance gene (R-genes), susceptibility genes (S-genes) and genes determining the target-insect effector interaction may also be used to develop immunity of the plant host to the insect-pests. To date, most researches are focused on optimizing and tweaking the components of the CRISPR/Cas molecules in individual crops.

Increasing knowledge of plant-insect interactions affords great possibilities for generating insect resistance by means of genome editing using the CRISPR/Cas9 system. Several examples of CRISPR/Cas9-based genome modifications in insects have been reported for control of pests and to counteract development of resistance to many insecticides as shown in Table 8.1. Supplanting alleles for insecticide resistance with their sensitive equivalents enable further application of otherwise redundant insecticides.

Pest control strategies also include targeting of insect species via gene drives to rapidly disseminate edited genes within the whole target insect species population. Gene drives based on CRISPR/Cas can induce sterility or lethality upon disruption of genes in the populations that are targeted. Such gene drives ultimately lead to population crash or even result in extinction due to the excessive burden of recessive lethal mutations.[132] Similarly, gene drives could skew the sex ratio toward males of a given insect population by specific disruption of the X-chromosome at the time of spermatogenesis. This ascertains the incidence of the Y chromosome in a majority of viable sperms

**TABLE 8.1** CRISPR/Cas9-based engineering of insects for pest/biotic stress management.

| Name of the insects | Target gene | Editing method | Method of transformation | References |
|---|---|---|---|---|
| *Drosophila melanogaster* | Nicotinic acetylcholine receptor α6 | Substitution | Embryo injection | Zimmer et al.[136] |
| *Anopheles gambiae* | rDNA | Knock-out | Embryo injection | Galizi et al.[137] |
| *Plutella xylostella* | Abdominal-A | Indels | Embryo microinjection | Huang et al.[138] |
| *Spodoptera exigua* | Ryanodine receptor | Substitution | Embryo microinjection | Zuo et al.[139] |
| *Culex quinquefasciatus* | Cytochrome P450 gene, CYP9M10 | Knock-out | Embryo injection | Itokawa et al.[140] |
| *Anopheles gambiae* | AGAP005958, AGAP011377, AGAP007280 | Knock-out, knock-in | Embryo microinjection | Hammond et al.[141] |
| *Tribolium castaneum* | E-cadherin gene, EGFP | Knock-out, knock-in | Embryo injection | Gilles et al.[142] |
| *Anopheles stephensi* | Kynurenine hydroxylase | Knock-in | Embryo microinjection | Gantz et al.[143] |
| Spodoptera *littoralis* | Olfactory receptor co-receptor (orco) gene | Knock-out | Egg injection | Koutroumpa et al.[144] |

Examples of CRISPR/Cas9-based engineering of insects for pest/biotic stress management.
From Bisht DS, Bhatia V, Bhattacharya R. Improving plant-resistance to insect-pests and pathogens: the new opportunities through targeted genome editing. *Seminars in Cell and Developmental Biology* 2019;**96**:65–76. https://doi.org/10.1016/j.semcdb.2019.04.008.

resulting in increase in the number of male progeny along with a concomitant decrease in the number of the insect females.

Therefore, curtailment of insect pest populations by genetic means via the release of strains of insects harboring deleterious traits such as biased sex ratio, infertility, lethality and insecticide sensitivity and introduction of genes conferring virus resistance in insect vectors to prevent viral disease transmission, are favorable strategies for biological control of insect-pests and their respective vectored diseases.

### Genome editing technologies for the engineering of plant disease resistance

Serious challenges are posed by plant pathogens to agriculture worldwide by greatly decreasing the annual productivity. Breeding plant varieties carrying durable and stable resistance effectual against wide-ranging classes of pathogens is formidable. Insight into the disease resistance genes and the complex molecular network involved in plant immunity can be used toward genome editing and genetic engineering for augmenting crop disease resistance.

## 8. Disease resistance breeding in the pre-editing times

Classical breeding of plants has been demonstrated to be beneficial in improving crop cultivars. There are many instances where significant measures by breeders have led to the generation and release of crop varieties harboring disease resistance.[145] The identification of DNA markers has further enabled breeders to shorten the breeding cycle of the developed varieties thereby augmenting the efficacy and precision of traditional plant breeding measures. Nevertheless, in spite of great success of the classical breeding measures, it becomes totally ineffectual in augmenting disease resistance traits in instances lacking natural allelic variations amid the crossable plant species.

Breeding measures toward generating mutations has met with some success in developing novel genetic traits. Nevertheless, breeding is a random procedure and evaluating and identifying desirable mutants is highly laborious and lengthy process. Additionally, upon identification of the desirable mutants, further breeding steps are necessary to establish homozygosity and to delete undesirable mutations.[146] Moreover, the longevity and endurance of resistance generated in crop varieties through breeding measures are mostly short-lived, primarily due to the emergence of new pathogenic determinants within the pathogen.

This impels the breeders to develop new varieties of crops regularly to generate adequate response to the evolution of new virulent pathogenic strains.[147] Production of genetically modified (GM) cultivars via recombinant DNA technology has opened a promising avenue complementary to traditional breeding measures as it enables the transmittance of beneficial genes across the species. Thus far, GM crops exhibiting resistance to viral diseases have been commercialized.[148]

However, GM crops have led to debates regarding their cultivation as they are engineered using foreign DNA, and several strong regulatory legislations have been formulated for their release into the environment in several countries. Hence, it has become highly challenging to use GM technology while striking a good balance between scientific requirements for assuring food security and dealing with socio-political concerns. Genome editing technologies afford novel possibilities for addressing the challenge of food security along with core concerns regarding GM regulations.

## 9. Prospects for genome editing–based engineering of disease resistance in plants

Genome editing techniques can generate precise point mutations in a stably inherited manner within the plant genome. Targeted mutational changes mediated by site-specific nucleases include gene elimination, modifying the sequences of regulatory elements determining gene expression, addition of desired genes at specific defined loci and assembling multiple genes into plant species with low risk of segregation while leaving no unwanted footprints. Genetically engineered crops free of markers such as antibiotic resistance genes[149] and any surplus genetic material such as plasmid remnants[150] have been generated using the CRISPR/Cas technology.

Such goals are unlikely to be achieved via conventional breeding methodologies or even through transgenesis. The major caveat of transgenesis is that the transgene expression is influenced by position effects due to the chromatin structure and regulatory sequences near the site of integration. Moreover, the transgenic locus is subject to silencing because of factors such as the occurrence of any homology between the transgene and any endogenous gene, copy number of the transgene, presence of truncated sequences and inverted repeats. Therefore, generating a comprehensive recipient line containing a specific locus of integration that ensures robust transgene expression without introducing unwanted detrimental effects on the plant could accelerate the construction of robust genetically modified plant lines.

The RNAi technology has been generally used to enable gene expression knockdown at the posttranscriptional level. Nevertheless, this technology has been demonstrated to result in incomplete transcript knockdowns, unforeseen off-targeting, and competition with gene regulation mediated via endogenous RNA.[151] On the other hand, the *Pyrococcus furiosus* type III-B CRISPR/Cas system is a more specific and unique RNA silencing system in which the homology-based degradation of the complementary RNA does not rely on host factors like dicer or RISC thereby reducing the prospects of silencing off-target genes.

In this light, CRISPR interference (CRISPRi) is a technique devised to inhibit multiple target genes at the same time by hindering transcriptional elongation, transcription factor, or RNA polymerase binding via the application of the dCas9–guide RNA complex. Transcriptional repression directed by the CRISPRi is greatly specific, and its reactions are reversible. RNAi silencing when used in gene silencing within the RNAi pathway itself (argonaute, dicer) results in difficult interpretation of the outcomes because of many direct and indirect effects. Additionally, the absence of CRISPR/Cas system within eukaryotes removes any possibility of competition with innate endogenous pathways indicating excellence of CRISPRi over the RNAi.

While the RNAi is predominantly restricted to attenuation of gene expression, the CRISPR/Cas system is more versatile by differentially targeting RNA or DNA in order to increase or decrease gene expression at the transcriptional and posttranscriptional stages. In the CRISPR/Cas9 technique, for target gene knockout at its genomic locus a single guide RNA and Cas9 enzyme are sufficient to introduce DSBs, while for the ZFNs and TALENS, two proteins are required to induce DSBs in addition to recoding of proteins using large (500–1500 bp) DNA segments specific for each individual target site as well as elaborate molecular design and assembly.[152]

Also, for multiplex gene editing ZFNs and TALENs are less suitable due to their large size. On the other hand, using CRISPR/Cas multiplex gene editing system several small-sized sgRNAs can be delivered along with the Cas9 in order to enable simultaneous editing of more than a single target gene. Targeted insertion of genes in addition to homologous recombination has been shown for several crops such as rice, maize, and tobacco.[152,153] Genome editing systems are distinguished from other current technologies by their simplicity, high efficacy and specificity features in addition to being time-saving, eco-friendly, amenable to multiplexing, and minimizing off target effects, which enable their wide-ranging applications. Single stranded nucleases (SSNs) afford several opportunities for engineering disease-resistant plants. Formerly, editing of disease susceptibility genetic elements for diminished pathogenesis was used as a direct approach for conferring overall disease resistance.[154] Nevertheless, with the emergence of the CRIPSR/Cas9 technology, the range of potential plant genome targets has also widened.

## 10. Targeting of susceptibility genes for generating disease resistance

Pathogens need host susceptibility genes (S) to enable successful infection, establishment, and propagation.[155] Based on the different steps involved in disease establishment, the S genes could be categorized as (a) genes concerned with aiding pathogen penetration into the host, (b) genes involved in supporting pathogen subsistence at the post-penetration stage, and (c) genes acting as negative determinants of immune signaling.[156]

Disruption of susceptibility genes could hence circumscribe the ability of the pathogen to induce disease. Upon mutation, disease resistance mediated by the S genes could be specific to the pathogen if the concerned mutation leads to disruption of molecules involved in prepenetration, penetration and post-penetration stages for a given class of pathogen.[154] Alternately, the S gene mutation could prove to evolve broad-spectrum resistance if the gene target generates constitutive or prolonged defense response.[157] S gene disruption has been used economically for generating plant disease resistance. Table 8.2 enlists some of the crops wherein disease susceptibility factors were edited using CRISPR/Cas9 system.

For example, the mutation generated barley *Mildew Locus* (*Mlo*) recessive alleles confer broad spectrum resistance to the fungus causing powdery mildew.[133] Also, homozygous mutations in the *Mlo* locus of tomato,[166] Arabidopsis[167] and pea[168] have been demonstrated to induce immune response to powdery mildew fungi. The Mlo gene codes for a membrane protein containing seven transmembrane domains broadly conserved among plants.[169] Efficacious use of CRISPR/Cas9 and TALEN to generate mutations in the three homeoalleles of the wheat *Mlo* susceptibility gene led to broad spectrum resistance to the fungus *Blumeria graminis* f. sp. *tritici* (*Bgt*).[134]

Such *Mlo* locus targeting by the CRISPR/Cas9 has also been shown in apple, grape vine, and tomato.[164,158] The S genes have been targeted also to induce immunity in plants against bacterial pathogens. A DMR6 gene frameshift deletion was induced by the Cas9/sgRNA, which displayed wide-ranging resistance against several pathogens such as *Phytophthora capsici, Pseudomonas syringae, and Xanthomonas* spp without causing significant impacts on plant growth and development.[159]

Targeting disease susceptibility factors has additionally been successful in inducing resistance against viral pathogens. The translation initiation factor eIF4e is responsible for allowing most of the positive sense single stranded RNA viruses to infect plants[170] due to which they have become potential genome editing targets leading to plants resistant to viruses. Disruption of the eIF4Es using the CRISPR/Cas9 system has been shown to induce potyvirus and ipomo virus resistance in cucumber[161] and Arabidopsis.[160] Similarly, recessive mutations in the rice translation initiation factor 4 gamma (eIF4G) has been shown to elicit resistance to rice tungro spherical virus (RTSV).[171] The eIF4G gene was edited by the Cas9/gRNA, which generated a novel eIF4G allele that conferred resistance against RTSV.[162] Also, these genome edited plants lacked any undesirable off-target mutations.

In addition to the targeting of coding regions, gene editing of the effector binding elements has also been shown to act as probable targets for engineering disease resistance.

The sugar transporter, *MeSWEET10a* has been shown to be the susceptible host gene in Cassava[172] during infection by *Xanthomonas axonopodis* pv. *manihotis* whose TAL (Transcription activation like) effector specifically

**TABLE 8.2** Plant host factors modified by gene-editing technology for engineering disease/biotic stress resistance.

| Plant species and its host factor | Gene function | Engineered resistance for pathogen | Mode of editing | References |
|---|---|---|---|---|
| *Solanum lycopersicum*; MLO -1 | Susceptibility gene associated with pathogenesis of Powdery mildew pathogen | Powdery mildew (*Oidium neolycopersici*) | CRISPR/Cas9 | Nekrasov et al.[158] |
| *Solanum lycopersicum*; DMR6 | This host factor belongs to the superfamily of 2-oxoglutarate Fe(II)-dependent oxygenases and is specifically overexpressed during pathogen infection | *Pseudomonas syringae*, *Phytophthora capsici*, and *Xanthomonas* spp. | | de Toledo Thomazella et al.[159] |
| *Arabidopsis thaliana*; eIF(iso)4E | Potyviral VPg (viral protein genome linked) interacts with eIF(iso)4E and enables translations of viral genes | Potyvirus | CRISPR/Cas9 | Pyott et al.[160] |
| *Cucumis sativus* L.; eIF4E (eukaryotic translation initiation factor 4E) | This host factor facilitates the transcription of uncapped viruses containing viral protein VPg that enables virus infectivity | Cucumber vein yellowing virus (Ipomovirus), Zucchini yellow mosaic virus, and Papaya ring spot mosaic virus-W | CRISPR/Cas9 | Chandrasekaran et al.[161] |

| | | | | |
|---|---|---|---|---|
| *Oryza sativa* var. *indica* cv. IR64; eIF4G, (eukaryotic translation initiation factor 4 gamma gene) | Functions in the assembly of basal translational initiation factors at the 5′ end of mRNA and in the cap-independent translation of viral RNA genomes | *Rice tungro spherical virus* (RTSV) | CRISPR/Cas9 | Macovei et al.[162] |
| *Oryza sativa;* SWEET14 | SWEET14 gene codes for the sucrose-efflux transporter whose activation by bacterial effectors is essential for the pathogen nutrition requirements | *Xanthomonas oryzae* pv. oryzae (Xoo) | TALEN-mediated disruption of the effector binding elements within the promoter of OsSWEET14 | Li et al.[163] |
| *Triticum aestivum;* MLO-A1 | Susceptibility gene associated with pathogenesis of Powdery mildew pathogen | Powdery mildew *(Blumeria graminis* f. sp. *tritici)* | CRISPR/Cas9 | Wang et al.[134] |
| *Vitis vinifera;* MLO-7 | Susceptibility gene associated with pathogenesis of Powdery mildew pathogen | Powdery mildew | CRISPR/Cas9 | Malnoy et al.[164] |

*Continued*

**TABLE 8.2** Plant host factors modified by gene-editing technology for engineering disease/biotic stress resistance.—cont'd

| Plant species and its host factor | Gene function | Engineered resistance for pathogen | Mode of editing | References |
|---|---|---|---|---|
| *Citrus sinensis;* CsLOBl (*Citrus sinensis* lateral organ boundaries) | CsLOB1 host factor is encoded by a susceptibility gene that is induced by the virulence factor of *Xanthomonas citri* | Citrus canker caused *by Xanthomonas citri subspecies citri (Xcc)* | CRISPR/Cas9-mediated disruption of effector binding elements within the promoter region of the CsLOB1 gene | Jia et al.[165] |

induces *MeSWEET10a* to promote pathogen virulence. Disruption of the effector binding element within the promoter using TALEN-based gene editing confers bacterial leaf blight resistance.[163] Likewise, disrupting the *OsSweet 13* gene using CRISPR/Cas9 prevents the TAL effector gene pthXo2 interaction leading to enhanced resistance against rice bacterial leaf blight.[173]

Cas9/sgRNA was used to edit the citrus canker susceptibility gene *CsLOB1* (CsLOB: Citrus *sinensis* lateral organ boundaries) effector binding element toward generating citrus varieties resistant to canker.[165] Many other plant disease susceptibility genes have been described in various crops.[157]

## 11. Gene editing–based engineering of immune receptors toward generating broad-spectrum resistance

Traditional breeding toward disease resistance is based on the identification of the R gene and its mutation within the respective cultivated plant varieties. Nevertheless, for the most part R gene-based disease resistance is specific to the strain and is circumvented by the pathogen, which evolves into new strains. Therefore, the primary challenge for breeders is the development of new varieties containing novel determinants of disease resistance that can be generated rapidly in order to keep up with the evolution of the pathogen.

Of importance are the conformational changes in the R protein upon interaction with its respective effector or an effector-modulated plant protein and its proper oligomerization and its interplay with other proteins.[174] Specific R gene residues are responsible for the recognition process and mutating them can alter defense activation and develop resistance to pathogens that are phylogenetically divergent. This could potentially help engineer durable disease resistance in crops.[175,176]

Also, transfer of the PRRs localized in the plasma membrane between different plant species can confer durable broad-spectrum resistance in crops thus abrogating taxonomic restriction.[177] In this context, expression of the Arabidopsis EFR pattern recognition receptor in rice facilitates the rice plant to sense the EFR bacterial ligand and stimulate an immune response.[178] The knowledge obtained on PRRs of diverse origins could potentially be used by genome editing tools to develop designer PRRs capable of sensing a wide range of plant pathogens. Therefore, modular R receptor possessing broad-spectrum specificities could be generated by leveraging immune receptor effector sensing capability and the resourceful programmability of CRISPR/Cas9–based gene editing.[25]

## 12. Spatial uncoupling of plant hormones implicated in defense response

Hormonal signaling tightly regulates plant defense response against invading pathogens.[179] Many plant hormones coordinate the immune response. Salicylic acid mediates immunity against hemi-biotrophic and biotrophic pathogens, while ethylene and jasmonic acid are implicated in defense response against necrotrophic pathogens wherein both of these hormones function antagonistically with respect to each other. Increased salicylic acid levels are correlated with enhanced susceptibility toward necrotrophic pathogens while decreased SA levels correlate with resistance.[179] Fine tuning molecular pathways of hormones resulting in spatio-temporal uncoupling of this antagonistic relationship of SA and JA/ET might provide broad spectrum pathogen resistance.

This interaction between hormones involved in defense regulation was used in the generation of tomato varieties resistant to black speck disease using the CRISPR/Cas9 system.[180] This disease is caused by the pathogen *Pseudomonas syringae pv. tomato* (Pto) DC3000 that secretes the hormone coronatine (COR) inciting opening of the stomata enabling pathogen penetrance and its subsequent colonization. It was observed that the COR co-receptor SlJAZ2 specifically accumulates within the guard cells and CRISPR/Cas9 editing of this SLJAZ2 led to the generation of a dominant repressor polypeptide that precluded COR-mediated stomatal reopening and conferred resistance to *PtoDC300*. Also, in the gene edited plants the immunity toward the necrotrophic fungus was not affected.

## 13. CRISPR/cas technology in engineering plant defense against biotic stresses

Intentional dissemination of gene elements in wild insect pest populations that change their sex ratio or generate lethal mutations is a targeted and eco-friendly strategy for controlling insect pests. Nevertheless, elicitation of insect resistance in response to gene drive based on CRISPR technology is an apparent risk at both experimental and theoretical levels.[181] Consensus regarding ethical and scientific issues toward insect resistance is required to practice this technology. Moreover, since CRISPR/Cas9-edited insects harboring gene drives have the probability to alter the entire insect population/eco-system, which might negatively impact pollinators and other beneficial insects, this could in turn adversely influence food chains that could result in unpredictable changes in the community structure.

Thus, strict assessment of risks due to nontarget effects should be considered prior to release of CRISPR/Cas9-based gene drives. Moreover, the prospect of gene flow between target insect species and its nontarget equivalents could aggravate the situation. Significant decrease in off-target outcomes could be obtained via the use of Cas9 variants that are engineered,[182] Cas9 orthologs containing definite DNA binding specificity, couple of gRNAs along with a Cas9 version harboring a mutated nickase,[183] short guide sequences,[184] dimeric RNA-guided FokI nucleases RFNs,[185] and guide sequences having no or tolerable nontarget effects.[186]

Targeted elimination of insect pests, insect vectors harboring viruses, or other invasive species of insects using gene drive technology looks greatly feasible provided the accompanying risks are assessed in view of unintended consequences on the eco-system. Toward this purpose, terminator genes that can program the life of the gene-edited insects along with the usage of tagged insects for monitoring gene flow could prove to be a necessary step toward the employment of gene drives while ensuring biosafety.

Knockout of genes determining susceptibility (S) using CRISPR/Cas9 has proved to be successful in generating disease resistance of plants against invading pathogens. Notwithstanding, a major problem regarding the knock out of the S gene is the elicitation of pleiotropic outcomes within the plants that could lead to accompanying negative impacts on plant fitness. In this context, the effector binding element can be edited instead of the gene itself, which could prove to be a viable, alternate strategy for obtaining disease resistance without impacting the performance of the plants.[163]

Aside from editing the S gene, R gene stacking could provide a useful strategy for engineering broad-spectrum, durable resistance in crop species. However, stacking and transfer of the R gene is highly time and labor consuming for tree species due to their prolonged generation cycle. For examples, more than 50 years was required for transmission of the *Malus floribunda* scab resistance gene, *Vf*, to economically important cultivars of apple.[187] Genes determining flowering could be ectopically expressed and key genes involved in flowering could be silenced toward reducing the juvenility period and elicit early flowering among tree species.[187,188] Genome editing can be used to silence juvenility causing genes and upon transfer of the favorable resistance genes the respective mutation can be segregated in succeeding generations.

Therefore, the CRISPR/Cas9 technology generating disease resistance among tree species will increasingly emerge as a robust tool for conveying genetic traits in economically significant cultivars within brief time periods.[132] Another option to obtain broad spectrum resistance among crops is to use endogenous PRRs and engineer them to be capable of recognizing diverse pathogenic determinants. For this, in-depth knowledge of the PRRs as well as their role in inducing immune responses is required. Generally, better discernment of the plant immune responses and tweaking the genome editing toolbox would enable more specific targeting of molecules involved in immune responses toward the generation of disease resistant cultivars ensuring more sustainable global agricultural practices.

## 14. Recent developments in engineering resistance against biotic stresses

### Use of RNAi technology to alleviate biotic stress

Antiviral RNA silencing is a host response induced by viral dsRNA involving an RNA-induced silencing complex (RISC) containing multiple proteins and one strand of siRNA (small interfering RNA). The RISC utilizes this RNA to lead the complex toward recognizing complementary viral mRNA. Consequent to recognition, the RNase part of the RISC known as Slicer is activated followed by cleavage of the viral mRNA. The initial RNA required for triggering this process is provided by Dicer, an RNase type III—like enzyme by recognition and cleavage of the dsRNA intermediates generated during the process of viral replication. This siRNA strategy has been found to be efficacious against both RNA and DNA viruses, but nevertheless, it is a relatively slow process and does not result in complete eradication of the virus.[189] This defense mechanism is called pathogen-derived resistance.

Tobacco plants were genetically modified to confer resistance to whiteflies by producing siRNA against the V-ATPaseA gene in *Bemisia tabaci*.[190] These transgenic plants were shown to synthesize the dsRNA precursor to form siRNA which knocked out the expression of V-ATPase A in *B. tabaci*[191] after the bugs fed on the transgenic plants and subsequently led to lethality.[192] The vector *Myzus persicae* transmits Potato Virus Y (PVY) across plants resulting in serious impacts on crop yield particularly in potatoes. The expression of two cuticle proteins in these vectors was reduced by feeding with gene-specific dsRNA which consequently led to 47% decrease in PVY transmission.[193] The αCop and vATPase were targeted in *Nezara viridula* by feeding dsRNA which led to 90% mortality in these insects.[194] Elicitation of RNAi for controlling insect pests is increasingly becoming a modern methodology for the management of these pests in major agricultural crop plants.

Augmented resistance to the blast fungus *Magnaporthe grisea* and the leaf blight bacteria *Xanthomonas oryzae* pv. *oryzae* was demonstrated by RNAi-mediated knock down of OsSSI2 fatty acid desaturase activity[195] in rice. Host elicited suppression of gene expression led to reduced fungal disease attachment in barley and wheat by transient RNAi—mediated gene regulation. Enhanced resistance to *Sclerotinia sclerotiorum* was observed by RNAi-mediated targeting of genes involved in lignin biosynthesis by lowering lignin levels.[196] In rice plants, osa-miR7695 negatively controlled macrophage protein 6 associated with resistance to the blast fungus, *Magnaporthe oryzae*, wherein overexpression of this protein led to better resistance to blast disease in these plants.[197]

Exogenous application of complementary siRNAs or dsRNAs to plant surfaces led to inhibition of endogenous post-transcriptional signaling and elicited RNAi-mediated resistance to pathogens upon diffusion of these molecules both locally and systemically.[198] RNAi technology has been widely employed in the development of transgenic plants conferring virus resistance.[190] This has involved the construction of RNAi vectors carrying viral target genes within the transgenic plants that resulted in silencing of virus multiplication in the cell. Antiviral defense mediated by RNAi is controlled by multigenic classes of Dicer-like enzymes that generate siRNAs from dsRNA at the time of transcription and/or replication of the RNA or DNA virus. During this process, argonaute enzymes interact with the viral siRNA and target the viral DNA that leads to silencing at the transcriptional and posttranscriptional levels.[199] Transgenic bean plants producing dsRNA targeting the Bean Golden Mosaic Virus (BGMV) sequences demonstrated BGMV resistance.[200] An equivalent strategy was employed to successfully develop resistance in barley crops against the Barley yellow dwarf virus (BYDV-PAV).[201] Papaya plants transgenically expressing the Papaya ringspot virus type W CP gene were found to be resistant to this virus.[202] *Solanum tuberosum* plants transgenic for the CP-encoding sequence from Potato virus X (PVX), PVY, and Potato virus S (PVS) exhibited resistance against these three viruses.[203] dsRNAs that target the coding sequences of the CP or the potyviral Nlb protein were topically delivered to confer resistance to mechanical BCMV infection in *Nicotiana benthamiana* and *Vigna unguiculata* plants.[204] The RNA silencing strategy can be applied to elicit resistance to both RNA and DNA viruses.

RNAi-based targeting of the genes 4G06, 10A06, 3B05, and 8H07 in *Heterodera schachtii*, the sugar beet cyst nematode, led to 23%–64% decrease in adult nematode females, thereby conferring resistance, albeit incomplete.[205] In another study, RNAi technology was used to inhibit the parasitism gene 16D10 of root-knot nematodes, wherein resistance to four major nematode species was reported in *Arabidopsis* that resulted in the in vivo 16D10 dsRNA expression.[206]

## Use of beneficial microorganisms to combat biotic stress

Under conditions of biotic stress, plants engage beneficial microbes from their surroundings to preclude pathogen invasion. When exposed to the fungal pathogen *Rhizoctonia solani*, sugar beets block the fungal pathogens by augmenting the chitinase content to elicit the growth of *Flavobacterium*.[207] Upon invasion by this fungus, species of *Streptomyces*, *Pseudomonas*, and *Burkholderia* within the soil aggregate together at the root system, thus functioning as the primary line of defense against this pathogen.[207,208] Beneficial microbes in the soil compete with infectious pathogens for living space and limited soil resources that circumscribe pathogen growth and virulence and consequently decrease plant stress via these interrelationships.[209,210] For instance, some of the resident soil bacteria compete for resources with the plant pathogen, *Ralstonia solanacearum*, which significantly decreases its rate of infection in tomato plants.[211] *Pseudomonas* and *Bacillus* secrete antibiotics with antiviral and insecticide functions and preclude or repress plant diseases via competition with the pathogens.[212] They also synthesize antipathogenic metabolites[213] or secrete iron carriers to deplete pathogens of iron,[214] thus circumscribing pathogen activities. In *Arabidopsis thaliana*, during infection caused by *Botrytis cinerea* and *Hyaloperonospora arabidopsidis*, the plant recruits synergistic bacteria to aid in defense against the infection and preclude pathogen invasion.[215]

## Application of transgenic technology to develop resistance against biotic stress

There are currently five different varieties of approved GM potatoes having antiviral resistance, inclusive of the Shepody NewLeaf Y potato, New Leaf Y Russet Burbank potato, New Leaf Plus Russet Burbank potato, Hi-Lite NewLeaf Y potato, and a new potato SPT TICAR.[216] Excepting potato SPT TICAR which was produced by Technoplant, Argentina, and approved for use in feed, food, and cultivation since 2018 in Argentina,[217] all the other GM potatoes were generated by the Monsanto Company, USA. The primary advantage of these GM potatoes is the reduction of cost for the farmers as lesser amounts of insecticides are necessary and the tubers do not have to be screened prior to planting for the next year. These GM potatoes have been approved for cultivation purposes only in Canada and the United States. Also, Japan, South Korea, the United States, Canada, Australia, New Zealand, and Mexico have approved these GM potatoes for food use.[218,219] All these approved GM potatoes use the strategy of gene silencing for pathogen resistance.[220] For PVY resistance, the GM potato plants encode PVY CP,[221] while for the PLRV (Potato Leaf Roll Virus) resistance, the plants have been engineered to encode and express the PLRV replicase.[222] These GM plants do not synthesize detectable levels of the transgenic proteins substantiating the functional mechanism involved in gene silencing and authenticating their safety for consumption.[223–225]

Albicidin (albD) gene was transformed into sugarcane which yielded plants rid of leaf scald disease.[226,227] Toward containing fungal infections, the *Trichoderma* spp. β-1,3-glucanase gene was introduced into sugarcane by transgenic technology and these transgenic plants displayed variable levels of resistance against the red rot fungus, *Colletotrichum falcatum*.[228] Transgenic lines of sugarcane that expressed a class II barley chitinase gene afforded protection to *C. falcatum* in vitro wherein crude protein extracts prepared from these transgenic plants blocked mycelial growth.[229]

Advancements in genetic engineering of plants have accomplished broad defense against pests by employing genes from various organisms such as bacteria, pests, and plants.[230–233] Protease inhibitors and lectins are insecticidal proteins generated by plants. Genes sourced from plants such as gna (*Galanthus nivalis* agglutinin gene), sbbi (Bowman-Birk inhibitor gene), skti

(Kunitz soybean trypsin inhibitor gene), and avac (*Amaranthus viridis* L. agglutinin gene) have been variously employed to generate transgenic sugarcane plants resistant to crop pests, wherein suppression of pest growth, reproduction, and development were observed.[234–236] Aprotinin gene was expressed in transgenic sugarcane plants which reduced 99.8% of the weight of *Scirpophaga excerptalis* larvae.[237] Subsequently, transgenic sugarcane lines overexpressing the CaneCPI-1 gene were produced that conferred resistance to *Sphenophorus levis* larvae.[230]

The major insecticidal proteins for engineering transgenic plants are expressed by the bacterium, *Bacillus thuringiensis* (Bt), which synthesizes proteins at the time of its vegetative growth (vip) as well as sporulation (cry). These proteins show high toxicity toward a broad range of insects upon entry into the alimentary tract.[238,239] As a result, these transgenic plants have been shown to be efficacious against target insects only but do not exhibit any toxicity toward nontarget insects.[240] Generation of crops that have been genetically transformed with the Bt genes have revolutionized the field of pest control involving Coleoptera and Lepidoptera pests in commercially important crops such as cotton, soybeans, and corn,[241] besides which they have reduced the negative impact of noxious chemical insecticides in agricultural practices.[242] Transgenic Bt sugarcane plants having resistance to stem borers have been produced successfully using transformation with various Bt genes such as cry1Aa3, cry1Ab, cry1Ac, m-cry1Ac, s-cry1Ac, vip3A, and cry2A under the control of the maize ubiquitin promoter and the CaMV 35S promoter. In 2017, the commercial GM sugarcane line resistant to *Diatraea saccharalis* was released for the first time. Subsequently, other transgenic sugarcane varieties expressing the cry gene family toxin were released for commercial use in Brazil, Canada, and the United States which conferred resistance to Lepidopteran insects.[243]

Transgenic rice plants expressing caffeine were generated by the introduction of 3 N-methyltransferase genes associated with the biosynthesis of caffeine in coffee plants.[244] These plants showed strongly augmented resistance to rice blast, bacterial leaf blight, and white-backed planthopper attack wherein caffeine served as a repellent factor against these rice pathogens. Additionally, caffeine elicits a panoply of $Ca^{2+}$ signaling-like events to express the plant resistance hormone, salicylic acid. Also, in these caffeine-producing rice plants, there was inhibition of phosphodiesterase accompanied by an increase in cGMP and cAMP, elevation of intracellular $Ca^{2+}$, activation of phenylalanine lyase, and salicylic acid. Thus, engineering of transgenic caffeine-producing plants could be adopted as a novel strategy to augment resistance to biotic stress factors. Caffeine-synthesizing genes derived from *Coffea arabica* were introduced into chrysanthemum and tobacco plants to induce enhanced pathogen resistance.[245–247]

## Recent progress in CRISPR/Cas9-based genome editing for developing disease resistance in crops

Several major crop species such as banana, tomato, cassava, wheat, rice, cucumber, citrus, and grapes have been engineered using the CRISPR/Cas9 system.[248] Broad-spectrum disease resistance is an efficacious strategy for management of diseases in crops as these gene loci confer resistance to a diverse array of pathogens. In rice plants, broad-spectrum resistance to *X. oryzae* pv. *oryzae* and *M. oryzae* was engineered by developing the broad-spectrum resistance Kitaake-1 mutant.[249] Besides, Oliva et al.[250] generated broad-spectrum resistance in two mega rice varieties Ciherang-Sub1 and IR64 as well as in the rice line Kitaake. Three homeoalleles of *Triticum aestivum* Mildew Resistance Locus (TaMLO) in hexaploid bread wheat were simultaneously mutated which conferred heritable and broad-spectrum powdery mildew resistance.[251] Likewise, three homeologs of *T. aestivum* Mildew Resistance Locus (TaEDR1) in wheat plants were simultaneously modified to generate edr1 plants which confirmed the negative role of this locus in resistance against powdery mildew.[252]

Knockout lines of tomato Powdery Mildew Resistance 4 (PMR4) elicited augmented resistance against *Oidium neolycopersici*, a powdery mildew pathogen.[253] Nevertheless, the antagonistic relationships between the JA and SA defense pathways circumscribe the genetic manipulation of these pathways. Tomato plants resistant to pathogen Pto DC3000 causing bacterial speck disease were engineered by the spatial uncoupling of JA-SA antagonism at the region of the stomata by editing the SlJAZ2 gene without impacting plant resistance to necrotrophic pathogens.[253,254] Also, genome-edited resistance characteristics were reported to a plethora of diseases such as citrus canker in Citrus plants, brown streak disease in Cassava, Fusarium wilt in banana, *Ipomovirus* in cucumber, *B. cinerea* in grapes.[255–260]

Tripathi et al.[261] observed augmented resistance to banana Xanthomonas wilt (BXW) disease in the edited banana cultivar called "Sukali Ndiizi," upon CRISPR/Cas9-mediated knockout of the gene Musa Downy mildew resistance 6 (Musa DMR6). The DMR6 gene codes for a 2-oxoglutarate Fe (II)-dependent oxygenase (2OGO) and has been determined as a negative regulator of defense mechanisms in plants as it causes hydrolysis of salicylic acid, the defense signaling molecule,[262] and its expression is enhanced upon pathogen infection.[263] CRISPR/Cas9 technology has been used to target and inactivate all three open reading frames of the Banana Streak Virus (BSV) that integrates into the genome of the host banana plant causing significant hurdle to germplasm movement and banana breeding.[264,265] The genome-edited "Gonja Manjaya" banana variety upon regeneration displayed targeted mutations within the endogenous BSV sequences integrated into the host genome.

Resistance to Cassava Bacterial Blight (CBB) disease has been engineered through CRISPR/Cas9-mediated editing of the promoter of the

MeSWEET10a gene of cassava.[266] Also, the CRISPR/Cas9 technology has been used in attempts to develop cassava plants resistant to two major viral diseases, Cassava Brown Streak Disease and Cassava Mosaic Disease.[267,268] *Striga* is a parasitic plant genus that greatly circumscribes agricultural productivity in Africa. Varieties of sorghum edited through CRISPR/Cas9 technology that targets the LOW GERMINATION LOCI 1 (LGS1) led to the generation of high-level resistance to the *Striga* parasite in the lsg1 mutants.[269] The DMR6 gene was edited using CRISPR/Cas9 technology in banana, sweet basil, and tomato which conferred broad-spectrum resistance to a panoply of bacterial and fungal pathogens.[261,270,271] A tomato variety showing resistance to the obligate root parasite, *Phelipanche aegyptiaca*, was developed by CRISPR/Cas9-based modification of the carotenoid cleavage dioxygenase 8 (CCD8) gene which plays a critical role in strigolactone biosynthesis.[272] Likewise, a rice variety with diminished *Striga hermonthica* germination was developed by CRISPR/Cas-based targeting of the OsCCD7 gene.[273] In wheat crops, Li et al.[274] showed that a mutant, Tamlo-R32, containing a targeted deletion within the MLO-B1 locus, displayed strong resistance to powdery mildew.

## 15. Conclusions and future directions

The current era has witnessed tremendous advances in genome editing platforms that have revolutionized plant research at both the laboratory and field levels. CRISPR/Cas9 technology is immensely popular as a tool for genome editing due to its easy applicability, affordability and versatility. Several CRISPR/Cas-edited plants show great promise in substantially increasing crop yields. The goal of developing designer plant species with multiple attributes and abilities to confront all of the abiotic and biotic stresses is not far away. Therefore, this technology has to be further explored and improved upon for generating crops immune to pathogens and insect pests.

This will help offset enormous losses in crop yields caused by biotic stress factors. Nevertheless, the future of CRISPR-modified genome edited plants in crop enhancement programs will mostly be decided by the respective regulatory agencies worldwide. The regulatory platforms for developing new varieties of crops follow practices that are based on the processes involved as well as the products developed. The magnitude of regulations exercised on CRISPR-edited crops will influence their development costs and determine their rates in the commercial markets. Imposition of product-based regulations on edited crops will mandate them to be categorized along with plant products generated via classical mutagenesis thereby clearing them from regulatory restrictions levied on genetically modified crop products. This will help improve public perception of the CRISPR technology and lead to approval by the general population worldwide.

## References


1. Mittler R. Abiotic stress, the field environment and stress combination. *Trends in Plant Science* 2006;**11**(1):15–9. https://doi.org/10.1016/j.tplants.2005.11.002.
2. Al-Whaibi MH. Plant heat-shock proteins: a mini review. *Journal of King Saud University Science* 2011;**23**(2):139–50. https://doi.org/10.1016/j.jksus.2010.06.022.
3. Guo M, Liu JH, Ma X, Luo DX, Gong ZH, Lu MH. The plant heat stress transcription factors (HSFS): structure, regulation, and function in response to abiotic stresses. *Frontiers in Plant Science* 2016;**7**(2016). https://doi.org/10.3389/fpls.2016.00114.
4. Swindell WR, Huebner M, Weber AP. Transcriptional profiling of Arabidopsis heat shock proteins and transcription factors reveals extensive overlap between heat and non-heat stress response pathways. *BMC Genomics* 2007;**8**. https://doi.org/10.1186/1471-2164-8-125.
5. Xu Y, Zhan C, Huang B. Heat shock proteins in association with heat tolerance in grasses. *International Journal of Proteomics* 2011;**2011**:1–11. https://doi.org/10.1155/2011/529648.
6. Dodds PN, Rathjen JP. Plant immunity: towards an integrated view of plant- pathogen interactions. *Nature Reviews Genetics* 2010;**11**(8):539–48. https://doi.org/10.1038/nrg2812.
7. Gorovits R, Moshe A, Amrani L, Kleinberger R, Anfoka G, Czosnek H. The six Tomato yellow leaf curl virus genes expressed individually in tomato induce different levels of plant stress response attenuation. *Cell Stress and Chaperones* 2017;**22**(3):345–55. https://doi.org/10.1007/s12192-017-0766-0.
8. Jones JT, Haegeman A, Danchin EGJ, et al. Top 10 plant-parasitic nematodes in molecular plant pathology. *Molecular Plant Pathology* 2013;**14**(9):946–61. https://doi.org/10.1111/mpp.12057.
9. Li R, Rashotte AM, Singh NK, Lawrence KS, Weaver DB, Locy RD. Transcriptome analysis of cotton (*Gossypium hirsutum* L.) genotypes that are susceptible, resistant, and hypersensitive to reniform nematode (Rotylenchulus reniformis). *PLoS One* 2015;**10**(11). https://doi.org/10.1371/journal.pone.0143261.
10. Rybicki EP. A Top Ten list for economically important plant viruses. *Archives of Virology* 2015;**160**(1):17–20. https://doi.org/10.1007/s00705-014-2295-9.
11. Srivastava S, Pathak AD, Gupta PS, Kumar A, Kumar A. Hydrogen peroxide-scavenging enzymes impart tolerance to high temperature induced oxidative stress in sugarcane. *Journal of Environmental Biology* 2012;**33**(3):657–61. http://www.jeb.co.in/journal_issues/201205_may12/paper_20.pdf.
12. Wahid A, Gelani S, Ashraf M, Foolad MR. Heat tolerance in plants: an overview. *Environmental and Experimental Botany* 2007;**61**(3):199–223. https://doi.org/10.1016/j.envexpbot.2007.05.011.
13. Demirevska-Kepova K, Holzer R, Simova-Stoilova L, Feller U. Heat stress effects on ribulose-1,5-bisphosphate carboxylase/oxygenase, Rubisco binding protein and Rubisco activase in wheat leaves. *Biologia Plantarum* 2005;**49**(4):521–5. https://doi.org/10.1007/s10535-005-0045-2.
14. Djanaguiraman M, Annie Sheeba J, Durga Devi D, Bangarusamy U. Cotton leaf senescence can be delayed by nitrophenolate spray through enhanced antioxidant defence system. *Journal of Agronomy and Crop Science* 2009;**195**(3):213–24. https://doi.org/10.1111/j.1439-037X.2009.00360.x.
15. Jiang C, Xu J, Zhang H, et al. A cytosolic class i small heat shock protein, RcHSP17.8, of Rosa chinensis confers resistance to a variety of stresses to *Escherichia coli*, yeast and *Arabidopsis thaliana*. *Plant, Cell and Environment* 2009;**32**(8):1046–59. https://doi.org/10.1111/j.1365-3040.2009.01987.x.

16. Tan W, wei MQ, Brestic M, Olsovska K, Yang X. Photosynthesis is improved by exogenous calcium in heat-stressed tobacco plants. *Journal of Plant Physiology* 2011;**168**(17):2063–71. https://doi.org/10.1016/j.jplph.2011.06.009.
17. Karuppanapandian T, Wang HW, Prabakaran N, et al. 2,4-dichlorophenoxyacetic acid-induced leaf senescence in mung bean (*Vigna radiata* L. Wilczek) and senescence inhibition by co-treatment with silver nanoparticles. *Plant Physiology and Biochemistry* 2011;**49**(2):168–77. https://doi.org/10.1016/j.plaphy.2010.11.007.
18. Møller IM, Jensen PE, Hansson A. Oxidative modifications to cellular components in plants. *Annual Review of Plant Biology* 2007;**58**:459–81. https://doi.org/10.1146/annurev.arplant.58.032806.103946.
19. Wang LJ, Li SH. Salicylic acid-induced heat or cold tolerance in relation to $Ca^{2+}$ homeostasis and antioxidant systems in young grape plants. *Plant Science* 2006;**170**(4):685–94. https://doi.org/10.1016/j.plantsci.2005.09.005.
20. Zhang B, Pan X, Cobb GP, Anderson TA. Plant microRNA: a small regulatory molecule with big impact. *Developmental Biology* 2006;**289**(1):3–16. https://doi.org/10.1016/j.ydbio.2005.10.036.
21. Savary S, Ficke A, Aubertot JN, Hollier C. Crop losses due to diseases and their implications for global food production losses and food security. *Food Security* 2012;**4**(4):519–37. https://doi.org/10.1007/s12571-012-0200-5.
22. Wang RR-C. Chromosomal distribution of genes conferring tolerance to abiotic stresses versus that of genes controlling resistance to biotic stresses in plants. *International Journal of Molecular Sciences* 2020;**21**(5):1820.
23. Broekgaarden C, Snoeren TAL, Dicke M, Vosman B. Exploiting natural variation to identify insect-resistance genes. *Plant Biotechnology Journal* 2011;**9**(8):819–25. https://doi.org/10.1111/j.1467-7652.2011.00635.x.
24. Chaerani R, Voorrips RE. Tomato early blight (*Alternaria solani*): the pathogen, genetics, and breeding for resistance. *Journal of General Plant Pathology* 2006;**72**(6):335–47. https://doi.org/10.1007/s10327-006-0299-3.
25. Andolfo G, Iovieno P, Frusciante L, Ercolano MR. Genome-editing technologies for enhancing plant disease resistance. *Frontiers in Plant Science* 2016;**7**(2016). https://doi.org/10.3389/fpls.2016.01813.
26. Taning CNT, Van Eynde B, Yu N, Ma S, Smagghe G. CRISPR/Cas9 in insects: applications, best practices and biosafety concerns. *Journal of Insect Physiology* 2017;**98**:245–57. https://doi.org/10.1016/j.jinsphys.2017.01.007.
27. Macho AP, Zipfel C. Plant PRRs and the activation of innate immune signaling. *Molecular Cell* 2014;**54**(2):263–72. https://doi.org/10.1016/j.molcel.2014.03.028.
28. Mithöfer A, Boland W. Recognition of herbivory-associated molecular patterns. *Plant Physiology* 2008;**146**(3):825–31. https://doi.org/10.1104/pp.107.113118.
29. Jones JDG, Dangl JL. The plant immune system. *Nature* 2006;**444**(7117):323–9. https://doi.org/10.1038/nature05286.
30. Monaghan J, Zipfel C. Plant pattern recognition receptor complexes at the plasma membrane. *Current Opinion in Plant Biology* 2012;**15**(4):349–57. https://doi.org/10.1016/j.pbi.2012.05.006.
31. Ronald PC, Beutler B. Plant and animal sensors of conserved microbial signatures. *Science* 2010;**330**(6007):1061–4. https://doi.org/10.1126/science.1189468.
32. Couto D, Zipfel C. Regulation of pattern recognition receptor signalling in plants. *Nature Reviews Immunology* 2016;**16**(9):537–52. https://doi.org/10.1038/nri.2016.77.

33. Rathinam M, Mishra P, Mahato AK, Singh NK, Rao U, Sreevathsa R. Comparative transcriptome analyses provide novel insights into the differential response of Pigeonpea (*Cajanus cajan* L.) and its wild relative (*Cajanus platycarpus* (Benth.) Maesen) to herbivory by *Helicoverpa armigera* (Hübner). *Plant Molecular Biology* 2019;**101**(1–2):163–82. https://doi.org/10.1007/s11103-019-00899-7.
34. Rathinam M, Mishra P, Vasudevan M, et al. Comparative transcriptome analysis of pigeonpea, *Cajanus cajan* (L.) and one of its wild relatives *Cajanus platycarpus* (Benth.) Maesen. *PLoS One* 2019;**14**(7):e0218731. https://doi.org/10.1371/journal.pone.0218731.
35. De Lorenzo G, D'Ovidio R, Cervone F. The role of polygalacturonase-inhibiting proteins (PGIPS) in defense against pathogenic fungi. *Annual Review of Phytopathology* 2001;**39**:313–35. https://doi.org/10.1146/annurev.phyto.39.1.313.
36. Kaewwongwal A, Chen J, Somta P, et al. Novel alleles of two tightly linked genes encoding polygalacturonase-inhibiting proteins (VrPGIP1 and VrPGIP2) associated with the Br locus that confer bruchid (*Callosobruchus* spp.) resistance to mungbean (*Vigna radiata*) accession V2709. *Frontiers in Plant Science* 2017;**8**. https://doi.org/10.3389/fpls.2017.01692.
37. Kalunke RM, Tundo S, Benedetti M, Cervone F, De Lorenzo G, D'Ovidio R. An update on polygalacturonase-inhibiting protein (PGIP), aleucine-rich repeat protein that protects crop plants against pathogens. *Frontiers in Plant Science* 2015;**6**. https://doi.org/10.3389/fpls.2015.00146.
38. Li H, Smigocki AC. Wound induced *Beta vulgaris* polygalacturonase-inhibiting protein genes encode a longer leucine-rich repeat domain and inhibit fungal polygalacturonases. *Physiological and Molecular Plant Pathology* 2016;**96**:8–18. https://doi.org/10.1016/j.pmpp.2016.06.004.
39. Veronico P, Melillo MT, Saponaro C, Leonetti P, Picardi E, Jones JT. A polygalacturonase-inhibiting protein with a role in pea defence against the cyst nematode Heterodera goettingiana. *Molecular Plant Pathology* 2011;**12**(3):275–87. https://doi.org/10.1111/j.1364-3703.2010.00671.x.
40. Li H, Smigocki AC. Sugar beet polygalacturonase-inhibiting proteins with 11 LRRs confer Rhizoctonia, Fusarium and Botrytis resistance in Nicotiana plants. *Physiological and Molecular Plant Pathology* 2018;**102**:200–8. https://doi.org/10.1016/j.pmpp.2018.03.001.
41. Broetto SG, Fabi JP, do Nascimento JRO. Cloning and expression analysis of two putative papaya genes encoding polygalacturonase-inhibiting proteins. *Postharvest Biology and Technology* 2015;**104**:48–56. https://doi.org/10.1016/j.postharvbio.2015.03.002.
42. Gui Z, Gao J, Xin N, et al. Association of polygalacturonase-inhibiting protein gene 2 (MsPGIP2) to common leaf spot resistance in alfalfa. *European Journal of Plant Pathology* 2016;**144**(2):245–56. https://doi.org/10.1007/s10658-015-0720-x.
43. Liu N, Zhang X, Sun Y, et al. Molecular evidence for the involvement of a polygalacturonase-inhibiting protein, GhPGIP1, in enhanced resistance to Verticillium and Fusarium wilts in cotton. *Scientific Reports* 2017;**7**. https://doi.org/10.1038/srep39840.
44. Ferrari S, Sella L, Janni M, De Lorenzo G, Favaron F, D'Ovidio R. Transgenic expression of polygalacturonase-inhibiting proteins in Arabidopsis and wheat increases resistance to the flower pathogen *Fusarium graminearum*. *Plant Biology* 2012;**14**(1):31–8. https://doi.org/10.1111/j.1438-8677.2011.00449.x.
45. Joubert DA, Kars I, Wagemakers L, et al. A polygalacturonase-inhibiting protein from grapevine reduces the symptoms of the endopolygalacturonase BcPG2 from *Botrytis cinerea* in *Nicotiana benthamiana* leaves without any evidence for in vitro interaction. *Molecular Plant-Microbe Interactions* 2007;**20**(4):392–402. https://doi.org/10.1094/MPMI-20-4-0392.

46. Zhang C, Feng C, Wang J, Kong F, Sun W, Wang F. Cloning, expression analysis and recombinant expression of a gene encoding a polygalacturonase-inhibiting protein from tobacco, *Nicotiana tabacum*. *Heliyon* 2016;**2**(5):e00110. https://doi.org/10.1016/j.heliyon.2016.e00110.
47. Abu-Goukh AA, Strand LL, Labavitch JM. Development-related changes in decay susceptibility and polygalacturonase inhibitor content of "Bartlett" pear fruit. *Physiological Plant Pathology* 1983;**23**(1):101–9. https://doi.org/10.1016/0048-4059(83)90037-1.
48. Bergmann CW, Ito Y, Singer D, et al. Polygalacturonase-inhibiting protein accumulates in *Phaseolus vulgaris* L. in response to wounding, elicitors and fungal infection. *The Plant Journal* 1994;**5**(5):625–34. https://doi.org/10.1111/j.1365-313X.1994.00625.x.
49. Johnston DJ, Ramanathan V, Williamson B. A protein from immature raspberry fruits which inhibits endopolygalacturonases from *Botrytis cinerea* and other micro-organisms. *Journal of Experimental Botany* 1993;**44**(5):971–6. https://doi.org/10.1093/jxb/44.5.971.
50. Mehli L. Temporal expression of a PGIP-gene in strawberry cultivars induced by wounding or by *Botrytis cinerea* infection. *Plant Protection Science* 2017;**38**:504–6. https://doi.org/10.17221/10536-pps. SI 2–6th Conf EFPP 2002.
51. Akhgari AB, Motallebi M, Zamani MR. Bean polygalacturonase-inhibiting protein expressed in transgenic *Brassica napus* inhibits polygalacturonase from its fungal pathogen Rhizoctonia solani. *Plant Protection Science* 2012;**48**(1):1–9. https://doi.org/10.17221/46/2009-pps.
52. Janni M, Sella L, Favaron F, Blechl AE, De Lorenzo G, D'Ovidio R. The expression of a bean PGIP in transgenic wheat confers increased resistance to the fungal pathogen *Bipolaris sorokiniana*. *Molecular Plant-Microbe Interactions* 2008;**21**(2):171–7. https://doi.org/10.1094/MPMI-21-2-0171.
53. Manfredini C, Sicilia F, Ferrari S, et al. Polygalacturonase-inhibiting protein 2 of *Phaseolus vulgaris* inhibits BcPG1, a polygalacturonase of *Botrytis cinerea* important for pathogenicity, and protects transgenic plants from infection. *Physiological and Molecular Plant Pathology* 2005;**67**(2):108–15. https://doi.org/10.1016/j.pmpp.2005.10.002.
54. Mohammadzadeh R, Zamani M, Motallebi M, et al. *Agrobacterium tumefaciens*'-mediated introduction of polygalacturonase inhibiting protein 2 gene (PvPGIP2) from "*Phaseolus vulgaris*" into sugar beet ("*Beta vulgaris*" L.). *Australian Journal of Crop Science* 2012;**6**(8).
55. Borras-Hidalgo O, Caprari C, Hernandez-Estevez I, De Lorenzo G, Cervone F. A gene for plant protection: expression of a bean polygalacturonase inhibitor in tobacco confers a strong resistance against *Rhizoctonia solani* and two oomycetes. *Frontiers in Plant Science* 2012;**3**. https://doi.org/10.3389/fpls.2012.00268.
56. Kulkarni KS, Zala HN, Bosamia TC, et al. De novo transcriptome sequencing to dissect candidate genes associated with pearl millet-downy mildew (*Sclerospora graminicola* sacc.) interaction. *Frontiers in Plant Science* 2016;**7**. https://doi.org/10.3389/fpls.2016.00847.
57. Prabhu SA, Kini KR, Raj SN, Moerschbacher BM, Shetty HS. Polygalacturonase-inhibitor proteins in pearl millet: possible involvement in resistance against downy mildew. *Acta Biochimica et Biophysica Sinica* 2012;**44**(5):415–23. https://doi.org/10.1093/abbs/gms015.
58. Feng C, Zhang X, Wu T, et al. The polygalacturonase-inhibiting protein 4 (OsPGIP4), a potential component of the qBlsr5a locus, confers resistance to bacterial leaf streak in rice. *Planta* 2016;**243**(5):1297–308. https://doi.org/10.1007/s00425-016-2480-z.
59. Hwang BH, Bae H, Lim HS, et al. Overexpression of polygalacturonase-inhibiting protein 2 (PGIP2) of Chinese cabbage (*Brassica rapa* ssp. pekinensis) increased resistance to the bacterial pathogen *Pectobacterium carotovorum* ssp. carotovorum. *Plant Cell, Tissue and Organ Culture* 2010;**103**(3):293–305. https://doi.org/10.1007/s11240-010-9779-4.

60. Schacht T, Unger C, Pich A, Wydra K. Endo- and exopolygalacturonases of *Ralstonia solanacearum* are inhibited by polygalacturonase-inhibiting protein (PGIP) activity in tomato stem extracts. *Plant Physiology and Biochemistry* 2011;**49**(4):377–87. https://doi.org/10.1016/j.plaphy.2011.02.001.
61. Wang R, Lu L, Pan X, et al. Functional analysis of OsPGIP1 in rice sheath blight resistance. *Plant molecular biology* 2015;**87**(1–2):181–91. https://doi.org/10.1007/s11103-014-0269-7.
62. Wang Z, Wan L, Xin Q, et al. Overexpression of OsPGIP2 confers *Sclerotinia sclerotiorum* resistance in *Brassica napus* through increased activation of defense mechanisms. *Journal of Experimental Botany* 2018;**69**(12):3141–55. https://doi.org/10.1093/jxb/ery138.
63. Agüero CB, Uratsu SL, Greve C, et al. Evaluation of tolerance to Pierce's disease and Botrytis in transgenic plants of *Vitis vinifera* L. expressing the pear PGIP gene. *Molecular Plant Pathology* 2005;**6**(1):43–51. https://doi.org/10.1111/J.1364-3703.2004.00262.X.
64. Shah S. *Analysis of the role of polygalacturonase inhibiting proteins in the interaction between plants and nematodes* (Doctoral Dissertation). 2018.
65. Li R, Rimmer R, Yu M, et al. Two *Brassica napus* polygalacturonase inhibitory protein genes are expressed at different levels in response to biotic and abiotic stresses. *Planta* 2003;**217**(2):299–308. https://doi.org/10.1007/s00425-003-0988-5.
66. Reymond P, Weber H, Damond M, Farmer EE. Differential gene expression in response to mechanical wounding and insect feeding in Arabidopsis. *Plant Cell* 2000;**12**(5):707–19. https://doi.org/10.1105/tpc.12.5.707.
67. Shelomi M, Danchin EGJ, Heckel D, et al. Horizontal gene transfer of pectinases from bacteria preceded the diversification of stick and leaf insects. *Scientific Reports* 2016;**6**. https://doi.org/10.1038/srep26388.
68. Frati F, Galletti R, De Lorenzo G, Salerno G, Conti E. Activity of endo-polygalacturonases in mirid bugs (Heteroptera: miridae) and their inhibition by plant cell wall proteins (PGIPs). *European Journal of Entomology* 2006;**103**(3):515–22. https://doi.org/10.14411/eje.2006.067.
69. D'Ovidio R, Raiola A, Capodicasa C, et al. Characterization of the complex locus of bean encoding polygalacturonase- inhibiting proteins reveals subfunctionalization for defense against fungi and insects. *Plant Physiology* 2004;**135**(4):2424–35. https://doi.org/10.1104/pp.104.044644.
70. Chotechung S, Somta P, Chen J, Yimram T, Chen X, Srinives P. A gene encoding a polygalacturonase-inhibiting protein (PGIP) is a candidate gene for bruchid (Coleoptera: bruchidae) resistance in mungbean (*Vigna radiata*). *Theoretical and Applied Genetics* 2016;**129**(9):1673–83. https://doi.org/10.1007/s00122-016-2731-1.
71. Casasoli M, Federici L, Spinelli F, et al. Integration of evolutionary and desolvation energy analysis identifies functional sites in a plant immunity protein. *Proceedings of the National Academy of Sciences of the United States of America* 2009;**106**(18):7666–71. https://doi.org/10.1073/pnas.0812625106.
72. Kosová K, Vítámvás P, Urban MO, Klíma M, Roy A, Tom Prášil I. Biological networks underlying abiotic stress tolerance in temperate crops-a proteomic perspective. *International Journal of Molecular Sciences* 2015;**16**(9):20913–42. https://doi.org/10.3390/ijms160920913.
73. Erpen L, Devi HS, Grosser JW, Dutt M. Potential use of the DREB/ERF, MYB, NAC and WRKY transcription factors to improve abiotic and biotic stress in transgenic plants. *Plant Cell, Tissue and Organ Culture* 2018;**132**(1). https://doi.org/10.1007/s11240-017-1320-6.

74. Hernandez-Garcia CM, Finer JJ. Identification and validation of promoters and cis-acting regulatory elements. *Plant Science* 2014;**217–218**:109–19. https://doi.org/10.1016/j.plantsci.2013.12.007.
75. Riechmann JL, Heard J, Martin G, et al. Arabidopsis transcription factors: genome-wide comparative analysis among eukaryotes. *Science* 2000;**290**(5499):2105–10. https://doi.org/10.1126/science.290.5499.2105.
76. Puranik S, Sahu PP, Srivastava PS, Prasad M. NAC proteins: regulation and role in stress tolerance. *Trends in Plant Science* 2012;**17**(6):369–81. https://doi.org/10.1016/j.tplants.2012.02.004.
77. Zhang H, Huang Y. Genome-wide survey and characterization of greenbug induced NAC transcription factors in sorghum (*Sorghum bicolor* (L.) Moench). In: *Proceedings of the XXI annual international plant & animal genome conference*; 2013.
78. Zhou W, Qian C, Li R, et al. TaNAC6s are involved in the basal and broad-spectrum resistance to powdery mildew in wheat. *Plant Science* 2018;**277**:218–28. https://doi.org/10.1016/j.plantsci.2018.09.014.
79. Chen YJ, Perera V, Christiansen MW, Holme IB, Gregersen PL, Grant MR, Collinge DB, Lyngkjær MF. The barley HvNAC6 transcription factor affects ABA accumulation and promotes basal resistance against powdery mildew. *Plant Molecular Biology* 2013;**83**(6):577–90. https://doi.org/10.1007/s11103-013-0109-1.
80. Liu Q, Yan S, Huang W, et al. NAC transcription factor ONAC066 positively regulates disease resistance by suppressing the ABA signaling pathway in rice. *Plant Molecular Biology* 2018;**98**(4–5):289–302. https://doi.org/10.1007/s11103-018-0768-z.
81. Baillo EH, Kimotho RN, Zhang Z, Xu P. Transcription factors associated with abiotic and biotic stress tolerance and their potential for crops improvement. *Genes* September 30, 2019;**10**(10):771. https://doi.org/10.3390/genes10100771.
82. Ibraheem F, Gaffoor I, Tan Q, Shyu CR, Chopra S. A sorghum MYB transcription factor induces 3-deoxyanthocyanidins and enhances resistance against leaf blights in maize. *Molecules* 2015;**20**(2):2388–404. https://doi.org/10.3390/molecules20022388.
83. Shan T, Rong W, Xu H, Du L, Liu X, Zhang Z. The wheat R2R3-MYB transcription factor TaRIM1 participates in resistance response against the pathogen *Rhizoctonia cerealis* infection through regulating defense genes. *Scientific Reports* 2016;**6**. https://doi.org/10.1038/srep28777.
84. Wang Y, Sheng L, Zhang H, et al. CmMYB19 over-expression improves aphid tolerance in Chrysanthemum by promoting lignin synthesis. *International Journal of Molecular Sciences* 2017;**18**(3):619. https://doi.org/10.3390/ijms18030619.
85. Peng X, Wang H, Jang JC, Xiao T, He H, Jiang D, Tang X. OsWRKY80-OsWRKY4 Module as a positive regulatory circuit in rice resistance against Rhizoctonia solani. *Rice (N Y)*. December 2016;**9**(1):63. https://doi.org/10.1186/s12284-016-0137-y.
86. Yang T, Zhu X, Yang J, Li X, Dong J, Zhang S, Zhao J, Yang T, Mao X, Zhu X, et al. OsWRKY67 positively regulates blast and bacteria blight resistance by direct activation of PR genes in rice. *BMC Plant Biology* 2018;**18**:1–13.
87. Wang J, Tao F, Tian W, et al. The wheat WRKY transcription factors TaWRKY49 and TaWRKY62 confer differential high-temperature seedling-plant resistance to *Puccinia striiformis* f. sp. tritici. *PLoS One* 2017;**12**(7):e0181963. https://doi.org/10.1371/journal.pone.0181963.

88. Chen L, Zhang L, Li D, Wang F, Yu D. WRKY8 transcription factor functions in the TMV-cg defense response by mediating both abscisic acid and ethylene signaling in Arabidopsis. *Proceedings of the National Academy of Sciences of the United States of America* 2013;**110**(21):E1963–71. https://doi.org/10.1073/pnas.1221347110.
89. Lee H, Cha J, Choi C, et al. Rice wrky11 plays a role in pathogen defense and drought tolerance. *Rice* 2018;**11**(1):1–12. https://doi.org/10.1186/s12284-018-0199-0.
90. Zhu Y, Li Y, Zhang S, Zhang X, Yao J, Luo Q, Wang X. Genome-wide identification and expression analysis reveal the potential function of ethylene responsive factor gene family in response to Botrytis cinerea infection and ovule development in grapes (*Vitis vinifera L.*). *Plant Biol (Stuttg)* 2019;**21**(4):571–84.
91. Wang M, Zhu Y, Han R, et al. Expression of vitis amurensis VaERF20 in *Arabidopsis thaliana* improves resistance to botrytis cinerea and pseudomonas syringae pv. Tomato DC3000. *International Journal of Molecular Sciences* 2018;**19**(3). https://doi.org/10.3390/ijms19030696.
92. Alves MS, Soares ZG, Vidigal PM, Barros EG, Poddanosqui AM, Aoyagi LN, Abdelnoor RV, Marcelino-Guimarães FC, Fietto LG. Differential expression of four soybean bZIP genes during Phakopsora pachyrhizi infection. *Functional Integrated Genomics* November 2015;**15**(6):685–96. https://doi.org/10.1007/s10142-015-0445-0.
93. Li X, Fan S, Hu W, Liu G, Wei Y, He C, Shi H. Two cassava basic leucine zipper (bZIP) transcription factors (MebZIP3 and MebZIP5) confer disease resistance against cassava bacterial blight. *Frontiers in Plant Science* 2017;**8**:1–11.
94. Lim CW, Baek W, Lim S, Han SW, Lee SC. Expression and functional roles of the pepper pathogen-induced bZIP transcription factor CabZIP2 in enhanced disease resistance to bacterial pathogen infection. *Molecular Plant Microbe Interaction* July 2015;**28**(7):825–33. https://doi.org/10.1094/MPMI-10-14-0313-R.
95. Feng H, Duan X, Zhang Q, et al. The target gene of tae-miR164, a novel NAC transcription factor from the NAM subfamily, negatively regulates resistance of wheat to stripe rust. *Molecular Plant Pathology* 2014;**15**(3):284–96. https://doi.org/10.1111/mpp.12089.
96. Samad AFA, Sajad M, Nazaruddin N, et al. MicroRNA and transcription factor: key players in plant regulatory network. *Frontiers in Plant Science* 2017;**8**. https://doi.org/10.3389/fpls.2017.00565.
97. Wang LL, Jin JJ, Li LH, Qu SH. Long non-coding RNAs responsive to blast. *Fungus Infection in Rice. Rice (N Y).* 2020;**13**(1):77. https://doi.org/10.1186/s12284-020-00437-w.
98. Zhang L, Wang M, Li N, Wang H, Qiu P, Pei L, Xu Z, Wang T, Gao E, Liu J, Liu S, Hu Q, Miao Y, Lindsey K, Tu L, Zhu L, Zhang X. Long noncoding RNAs involve in resistance to Verticillium dahliae, a fungal disease in cotton. *Plant Biotechnology Journal* 2018;**16**(6):1172–85. https://doi.org/10.1111/pbi.12861.
99. Nakashima K, Tran LS, Van Nguyen D, Fujita M, Maruyama K, Todaka D, Ito Y, Hayashi N, Shinozaki K, Yamaguchi-Shinozaki K. Functional analysis of a NAC-type transcription factor OsNAC6 involved in abiotic and biotic stress-responsive gene expression in rice. *The Plant Journal* 2007;**51**(4):617–30. https://doi.org/10.1111/j.1365-313X.2007.03168.x.
100. Bari R, Jones JDG. Role of plant hormones in plant defence responses. *Plant Molecular Biology* 2009;**69**(4):473–88. https://doi.org/10.1007/s11103-008-9435-0.
101. Loake G, Grant M. Salicylic acid in plant defence-the players and protagonists. *Current Opinion in Plant Biology* 2007;**10**(5):466–72. https://doi.org/10.1016/j.pbi.2007.08.008.

102. Müller M, Munné-Bosch S. Ethylene response factors: a key regulatory hub in hormone and stress signaling. *Plant Physiology* 2015;**169**(1):32–41. https://doi.org/10.1104/pp.15.00677.
103. Wasternack C, Hause B. Jasmonates: biosynthesis, perception, signal transduction and action in plant stress response, growth and development. An update to the 2007 review in Annals of Botany. *Annals of Botany* 2013;**111**(6):1021–58. https://doi.org/10.1093/aob/mct067.
104. Van Loon LC, Rep M, Pieterse CMJ. Significance of inducible defense-related proteins in infected plants. *Annual Review of Phytopathology* 2006;**44**:135–62. https://doi.org/10.1146/annurev.phyto.44.070505.143425.
105. Wu Y, Zhang D, Chu JY, et al. The arabidopsis NPR1 protein is a receptor for the plant defense hormone salicylic acid. *Cell Reports* 2012;**1**(6):639–47. https://doi.org/10.1016/j.celrep.2012.05.008.
106. Eulgem T, Somssich IE. Networks of WRKY transcription factors in defense signaling. *Current Opinion in Plant Biology* 2007;**10**(4):366–71. https://doi.org/10.1016/j.pbi.2007.04.020.
107. Kim KC, Lai Z, Fan B, Chen Z. Arabidopsis WRKY38 and WRKY62 transcription factors interact with histone deacetylase 19 in basal defense. *Plant Cell* 2008;**20**(9):2357–71. https://doi.org/10.1105/tpc.107.055566.
108. Lorenzo O, Piqueras R, Sánchez-Serrano JJ, Solano R. ETHYLENE RESPONSE FACTOR1 integrates signals from ethylene and jasmonate pathways in plant defense. *Plant Cell* 2003;**15**(1):165–78. https://doi.org/10.1105/tpc.007468.
109. Berrocal-Lobo M, Molina A, Solano R. Constitutive expression of ETHYLENE-RESPONSE-FACTOR1 in Arabidopsis confers resistance to several necrotrophic fungi. *The Plant Journal* 2002;**29**(1):23–32. https://doi.org/10.1046/j.1365-313x.2002.01191.x.
110. Brown RL, Kazan K, McGrath KC, Maclean DJ, Manners JM. A role for the GCC-box in jasmonate-mediated activation of the PDF1.2 gene of Arabidopsis. *Plant Physiology* 2003;**132**(2):1020–32. https://doi.org/10.1104/pp.102.017814.
111. Moffat CS, Ingle RA, Wathugala DL, Saunders NJ, Knight H, Knight MR. ERF5 and ERF6 play redundant roles as positive regulators of JA/Et-mediated defense against botrytis cinerea in arabidopsis. *PLoS One* 2012;**7**(4). https://doi.org/10.1371/journal.pone.0035995.
112. Chini A, Fonseca S, Fernández G, et al. The JAZ family of repressors is the missing link in jasmonate signalling. *Nature* 2007;**448**(7154):666–71. https://doi.org/10.1038/nature06006.
113. Fonseca S, Chini A, Hamberg M, et al. (+)-7-iso-Jasmonoyl-L-isoleucine is the endogenous bioactive jasmonate. *Nature Chemical Biology* 2009;**5**(5):344–50. https://doi.org/10.1038/nchembio.161.
114. Zhai Q, Yan L, Tan D, et al. Phosphorylation-coupled proteolysis of the transcription factor MYC2 is important for jasmonate-signaled plant immunity. *PLoS Genetics* 2013;**9**(4):e1003422. https://doi.org/10.1371/journal.pgen.1003422.
115. Jung C, Zhao P, Seo JS, Mitsuda N, Deng S, Chua NH. PLANT U-BOX PROTEIN10 regulates MYC2 stability in arabidopsis. *Plant Cell* 2015;**27**(7):2016–31. https://doi.org/10.1105/tpc.15.00385.
116. Solano R, Stepanova A, Chao Q, Ecker JR. Nuclear events in ethylene signaling: a transcriptional cascade mediated by ETHYLENE-INSENSITIVE3 and ETHYLENE-RESPONSE-FACTOR1. *Genes and Development* 1998;**12**(23):3703–14. https://doi.org/10.1101/gad.12.23.3703.

117. Ju C, Yoon GM, Shemansky JM, et al. CTR1 phosphorylates the central regulator EIN2 to control ethylene hormone signaling from the ER membrane to the nucleus in Arabidopsis. *Proceedings of the National Academy of Sciences of the United States of America* 2012;**109**(47):19486–91. https://doi.org/10.1073/pnas.1214848109.
118. Spoel SH, Koornneef A, Claessens SMC, et al. NPR1 modulates cross-talk between salicylate- and jasmonate-dependent defense pathways through a novel function in the cytosol. *Plant Cell* 2003;**15**(3):760–70. https://doi.org/10.1105/tpc.009159.
119. Li J, Brader G, Palva ET. The WRKY70 transcription factor: a node of convergence for jasmonate-mediated and salicylate-mediated signals in plant defense. *Plant Cell* 2004;**16**(2):319–31. https://doi.org/10.1105/tpc.016980.
120. Petersen M, Brodersen P, Naested H, et al. Arabidopsis MAP kinase 4 negatively regulates systemic acquired resistance. *Cell* 2000;**103**(7):1111–20. https://doi.org/10.1016/S0092-8674(00)00213-0.
121. Zhu Z, An F, Feng Y, et al. Derepression of ethylene-stabilized transcription factors (EIN3/EIL1) mediates jasmonate and ethylene signaling synergy in Arabidopsis. *Proceedings of the National Academy of Sciences of the United States of America* 2011;**108**(30):12539–44. https://doi.org/10.1073/pnas.1103959108.
122. O'Donnell PJ, Calvert C, Atzorn R, Wasternack C, Leyser HMO, Bowles DJ. Ethylene as a signal mediating the wound response of tomato plants. *Science* 1996;**274**(5294):1914–7. https://doi.org/10.1126/science.274.5294.1914.
123. Song S, Huang H, Gao H, et al. Interaction between MYC2 and ETHYLENE INSENSITIVE3 modulates antagonism between jasmonate and ethylene signaling in Arabidopsis. *Plant Cell* 2014;**26**(1):263–79. https://doi.org/10.1105/tpc.113.120394.
124. Choi J, Huh SU, Kojima M, Sakakibara H, Paek KH, Hwang I. The cytokinin-activated transcription factor ARR2 promotes plant immunity via TGA3/NPR1-dependent salicylic acid signaling in arabidopsis. *Developmental Cell* 2010;**19**(2):284–95. https://doi.org/10.1016/j.devcel.2010.07.011.
125. Kohli A, Sreenivasulu N, Lakshmanan P, Kumar PP. The phytohormone crosstalk paradigm takes center stage in understanding how plants respond to abiotic stresses. *Plant Cell Reports* 2013;**32**(7):945–57. https://doi.org/10.1007/s00299-013-1461-y.
126. Reusche M, Klásková J, Thole K, et al. Stabilization of cytokinin levels enhances Arabidopsis resistance against *Verticillium longisporum. Molecular Plant-Microbe Interactions* 2013;**26**(8):850–60. https://doi.org/10.1094/MPMI-12-12-0287-R.
127. Jiang CJ, Shimono M, Sugano S, et al. Cytokinins act synergistically with salicylic acid to activate defense gene expression in rice. *Molecular Plant-Microbe Interactions* 2013;**26**(3):287–96. https://doi.org/10.1094/MPMI-06-12-0152-R.
128. Dangl JL, Horvath DM, Staskawicz BJ. Pivoting the plant immune system from dissection to deployment. *Science* 2013;**341**(6147):746–51. https://doi.org/10.1126/science.1236011.
129. Gurr SJ, Rushton PJ. Engineering plants with increased disease resistance: what are we going to express? *Trends in Biotechnology* 2005;**23**(6):275–82. https://doi.org/10.1016/j.tibtech.2005.04.007.
130. Collinge DB, Jørgensen HJL, Lund OS, Lyngkjær MF. Engineering pathogen resistance in crop plants: current trends and future prospects. *Annual Review of Phytopathology* 2010;**48**:269–91. https://doi.org/10.1146/annurev-phyto-073009-114430.
131. Wally O, Punja ZK. Genetic engineering for increasing fungal and bacterial disease resistance in crop plants. *GM crops* 2010;**1**(4):199–206. https://doi.org/10.4161/gmcr.1.4.13225.

132. Bisht DS, Bhatia V, Bhattacharya R. Improving plant-resistance to insect-pests and pathogens: the new opportunities through targeted genome editing. *Seminars in Cell and Developmental Biology* 2019;**96**:65–76. https://doi.org/10.1016/j.semcdb.2019.04.008.
133. Büschges R, Hollricher K, Panstruga R, et al. The barley Mlo gene: a novel control element of plant pathogen resistance. *Cell* 1997;**88**(5):695–705. https://doi.org/10.1016/S0092-8674(00)81912-1.
134. Wang Y, Cheng X, Shan Q, et al. Simultaneous editing of three homoeoalleles in hexaploid bread wheat confers heritable resistance to powdery mildew. *Nature Biotechnology* 2014;**32**(9):947–51. https://doi.org/10.1038/nbt.2969.
135. Kolmer JA, Bernardo A, Bai G, Hayden MJ, Chao S. Adult plant leaf rust resistance derived from toropi wheat is conditioned by Lr78 and three minor QTL. *Phytopathology* 2018;**108**(2):246–53. https://doi.org/10.1094/PHYTO-07-17-0254-R.
136. Zimmer CT, Garrood WT, Puinean AM, Eckel-Zimmer M, Williamson MS, Davies TGE, Bass C. A CRISPR/Cas9 mediated point mutation in the alpha 6 subunit of the nicotinic acetylcholine receptor confers resistance to spinosad in Drosophila melanogaster. *Insect Biochemistry and Molecular Biology* 2016;**73**:62–9.
137. Galizi R, Hammond A, Kyrou K, Taxiarchi C, Bernardini F, O'Loughlin SM, Papathanos P-A, Nolan T, Windbichler N, Crisanti A. A CRISPR-Cas9 sex-ratio distortion system for genetic control. *Science Reports* 2016;**6**:31139.
138. Huang Y, Chen Y, Zeng B, Wang Y, James AA, Gurr GM, Yang G, Lin X, Huang Y, You M. CRISPR/Cas9 mediated knockout of the abdominal-A homeotic gene in the global pest, diamondback moth (*Plutella xylostella*). *Insect Biochemistry and Molecular Biology* 2016;**75**:98–106.
139. Zuo Y, Wang H, Xu Y, Huang J, Wu S, Wu Y, Yang Y. CRISPR/Cas9 mediated G4946E substitution in the ryanodine receptor of *Spodoptera exigua* confers high levels of resistance to diamide insecticides. *Insect Biochemistry and Molecular Biology* 2017;**89**:79–85.
140. Itokawa K, Komagata O, Kasai S, Ogawa K, Tomita T. Testing the causality between CYP9M10 and pyrethroid resistance using the TALEN and CRISPR/Cas9 technologies. *Science Reports* 2016;**6**:24652.
141. Hammond A, Galizi R, Kyrou K, Simoni A, Siniscalchi C, Katsanos D, Gribble M, Baker D, Marois E, Russell S, Burt A, Windbichler N, Crisanti A, Nolan T. A CRISPR-Cas9 gene drive system targeting female reproduction in the malaria mosquito vector Anopheles gambiae. *Nature Biotechnology* 2016;**34**:78–83.
142. Gilles AF, Schinko JB, Averof M. Efficient CRISPR-mediated gene targeting and transgene replacement in the beetle *Tribolium castaneum. Development* 2015;**142**:2832–9.
143. Gantz VM, Jasinskiene N, Tatarenkova O, Fazekas A, Macias VM, Bier E, James AA. Highly efficient Cas9-mediated gene drive for population modification of the malaria vector mosquito *Anopheles stephensi. Proceedings of the National Academy of Sciences United States of America* 2015;**112**:E6736–43.
144. Koutroumpa FA, Monsempes C, Francois M-C, de Cian A, Royer C, Concordet J-P, Jacquin-Joly E. Heritable genome editing with CRISPR/Cas9 induces anosmia in a crop pest moth. *Science Reports* 2016;**6**:29620.
145. Johnson R. Classical plant breeding for durable resistance to diseases. *Journal of Plant Pathology* 2000;**82**(1):3–7.
146. Schaart JG, van de Wiel CCM, Lotz LAP, Smulders MJM. Opportunities for products of new plant breeding techniques. *Trends in Plant Science* 2016;**21**(5):438–49. https://doi.org/10.1016/j.tplants.2015.11.006.

147. McDonald BA, Linde C. Pathogen population genetics, evolutionary potential, and durable resistance. *Annual Review of Phytopathology* 2002;**40**:349–79. https://doi.org/10.1146/annurev.phyto.40.120501.101443.
148. Aldemita RR, Reaño IMaE, Solis RO, Hautea RA. Trends in global approvals of biotech crops (1992-2014). *GM Crops and Food* 2015;**6**(3):150–66. https://doi.org/10.1080/21645698.2015.1056972.
149. Wang J, Wang C, Wang K. Generation of marker-free transgenic rice using CRISPR/Cas9 system controlled by floral specific promoters. *Journal of Genetics and Genomics* 2019;**46**(1):61–4. https://doi.org/10.1016/j.jgg.2018.11.005.
150. Zhang Y, Liang Z, Zong Y, et al. Efficient and transgene-free genome editing in wheat through transient expression of CRISPR/Cas9 DNA or RNA. *Nature Communications* 2016;**7**. https://doi.org/10.1038/ncomms12617.
151. Mushtaq M, Bhat JA, Mir ZA, et al. CRISPR/Cas approach: a new way of looking at plant-abiotic interactions. *Journal of Plant Physiology* 2018;**224–225**:156–62. https://doi.org/10.1016/j.jplph.2018.04.001.
152. Shan Q, Wang Y, Li J, et al. Targeted genome modification of crop plants using a CRISPR-Cas system. *Nature Biotechnology* 2013;**31**(8):686–8. https://doi.org/10.1038/nbt.2650.
153. Svitashev S, Young JK, Schwartz C, Gao H, Falco SC, Cigan AM. Targeted mutagenesis, precise gene editing, and site-specific gene insertion in maize using Cas9 and guide RNA. *Plant Physiology* 2015;**169**(2):931–45. https://doi.org/10.1104/pp.15.00793.
154. Zaidi SS, Mukhtar MS, Mansoor S. Genome editing: targeting susceptibility genes for plant disease resistance. *Trends in Biotechnology* 2018;**36**(9):898–906. https://doi.org/10.1016/j.tibtech.2018.04.005.
155. Pavan S, Jacobsen E, Visser RGF, Bai Y. Loss of susceptibility as a novel breeding strategy for durable and broad-spectrum resistance. *Molecular Breeding* 2010;**25**(1):1–12. https://doi.org/10.1007/s11032-009-9323-6.
156. Van Schie CCN, Takken FLW. Susceptibility genes 101: how to be a good host. *Annual Review of Phytopathology* 2014;**52**:551–81. https://doi.org/10.1146/annurev-phyto-102313-045854.
157. Lapin D, Van den Ackerveken G. Susceptibility to plant disease: more than a failure of host immunity. *Trends in Plant Science* 2013;**18**(10):546–54. https://doi.org/10.1016/j.tplants.2013.05.005.
158. Nekrasov V, Wang C, Win J, Lanz C, Weigel D, Kamoun S. Rapid generation of a transgene-free powdery mildew resistant tomato by genome deletion. *Scientific Reports* 2017;**7**(1). https://doi.org/10.1038/s41598-017-00578-x.
159. Thomazella DPT, Seong K, Mackelprang R, Dahlbeck D, Geng Y, Gill US, Qi T, Pham J, Giuseppe P, Lee CY, Ortega A, Cho MJ, Hutton SF, Staskawicz B. Loss of function of a DMR6 ortholog in tomato confers broad-spectrum disease resistance. *Proceedings of the National Academy of Science U S A* 2021;**118**(27). e2026152118. https://doi.org/10.1073/pnas.2026152118.
160. Pyott DE, Sheehan E, Molnar A. Engineering of CRISPR/Cas9-mediated potyvirus resistance in transgene-free Arabidopsis plants. *Molecular plant pathology* 2016;**17**(8):1276–88. https://doi.org/10.1111/mpp.12417.
161. Chandrasekaran J, Brumin M, Wolf D, et al. Development of broad virus resistance in non-transgenic cucumber using CRISPR/Cas9 technology. *Molecular plant pathology* 2016;**17**(7):1140–53. https://doi.org/10.1111/mpp.12375.

162. Macovei A, Sevilla NR, Cantos C, et al. Novel alleles of rice eIF4G generated by CRISPR/Cas9-targeted mutagenesis confer resistance to Rice tungro spherical virus. *Plant Biotechnology Journal* 2018;**16**(11):1918–27. https://doi.org/10.1111/pbi.12927.
163. Li T, Liu B, Spalding MH, Weeks DP, Yang B. High-efficiency TALEN-based gene editing produces disease-resistant rice. *Nature Biotechnology* 2012;**30**(5):390–2. https://doi.org/10.1038/nbt.2199.
164. Malnoy M, Viola R, Jung M-H, et al. DNA-free genetically edited grapevine and apple protoplast using CRISPR/Cas9 ribonucleoproteins. *Frontiers of Plant Science* 2016;**7**:1904.
165. Jia H, Orbovic V, Jones JB, Wang N. Modification of the PthA4 effector binding elements in Type I CsLOB1 promoter using Cas9/sgRNA to produce transgenic Duncan grapefruit alleviating XccΔpthA4: DCsLOB1.3 infection. *Plant Biotechnology Journal* 2016;**14**(5):1291–301. https://doi.org/10.1111/pbi.12495.
166. Bai Y, Pavan S, Zheng Z, et al. Naturally occurring broad-spectrum powdery mildew resistance in a Central American tomato accession is caused by loss of Mlo function. *Molecular Plant-Microbe Interactions* 2008;**21**(1):30–9. https://doi.org/10.1094/MPMI-21-1-0030.
167. Consonni C, Humphry ME, Hartmann HA, et al. Conserved requirement for a plant host cell protein in powdery mildew pathogenesis. *Nature Genetics* 2006;**38**(6):716–20. https://doi.org/10.1038/ng1806.
168. Pavan S, Schiavulli A, Appiano M, et al. Pea powdery mildew er1 resistance is associated to loss-of-function mutations at a MLO homologous locus. *Theoretical and Applied Genetics* 2011;**123**(8):1425–31. https://doi.org/10.1007/s00122-011-1677-6.
169. Devoto A, Piffanelli P, Nilsson IM, et al. Topology, subcellular localization, and sequence diversity of the Mlo family in plants. *Journal of Biological Chemistry* 1999;**274**(49):34993–5004. https://doi.org/10.1074/jbc.274.49.34993.
170. Bastet A, Robaglia C, Gallois JL. eIF4E resistance: natural variation should guide gene editing. *Trends in Plant Science* 2017;**22**(5):411–9. https://doi.org/10.1016/j.tplants.2017.01.008.
171. Lee JH, Muhsin M, Atienza GA, et al. Single nucleotide polymorphisms in a gene for translation initiation factor (elF4G) of rice (*Oryza sativa*) associated with resistance to rice tungro spherical virus. *Molecular Plant-Microbe Interactions* 2010;**23**(1):29–38. https://doi.org/10.1094/MPMI-23-1-0029.
172. Cohn M, Bart RS, Shybut M, et al. *Xanthomonas axonopodis* virulence is promoted by a transcription activator-like effector - mediated induction of a SWEET sugar transporter in Cassava. *Molecular Plant-Microbe Interactions* 2014;**27**(11):1186–98. https://doi.org/10.1094/MPMI-06-14-0161-R.
173. Zhou J, Peng Z, Long J, et al. Gene targeting by the TAL effector PthXo2 reveals cryptic resistance gene for bacterial blight of rice. *Plant Journal* 2015;**82**(4):632–43. https://doi.org/10.1111/tpj.12838.
174. Bonardi V, Cherkis K, Nishimura MT, Dangl JL. A new eye on NLR proteins: focused on clarity or diffused by complexity? *Current Opinion in Immunology* 2012;**24**(1):41–50. https://doi.org/10.1016/j.coi.2011.12.006.
175. Farnham G, Baulcombe DC. Artificial evolution extends the spectrum of viruses that are targeted by a disease-resistance gene from potato. *Proceedings of the National Academy of Sciences of The United States of America* 2006;**103**(49):18828–33. https://doi.org/10.1073/pnas.0605777103.

176. Giannakopoulou A, Steele JFC, Segretin ME, et al. Tomato I2 immune receptor can be engineered to confer partial resistance to the oomycete phytophthora infestans in addition to the fungus fusarium oxysporum. *Molecular Plant-Microbe Interactions* 2015;**28**(12):1316–29. https://doi.org/10.1094/MPMI-07-15-0147-R.
177. Lacombe S, Rougon-Cardoso A, Sherwood E, et al. Interfamily transfer of a plant pattern-recognition receptor confers broad-spectrum bacterial resistance. *Nature Biotechnology* 2010;**28**(4):365–9. https://doi.org/10.1038/nbt.1613.
178. Schwessinger B, Bahar O, Thomas N, et al. Transgenic expression of the dicotyledonous pattern recognition receptor EFR in rice leads to ligand-dependent activation of defense responses. *PLoS Pathogens* 2015;**11**(3):1–34. https://doi.org/10.1371/journal.ppat.1004809.
179. Glazebrook J. Contrasting mechanisms of defense against biotrophic and necrotrophic pathogens. *Annual Review of Phytopathology* 2005;**43**:205–27. https://doi.org/10.1146/annurev.phyto.43.040204.135923.
180. Ortigosa A, Gimenez-Ibanez S, Leonhardt N, Solano R. Design of a bacterial speck resistant tomato by CRISPR/Cas9-mediated editing of SlJAZ2. *Plant Biotechnology Journal* 2019;**17**(3):665–73. https://doi.org/10.1111/pbi.13006.
181. Hammond AM, Kyrou K, Bruttini M, et al. The creation and selection of mutations resistant to a gene drive over multiple generations in the malaria mosquito. *PLoS Genetics* 2017;**13**(10):e1007039. https://doi.org/10.1371/journal.pgen.1007039.
182. Kleinstiver BP, Pattanayak V, Prew MS, et al. High-fidelity CRISPR-Cas9 nucleases with no detectable genome-wide off-target effects. *Nature* 2016;**529**(7587):490–5. https://doi.org/10.1038/nature16526.
183. Ran FA, Hsu PD, Lin C-Y, et al. Double nicking by RNA-guided CRISPR Cas9 for enhanced genome editing specificity. *Cell* 2013;**154**(6):1380–9. https://doi.org/10.1016/j.cell.2013.08.021.
184. Fu Y, Sander JD, Reyon D, Cascio VM, Joung JK. Improving CRISPR-Cas nuclease specificity using truncated guide RNAs. *Nature Biotechnology* 2014;**32**(3):279–84. https://doi.org/10.1038/nbt.2808.
185. Tsai SQ, Wyvekens N, Khayter C, et al. Dimeric CRISPR RNA-guided FokI nucleases for highly specific genome editing. *Nature Biotechnology* 2014;**32**(6):569–76. https://doi.org/10.1038/nbt.2908.
186. Bae S, Park J, Kim JS. Cas-OFFinder: a fast and versatile algorithm that searches for potential off-target sites of Cas9 RNA-guided endonucleases. *Bioinformatics* 2014;**30**(10):1473–5. https://doi.org/10.1093/bioinformatics/btu048.
187. Flachowsky H, Le Roux PM, Peil A, Patocchi A, Richter K, Hanke MV. Application of a high-speed breeding technology to apple (Malus×domestica) based on transgenic early flowering plants and marker-assisted selection. *New Phytologist* 2011;**192**(2):364–77. https://doi.org/10.1111/j.1469-8137.2011.03813.x.
188. Yamagishi N, Kishigami R, Yoshikawa N. Reduced generation time of apple seedlings to within a year by means of a plant virus vector: a new plant-breeding technique with no transmission of genetic modification to the next generation. *Plant Biotechnology Journal* 2014;**12**(1):60–8. https://doi.org/10.1111/pbi.12116.
189. de Ronde D, Butterbach P, Kormelink R. Dominant resistance against plant viruses. *Frontiers in Plant Science* 2014;**5**:307.
190. Bharathi JK, Anandan R, Benjamin LK, Muneer S, Prakash MAS. Recent trends and advances of RNA interference (RNAi) to improve agricultural crops and enhance their resilience to biotic and abiotic stresses. *Plant Physiology and Biochemistry* January 2023;**194**:600–18.

191. Tsuji H, Aya K, Ueguchi-Tanaka M, Shimada Y, Nakazono M, Watanabe R, Nishizawa NK, Gomi K, Shimada A, Kitano H, Ashikari M, Matsuoka M. GAMYB controls different sets of genes and is differentially regulated by microRNA in aleurone cells and anthers. *The Plant Journal* August 2006;**47**(3):427–44.
192. Thakur N, Upadhyay SK, Verma PC, Chandrashekar K, Tuli R, Singh PK. Enhanced whitefly resistance in transgenic tobacco plants expressing double stranded RNA of v-ATPase A gene. *PLoS One* March 3, 2014;**9**(3):e87235.
193. Bahrami Kamangar S, Christiaens O, Taning CNT, et al. The cuticle protein MPCP2 is involved in *Potato virus Y* transmission in the green peach aphid *Myzus persicae*. *Journal of Plant Diseases and Protection* 2019;**126**:351–7.
194. Sharma R, Christiaens O, Taning CN, Smagghe G. RNAi-mediated mortality in southern green stinkbug Nezara viridula by oral delivery of dsRNA. *Pest Management Science* January 2021;**77**(1):77–84.
195. Jiang CJ, Shimono M, Maeda S, Inoue H, Mori M, Hasegawa M, Sugano S, Takatsuji H. Suppression of the rice fatty-acid desaturase gene OsSSI2 enhances resistance to blast and leaf blight diseases in rice. *Molecular Plant Microbe Interaction* July 2009;**22**(7):820–9.
196. Peltier AJ, Hatfield RD, Grau CR. Soybean stem lignin Concentration Relates to resistance to Sclerotinia sclerotiorum. *The Plant Disease* February 2009;**93**(2):149–54.
197. Campo S, Peris-Peris C, Siré C, Moreno AB, Donaire L, Zytnicki M, Notredame C, Llave C, San Segundo B. Identification of a novel microRNA (miRNA) from rice that targets an alternatively spliced transcript of the Nramp6 (Natural resistance-associated macrophage protein 6) gene involved in pathogen resistance. *New Phytology* July 2013;**199**(1):212–27.
198. Dubrovina AS, Kiselev KV. Exogenous RNAs for gene regulation and plant resistance. *International Journal of Molecular Science* May 8, 2019;**20**(9):2282.
199. Pooggin MM. RNAi-mediated resistance to viruses: a critical assessment of methodologies. *Current Opinion in Virology* October 2017;**26**:28–35. https://doi.org/10.1016/j.coviro.2017.07.010.
200. Bonfim K, Faria JC, Nogueira EO, Mendes EA, Aragão FJ. RNAi-mediated resistance to Bean golden mosaic virus in genetically engineered common bean (Phaseolus vulgaris). *Molecular Plant Microbe Interaction* June 2007;**20**(6):717–26. https://doi.org/10.1094/MPMI-20-6-0717.
201. Wang MB, Abbott DC, Waterhouse PM. A single copy of a virus-derived transgene encoding hairpin RNA gives immunity to barley yellow dwarf virus. *Molecular Plant Pathology* November 1, 2000;**1**(6):347–56.
202. Krubphachaya P, Juríček M, Kertbundit S. Induction of RNA-mediated resistance to papaya ringspot virus type W. *Journal of Biochemistry and Molecular Biology* May 31, 2007;**40**(3):404–11.
203. Hameed A, Tahir MN, Asad S, Bilal R, Van Eck J, Jander G, Mansoor S. RNAi-mediated simultaneous resistance against three RNA viruses in potato. *Molecular Biotechnology* March 2017;**59**(2–3):73–83.
204. Worrall EA, Bravo-Cazar A, Nilon AT, Fletcher SJ, Robinson KE, Carr JP, Mitter N. Exogenous application of RNAi-Inducing double-stranded RNA inhibits aphid-mediated transmission of a plant virus. *Frontiers in Plant Science* March 15, 2019;**10**:265.
205. Sindhu AS, Maier TR, Mitchum MG, Hussey RS, Davis EL, Baum TJ. Effective and specific in planta RNAi in cyst nematodes: expression interference of four parasitism genes reduces parasitic success. *Journal of Experimental Biology* 2009;**60**(1):315–24.

206. Huang J, Li Z, Zhao D. Deregulation of the OsmiR160 target gene OsARF18 causes growth and developmental defects with an alteration of auxin signaling in rice. *Scientific Reports* July 21, 2016;**6**:29938.
207. Carrión VJ, Perez-Jaramillo J, Cordovez V, Tracanna V, de Hollander M, Ruiz-Buck D, Mendes LW, van Ijcken WFJ, Gomez-Exposito R, Elsayed SS, Mohanraju P, Arifah A, van der Oost J, Paulson JN, Mendes R, van Wezel GP, Medema MH, Raaijmakers JM. Pathogen-induced activation of disease-suppressive functions in the endophytic root microbiome. *Science* November 1, 2019;**366**(6465):606–12.
208. Mendes R, Kruijt M, de Bruijn I, Dekkers E, van der Voort M, Schneider JHM, Piceno YM, DeSantis TZ, Andersen GL, Bakker PAHM, Raaijmakers JM. Deciphering the rhizosphere microbiome for disease-suppressive bacteria. *Science* 2011;**332**:1097–100.
209. Frey-Klett P, Burlinson P, Deveau A, Barret M, Tarkka M, Sarniguet A. Bacterial-fungal interactions: hyphens between agricultural, clinical, environmental, and food microbiologists. *Microbiology Molecular Biology and Review* 2011;**75**:583–609.
210. Raaijmakers JM, Mazzola M. Diversity and natural functions of antibiotics produced by beneficial and plant pathogenic bacteria. *Annual Review of Phytopathology* 2012;**50**:403–24.
211. Wei Z, Yang T, Friman V, Xu Y, Shen Q, Jousset A. Trophic network architecture of root-associated bacterial communities determines pathogen invasion and plant health. *Nature Communication* 2015;**6**.
212. Zelezniak A, Andrejev S, Ponomarova O, Mende DR, Bork P, Patil KR. Metabolic dependencies drive species co-occurrence in diverse microbial communities. *Proceedings of the National Academy of Science USA* 2015;**112**:6449–54.
213. Koprivova A, Schuck S, Jacoby RP, Klinkhammer I, Welter B, Leson L, Martyn A, Nauen J, Grabenhorst N, Mandelkow JF, Zuccaro A, Zeier J, Kopriva S. Root-specific camalexin biosynthesis controls the plant growth-promoting effects of multiple bacterial strains. *Proceedings of the National Academy of Science USA* 2019;**116**:15735–44.
214. Souza RD, Ambrosini A, Passaglia LMP. Plant growth-promoting bacteria as inoculants in agricultural soils. *Genetic Molecular Biology* 2015;**38**:401–19.
215. Berendsen RL, Vismans G, Yu K, Song Y, de Jonge R, Burgman WP, Burmølle M, Herschend J, Bakker PAHM, Pieterse CMJ. Disease-induced assemblage of a plant-beneficial bacterial consortium. *ISME Journal* 2018;**12**:1496–507.
216. Křížkovská B, Viktorová J, Lipov J. Approved genetically modified potatoes (*Solanum tuberosum*) for Improved stress resistance and food safety. *Journal of Agricultural Food Chemistry* September 28, 2022;**70**(38):11833–43.
217. *International Service for the Acquisition of Agri-biotech applications (ISAAA). Event Name: TIC-AR233-5.* Ithaca, NY: ISAAA; 2018.
218. Australia New Zealand Food Authority (ANZFA). *Food derived from insect and potato virus Y-protected potato lines RBMT15-101, SEMT15-02 and SEMT15-15.* Kingston, Australia and Wellington, New Zealand: ANZFA; 2001.
219. Canadian Food Inspection Agency (CFIA). *Determination of Environmental safety of SEMT15-02, SEMT15-15, and RBMT15-101 Colorado potato Beetle and potato virus Y resistant potato lines developed by Monsanto.* Ottawa, Ontario, Canada: CFIA; 2017.
220. Khalid A, Zhang Q, Yasir M, Li F. Small RNA based genetic engineering for plant viral resistance: application in crop protection. *Frontiers in Microbiology* 2017;**8**:43.
221. Bilal M, Tabassum B, Ali Q, Nasir IA. Down regulation of potato virus Y (PVY) Coat protein (CP) expression by iberis gibraltarica protein extract. *Cytology and Genetics* 2021;**55**:80–6.

222. Lawson EC, Weiss JD, Thomas PE, Kaniewski WK. NewLeaf Plus® Russet Burbank potatoes: replicase-mediated resistance to potato leafroll virus. *Molecular Breeding* 2001;**7**:1−12.
223. Ehrenfeld N, Romano E, Serrano C, Arce-Johnson P. Replicase mediated resistance against potato leafroll virus in potato desireé plants. *Biological Research* 2004;**37**:71−82.
224. Collinge DB, Jørgensen HJL, Lund OS, Lyngkjær MF. Engineering pathogen resistance in crop plants: current trends and future prospects. *Annual Review of Phytopathology* 2010;**48**:269−91.
225. Rashid M-O, Zhang X-Y, Wang Y, Li D-W, Yu J-L, Han C-G. The three Essential Motifs in P0 for suppression of RNA silencing activity of potato leafroll virus are required for virus systemic infection. *Viruses* 2019;**11**:170.
226. Zhang L, Xu J, Birch RG. Engineered detoxification confers resistance against a pathogenic bacterium. *Nature Biotechnology* 1999;**17**(10):1021.
227. Verma KK, Song XP, Budeguer F, Nikpay A, Enrique R, Singh M, Zhang BQ, Wu JM, Li YR. Genetic engineering: an efficient approach to mitigating biotic and abiotic stresses in sugarcane cultivation. *Plant Signal Behaviour* December 31, 2022;**17**(1):2108253.
228. Nayyar S, Sharma BK, Kaur A, Kalia A, Sanghera GS, Thind KS, et al. Red rot resistant transgenic sugarcane developed through expression of β-1,3-glucanase gene. *PLoS One* 2017;**12**(6):e0179723.
229. Tariq M, Khan A, Tabassum B, Toufiq N, Bhatti MU, Riaz S, et al. Antifungal activity of chitinase II against Colletotrichum falcatum Went. causing red rot disease in transgenic sugarcane. *Turkish Journal of Biology* 2018;**42**:45. https://doi.org/10.3906/biy-1709-17.
230. Schneider VK, Soares-Costa A, Chakravarthi M, Ribeiro C, Chabregas SM, Falco MC, et al. Transgenic sugarcane overexpressing CaneCPI1 negatively affects the growth and development of the sugarcane weevil Sphenophorus levis. *Plant Cell Reports* 2017;**36**(1):193−201.
231. Gill R, Malhotra PK, Gosal SS. Direct plant regeneration from cultured young leaf segments of sugarcane. *Plant Cell Tissue and Organ Culture* 2006;**84**(2):227−31.
232. Romeis J, Meissle M, Bigler F. Transgenic crops expressing Bacillus thuringiensis toxins and biological control. *Nature Biotechnology* 2006;**24**(1):63−71.
233. Gosal SS, Wani SH. Plant genetic transformation and transgenic crops: methods and applications. In: Gosal S, Wani S, editors. *Biotechnologies of crop Improvement.* vol 2. Cham: Springer; 2018. pp. 1−23.
234. Falco MC, Silva-Filho MC. Expression of soybean proteinase inhibitors in transgenic sugarcane plants: effects on natural defense against Diatraea saccharalis. *Plant Physiology and Biochemistry* 2003;**41**(8):761−6.
235. Deng ZN, Wei YW, WL L, Li YR. Fusion insect-resistant gene mediated by matrix attachment region (MAR) sequence in transgenic sugarcane. *Sugar Technology* 2008;**10**(1):87−90.
236. Sétamou M, Bernal JS, Legaspi JC, Mirkov TE, Legaspi Jr BC. Evaluation of lectin-expressing transgenic sugarcane against stalk borers (Lepidoptera: pyralidae): effects on life history parameters. *Journal of Economic Entomology* 2002;**2**(2):469−77.
237. Christy LA, Arvinth S, Saravanakumar M, Kanchana M, Mukunthan N, Srikanth J, et al. Engineering sugarcane cultivars with bovine pancreatic trypsin inhibitor (aprotinin) gene for protection against top borer (Scirpophaga excerptalis Walker). *Plant Cell Reports* 2009;**28**(2):175−84.

238. Knowles BH, Ellar DJ. Characterization and partial purification of a plasma membrane receptor for Bacillus thuringiensis var. kurstaki lepidopteran-specificδ-endotoxin. *Journal of Cell Science* 1986;**83**(1):89–101.
239. Chakroun M, Bel Y, Escriche B. Bacterial vegetative insecticidal proteins (Vip) from entomopathogenic bacteria. *Microbiology and Molecular Biology Review* 2016;**80**(2):329–50.
240. Baranek J, Konecka E, Kaznowski A. Interaction between toxin crystals and vegetative insecticidal proteins of Bacillus thuringiensis in lepidopteran larvae. *BioControl* 2017;**62**:649–58.
241. Tabashnik BE, Carrière Y, Brewer M. Evaluating cross-resistance between vip and cry toxins of Bacillus thuringiensis. *Journal of Economic Entomology* 2020;**113**(2):553–61.
242. Qamar Z, Riaz S, Nasir I, Ali Q, Husnain T. Transformation and transgenic expression studies of glyphosate tolerant and cane borer resistance genes in sugarcane (Saccharum officinarum L.). *Molecular Plant Breeding* 2015;**6**:1–17.
243. ISAAA. *Crop Biotech Update*. 2022. https://www.isaaa.org/gmap provaldatabase/crop/default.asp?CropID=27&Crop=Sugarcane. [Accessed 21 April 2022].
244. Park JC, Yoo Y, Lim H, Yun S, Win KTYS, Kim KM, Lee GS, Cho MH, Lee TH, Sano H, Lee SW. Intracellular $Ca^{2+}$ accumulation triggered by caffeine provokes resistance against a broad range of biotic stress in rice. *Plant Cell Environment* April 2022;**45**(4):1049–64.
245. Kim YS, Uefuji H, Ogita S, Sano H. Transgenic tobacco plants producing caffeine: a potential new strategy for insect pest control. *Transgenic Research* 2006;**15**(6):667–72.
246. Kim YS, Lim S, Yoda H, Choi YE, Sano H. Simultaneous activation of salicylate production and fungal resistance in transgenic Chrysanthemum producing caffeine. *Plant Signal Behaviour* 2011;**6**(3):409–12.
247. Kim YS, Sano H. Pathogen resistance of transgenic tobacco plants producing caffeine. *Phytochemistry* 2008;**69**(4):882–8.
248. Nerkar G, Devarumath S, Purankar M, Kumar A, Valarmathi R, Devarumath R, Appunu C. Advances in crop breeding through precision genome editing. *Frontiers in Genetics* 2022;**13**:880195.
249. Zhou X, Liao H, Chern M, Yin J, Chen Y, Wang J, et al. Loss of function of a rice TPR-domain RNA-binding protein confers Broad-Spectrum disease resistance. *Proceeding of the National Academy of Science U. S. A.* 2018;**115**:3174–9.
250. Oliva R, Ji C, Atienza-Grande G, Huguet-Tapia JC, Perez-Quintero A, Li T, et al. Broad-spectrum resistance to bacterial blight in rice using genome editing. *Nature Biotechnology* 2019;**37**(11):1344–50.
251. Wang Y, Cheng X, Shan Q, Zhang Y, Liu J, Gao C, et al. Simultaneous editing of three homoeoalleles in hexaploid bread wheat confers heritable resistance to powdery mildew. *Nature Biotechnology* 2014;**32**(9):947–51.
252. Zhang Y, Bai Y, Wu G, Zou S, Chen Y, Gao C, et al. Simultaneous modification of three homoeologs of TaEDR1 by genome editing enhances powdery mildew resistance in wheat. *The Plant Journal* 2017;**91**(4):714–24.
253. Santillan Martinez MI, Bracuto V, Koseoglou E, Appiano M, Jacobsen E, Visser RGF, et al. CRISPR/Cas9-targeted Mutagenesis of the tomato Susceptibility gene PMR4 for resistance against powdery mildew. *BMC Plant Biology* 2020;**20**(1):284.
254. Ortigosa A, Gimenez-Ibanez S, Solano R. Design of a bacterial speck resistant tomato by CRISPR/Cas9-mediated editing of SlJAZ2. *Plant Biotechnology Journal* 2019;**17**(3):665–73.

255. Dale J, James A, Paul J-Y, Khanna H, Smith M, Peraza-Echeverria S, et al. Transgenic cavendish bananas with resistance to Fusarium wilt tropical race 4. *Nature Communication* 2017;**8**(1):1496.
256. Jia H, Zhang Y, Orbović V, Xu J, White FF, Jones JB, et al. Genome editing of the disease Susceptibility gene CsLOB1 in citrus confers resistance to citrus canker. *Plant Biotechnology Journal* 2017;**15**(7):817–23.
257. Peng A, Chen S, He Y, Xu L, He Y, Wu L, et al. Engineering canker-resistant plants through CRISPR/Cas9-targeted editing of the susceptibility gene CsLOB1 promoter in citrus. *Plant Biotechnology Journal* 2017;**15**(12):1509–19.
258. Wang X, Tu M, Wang D, Liu J, Li Y, Li Z. CRISPR/Cas9- mediated efficient targeted Mutagenesis in grape in the first generation. *Plant Biotechnology Journal* 2017;**16**(4):844–55.
259. Gomez MA, Lin DZ, Moll T, Chauhan RD, Hayden L, Renninger K, et al. Simultaneous CRISPR/Cas9-Mediated editing of cassava eIF4E Isoforms nCBP-1 and nCBP-2 reduces cassava brown streak disease symptom severity and incidence. *Plant Biotechnology Journal* 2019;**17**(2):421–34.
260. Chandrasekaran J, Brumin M, Wolf D, Leibman D, Klap C, Pearlsman M, et al. Development of broad virus resistance in non-transgenic cucumber using CRISPR/Cas9 technology. *Molecular Plant Pathology* 2016;**17**(7):1140–53.
261. Tripathi JN, Ntui VO, Shah T, Tripathi L. CRISPR/Cas9-mediated editing of DMR6 orthologue in banana (Musa spp.) confers enhanced resistance to bacterial disease. *Plant Biotechnology Journal* July 2021;**19**(7):1291–3. https://doi.org/10.1111/pbi.13614.
262. Zhang Y, Zhao L, Zhao J, Li Y, Wang J, Guo R, et al. S5H/DMR6 encodes a salicylic acid 5-hydroxylase that fine-tunes salicylic acid homeostasis. *Plant Physiology* 2017;**175**(3):1082–93.
263. Low YC, Lawton MA, Di R. Validation of barley 2OGO gene as a functional orthologue of Arabidopsis DMR6 gene in Fusarium head blight susceptibility. *Scientific Reports* 2020;**10**:9935.
264. Tripathi JN, Ntui VO, Ron M, Muiruri SK, Britt A, Tripathi L. CRISPR/Cas9 editing of endogenous banana streak virus in the B genome of Musa spp. Overcomes a major Challenge in banana breeding. *Communication Biology* 2019;**2**(1):1–11.
265. Tripathi L, Dhugga KS, Ntui VO, Runo S, Syombua ED, Muiruri S, Wen Z, Tripathi JN. Genome editing for sustainable agriculture in Africa. *Frontiers in Genome Editing* 2022;**4**:876697.
266. Wang Y, Geng M, Pan R, Zhang T, Zhen X, Che Y. Engineering bacterial blight-resistant plants through2 CRISPR/Cas9-Targeted editing of the MeSWEET10a promoter in cassava. *bioRxiv preprint* 2022:1–15.
267. Gomez MA, Lin ZD, Moll T, Chauhan RD, Hayden L, Renninger K, et al. Simultaneous CRISPR/Cas9-Mediated editing of cassava eIF4E isoforms nCBP-1 and nCBP-2 reduces cassava brown streak disease symptom severity and incidence. *Plant Biotechnology Journal* 2019;**17**:421–34.
268. Mehta D, Stürchler A, Anjanappa RB, Zaidi SS-E-A, Hirsch-Hoffmann M, Gruissem W, et al. Linking CRISPR/Cas9 interference in cassava to the Evolution of editing-resistant Geminiviruses. *Genome Biology* 2019;**20**:80.
269. Bellis ES, Kelly EA, Lorts CM, Gao H, Deleo VL, Rouhan G, et al. Genomics of sorghum local adaptation to a parasitic plant. *Proceedings of the National Academy of Science U. S. A.* 2020;**117**:4243–51.

270. de Toledo Thomazella DP, Seong K, Mackelprang R, Dahlbeck D, Geng Y, Gill US, et al. Loss of function of a DMR6 ortholog in tomato confers broad-spectrum disease resistance. *Proceedings of the National Academy of Science U. S. A.* 2021;**118**(27). e2026152118.
271. Hasley JAR, Navet N, Tian M. CRISPR/Cas9-mediated mutagenesis of sweet basil candidate susceptibility gene ObDMR6 enhances downy mildew resistance. *PLoS One* 2021;**16**:e0253245.
272. Bari VK, Nassar JA, Kheredin SM, Gal-On A, Ron M, Britt A, et al. CRISPR/Cas9-mediated mutagenesis of CAROTENOID CLEAVAGE DIOXYGENASE 8 in tomato Provides resistance against the parasitic weed phelipanche aegyptiaca. *Scientific Reports* 2019;**9**:11438.
273. Butt H, Jamil M, Wang JY, Al-Babili S, Mahfouz M. Engineering plant architecture via CRISPR/Cas9-mediated alteration of strigolactone biosynthesis. *BMC Plant Biology* 2018;**18**:174.
274. Li S, Lin D, Zhang Y, Deng M, Chen Y, Lv B, et al. Genome-edited powdery mildew resistance in wheat without growth penalties. *Nature* 2022;**602**.

Chapter 9

# Strategies to address climate change: novel ways to generate food proteins using biotechnology to reduce greenhouse gas emissions

## 1. Introduction

As environmental sustainability becomes increasingly important, the utility of animal-sourced food products has been under extensive evaluation. Livestock feedstock alone represents almost 40% of the world's cropland, and the demand for meat has become even greater as low- and middle-income countries increase their economic status.[1] The increased need for arable land to produce animal protein has been a significant driver of biodiversity loss, pollution, and eutrophication through the overuse of fertilizers. At the same time, over 800 million people suffer from undernutrition and over two billion from malnutrition (World Health Organization, 2022). Improving the nutritional status of the world's population will require a doubling of food production, and if continuing to follow conventional patterns of agriculture, this will result in a decrease in environmental sustainability overall, culminating in an 80% increase in greenhouse gas emissions.[2]

In addition to sustainability issues, the overconsumption of meat has been linked to several health concerns, such as heart disease, cardiovascular disease, and colorectal cancer.[3] Food-borne illnesses such as *E. coli*, *Salmonella,* and *Campylobacter* are also associated with meat consumption, as are concerns of the overuse of antibiotics used in livestock.[4,5]

To preserve biodiversity and reduce our increasing carbon footprint, research and development into animal protein replacements has resulted in an explosion of innovation. The alternative meat and dairy industries are currently growing at a rate of 15.8%, with increased appeal to mainstream consumers rather than merely to "niche" markets.[6] Meat alternatives produced today have appealed to the fast-growing segment of "flexitarian" consumers.

Agricultural Biotechnology. https://doi.org/10.1016/B978-0-323-88476-1.00008-4

Over the past 10 years, increased demand has initiated an assortment of alternative protein products from diverse sources to enter the marketplace. Alternative proteins can simulate the appearance, mouthfeel, and flavor, and even nutritional profiles of animal products. Furthermore, alternative proteins can greatly reduce greenhouse gas emissions and issues concerning land usage and loss of diversity in wild ecosystems.[1] Finally, alternative protein can directly address animal welfare and other issues associated with the ethics of eating meat.[7] Thus, alternatives to meat-sourced food protein are a response to a market-driven narrative that addresses environmental sustainability, human nutrition, and ethical challenges that face the world today.

The way forward for low- and middle-income countries with respect to animal protein alternatives remains unclear. Nations which have increased economic status over the years embrace meat as a new attribute of wealth, and the maintenance of livestock for millennia throughout a number of rural cultures is likely to remain steadfast. In contrast, consumption of meat in Western countries has stagnated or in some regions has even declined.[8] The market for alternative protein could disrupt conventional trading partners and impact communities of the Global South in ways that are difficult to predict.

Alternative protein development can be divided into three main innovations: plant-based protein as meat substitutes, microbial fermentation to produce animal ingredient proteins, and cell-based or cultivated meat. These new technologies in food and agriculture will disrupt the global market significantly. The following review describes the history and status of these three innovations and then explores how these disruptors in the agricultural market will impact food sovereignty, inequity, and social justice. The review discusses consumer acceptance to meat alternatives on a variety of fronts and concludes with a prediction of future challenges and directions for this rapidly accelerating market.

## 2. History and current status of alternative proteins

Original plant-based protein such as tofu or tempeh from soy, or seitan from wheat, were produced and consumed in Asian countries for centuries and then became popular in Western countries in the late-20th Century. Today, there is interest in novel plant-based foods with enhanced nutrition as well as improved sensory characteristics. Thus, we can refer to plant-based protein as either traditional or novel. Novel foods often are characterized as having the texture and taste that makes them satisfying alternatives to animal-based products for consumers.

More recently, a renewed interest in high protein content, from energy bars to powders, has emerged as a popular food trend.[9] Plant protein on its own can also be used to transform meat-centric meals into healthy, nutrient-rich alternatives.[10] Thus, the development of alternative protein products that have the same look, mouthfeel, and taste as animal-sourced foods is the latest development, as it allows the consumer to not give up the sensory pleasures of meat and dairy that they have come to know and love.

Today, plant-based products that have the same mouthfeel as hamburger and sausage are being produced using soy and pea protein, by companies such as Beyond Meat and Impossible Food. Plant proteins are extracted from these crops and then mixed with other ingredients and processed into a meaty texture.[1]

To produce plant-based replacements for animal proteins, functional analogs in the plant world are sought. This requires the processing of crops to find the necessary ingredients (proteins, fats, and starch) and arrange them into a muscle-like texture that closely resembles meat. Product formulation to get the necessary texture and taste, yet retain or enhance the desired nutritional qualities, can be complex and may consist of a variety of manufacturing processes such as kneading, pressing, 3D printing, and extrusion, among others.[6]

## History and current status of fermentation

Fermentation has been used in food production for thousands of years, from the dawn of civilization. From bread to kimchi, microbial cultures are used to process food products for preservation, taste, and nutritional content. Over the years, fermentation as a technology has been used for a number of disciplines ranging from pharmaceuticals to biofuel production.[11] Today, fermentation can be used to generate several novel foods, including alternative proteins. Examples of traditional fermentation include fungi to ferment soy into tempeh and lactic acid bacteria to produce cheese and yogurt. In certain instances, microbial biomass itself can be used to make the food product itself. For example, the mycelium used in the products of the company Quorn, a company that has been in operation since 1985, are based on filamentous fungi. The fungi can be grown in microbial bioreactors. These products are highly nutritious and require minimal processing. The company can use food waste to produce its mycelium products, using large fermenters, and makes a carbon footprint that is 10 times lower than that of beef. Currently, Quorn produces 25,000 tonnes (dry mass) per year with global retail sales of approximately 200 million Euros. Mycoprotein as a source of protein is projected to increase annually by 20%.[12]

Precision fermentation is another form of fermentation that incorporates modern biotechnology to generate ingredient proteins, flavorings, and natural pigments using microbial bioreactors (Fig. 9.1). For example, Impossible Foods produces heme protein for its Impossible Burger by using precision fermentation technology[11,13] Fig. 9.1.

In general, the use of controlled bioreactors to cultivate microbes such as bacteria or fungi to produce either biomass or specific food ingredients is much more efficient than growing crops in an open field. Production in

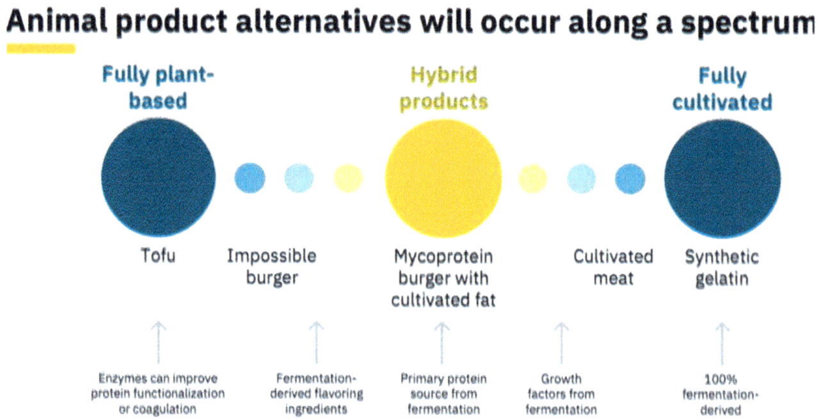


**FIGURE 9.1** **Animal product alternatives.** Alternative animal products include a range of various foods from fully plant based to hybrid products to fully cultivated or fermentation based. *Image from The Good Food Institute.*

bioreactors do not require soil, climate, water usage, pesticides, or fertilizers. The carbon footprint of bioreactors is low, and the fact that they can be located on nonagricultural land or even within urban centers enables them to not compete for arable land. In certain instances, production could even take place in industrial areas, where inorganic carbon sources such as $CO_2$ and $H_2$ could be utilized.[13]

## History and current status of cultivated meat production

Cultivated/cell-based meat now has a near two-decade history of laboratory research and finds its origins in the biomedical and pharmaceutical industries. It is interesting to reflect on how socio-cultural framings have served to move the trajectory of cultivated meat into the global movement that it is today. This trajectory can be thought of in terms of two critical movements, the first initiated by university-based projects that were initially driven by environmental and animal ethics concerns that culminated in a slow and steadily growing knowledge base in the field and the second by philanthropists with environmental sustainability concerns.

In 2011, an article for the New Yorker reported that the technology for cultivated meat was readily available, if only there was sufficient funding for it.[14] Shortly afterward, philanthropic donors offered to finance a program at Maastricht University to produce the first lab grown burger in a public event.[15] The cultivated meat space soon became associated with high tech investment streams, often funded by wealthy individuals who focused on breakthrough technologies to address humanity's global challenges. This became the second phase of this movement. The industry shifted again when investments from multinational (food) corporations such as Cargill, Nestle, and Tyson Foods become available, beginning in 2017. Today, many startup companies are in this space, such as Memphis Meats (who raised 17 million in Series A funding). Just Eat was the first cultivated meat startup to gain regulatory approval and sells cultured chicken meat in Singapore (Just Eat.com). Israeli startup AlephFarms has also moved forward, planning to create whole-meat cuts or slices.

To produce cultivated meat products that are authentic in terms of texture and taste, multiple types of cells are required, including muscle precursor cells, fat cells, and connective tissue.[16] The use of media can also impact meat quality and taste. The cultivated meat industry has focused on two major products, one that is unstructured, such as hamburger or sausage, versus one that is highly structured, such as chicken breasts or beefsteaks. In order to achieve the latter, stem cells must be grown in a bioreactor for 40–50 generations, and then the media must be changed to promote proliferation into differentiated cell types, including muscle, fat, and connective tissue (Fig. 9.2). Since these cells are adherent and need to be attached to a scaffold of some sort, new biomaterials have been developed, which could act as microcarriers. The media used must be animal product-free and of food grade quality[17] Fig. 9.2.

Animal cell culture has been well-established in the disciplines of biomedical research and requires carbon and nitrogen sources, salts, sugars, amino acids, vitamins, buffers, and signaling molecules such as cytokines and growth factors.[18] Some of these constituents can influence sensory qualities, for example, glutamic acid and asparagine can create an umami flavor.[19] Changes in media preparation can also impact the proliferation and differentiation of muscle cells in culture. For example, myoblasts require different growth factors and cytokines to proliferate, including fibroblast growth factor, insulin growth factor, hepatocyte growth factor, transforming growth factor-β, and cytokines such as tumor necrosis factor-α.[20] These signaling molecules stimulate myogenesis through a variety of metabolic pathways and are prohibitively expensive. Due to the cost of growth factors, less expensive methods are critical. One way to reduce the need and cost of animal-derived growth factors is to screen for amino acid homology of animal growth factors with analogous proteins in plants and fungi. For example, chickpea peptide extracts have been shown to stimulate insulin associated cell signaling.[21] It will be necessary to effectively replace animal serum with serum-free media. It is also possible to make use of complex ingredients to produce some of the constituent of media, such as molasses, in place of purified glucose.[22]

To create cultivated meat products such as chicken breasts and steaks in vitro, a natural, edible scaffold framework must be established, that recreates the microenvironment that cells adhere to.[23] The scaffold will be critical for the duration of cell cultivation and must be biocompatible with the cells so that they can proliferate and differentiate, yet enable the free flow of oxygen,

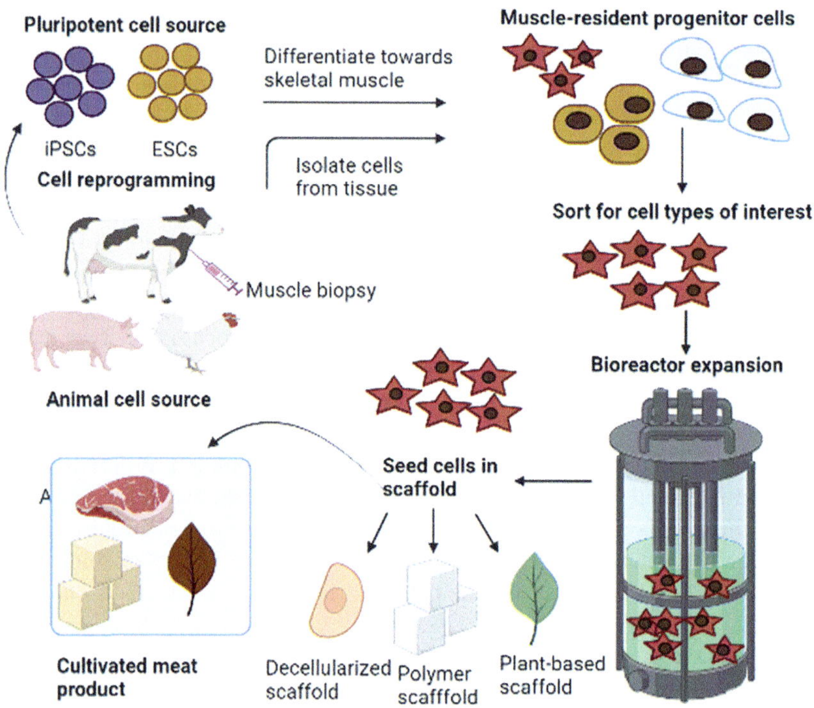


**FIGURE 9.2 General workflow for cultivated meat production.** The first step is cell sourcing, which may be either an animal cell source or pluripotent cell source. Relevant cell types from the chosen cell source are then isolated and expanded in a bioreactor to yield a large quantity of cells. Finally, the cells are matured in a biocompatible tissue scaffold that supports cell development and provides the cultivated meat product with a specific structure. *From Reiss J, Robertson S, Suzuki M. Cell sources for cultivated meat: applications and considerations throughout the production workflow.* International Journal of Molecular Sciences *2021;***22***(14):513. https://doi.org/10.3390/ijms22147513.*

nutrients, and waste. Scaffolds can be made by electrospinning, mold cast and injectable systems, and 3D extrusion printing.[22]

In a conventional lab setting, a 2D system composed of Petri dishes and/or tissue culture flasks is employed to offer mechanical support for cells. Under these conditions, however, muscle cells change their phenotypic properties and behaviors, due to altered gene expression. Eventually, the cells form a monolayer and become unsustainable.[22] On the other hand, a 3D scaffold matrix comprised a hydrogel of crosslinked hydrophilic polymers and possibly with growth factors and cell adhesive molecules incorporated within, can better functionalize muscle cells for cultivated meat.[24] For scale up, cell suspensions supported by microcarriers will be necessary to reach the volumes necessary for market demand. These microcarriers could be edible, dissolve over time, or be readily extractable during processing.[22] Microcarriers could also encapsulate and release growth factors to promote cell proliferation or differentiation over time. For example, one company (Mosa Meats) has developed a technology to grow cultured meat using pillars, which scaffold the formation of a hydrogel containing muscle cells.[25] These muscle cells then self-assemble to form contractable rings, which, in turn, promote skeletal muscle maturation.

A challenge, which needs to be overcome, is the fact that there is a limit to cell division; cells will eventually enter senescence after dividing to a certain number. By adding telomerase, the number of cell divisions would be extended.[26] Another challenge is the need to use animal free media. Prior to the development of the cultivated meat industry, animal-derived serum has been used, such as fetal calf serum. To move away from issues such as animal welfare, serum free media, which contains replacements for animal serum, will be necessary.[27] For example, serum contains insulin, and the animal-derived version can be replaced with a recombinant one produced in microbial bio-reactors. In the future, animal proteins including albumin, insulin-like growth factor 1 (IGF-1), and others can possibly be replaced with proteins from plant sources, such as albumin storage proteins.[28]

The research and development of cultivated meat will spur a pronounced interest over a variety of disciplines, such as sourcing animal stem cells, scaling up processes, enhancing cell culture proliferation and differentiation, scaffold development, and the production of fetal bovine serum−free growth media using other sources from plants and fungi. Scaling up would include additional, more physical challenges, such as helping animal cell culture to withstand shear stresses in a large bioreactor setting.[29]

## Other sources of protein

Other, unconventional crops can be used for both food and feed. Ruminants, such as cattle and sheep, can feed on microalgae and duckweed. These crops are high yielding and economically competitive forms of protein and fiber, which do not compete with land required for human food. They have increased nutrient availability, improve soil fertility, and can even provide nitrogen fixation for other crops.[30] Algae is frequently used in the food industry, both as food supplements and as a functional food.[31] More specifically, food products based on the cyanobacteria Spirulina are beginning to emerge.[32]

Insects are also being considered as alternative sources of protein. Edible insects can lower agricultural land use and greenhouse gas emissions, yet are high in nutritional composition.[33] The practice of entomophagy has been part of the early history of mankind. While followed in China for over 3000 years; only recently has it begun to gain momentum in Western culture, with the greatest barrier to consumption being cultural. Over 100 countries around the world, about two billion people, practice entomophagy today.[34,35] Consumer behavior reports suggest that there is a greater willingness to consume products composed of shredded insects or powders rather than whole insects. Insects are high in protein content, making them a suitable alternative for human consumption. Other studies have illustrated that a number of insect peptides found in food products contain antihypertensive, antimicrobial, and http://about:blank.[36,37] Currently, the building of large-scale production facilities for edible insects is a challenge being addressed.

## 3. Circular food systems

Land usage for livestock feed has been well characterized, both in terms of acreage for grazing and acreage to produce crops for feed. For alternative protein, land usage may be required to generate feedstock. In comparison, the land usage required to produce cultivated meat could be reduced even further if feedstock were made from microalgae instead of crops.[38,39] Using similar measurements, protein sourced from insects can be derived from biomass that is composed of food waste residue or leftover streams that is unwanted for human consumption.[34] Thus, many forms of alternative protein facilitate a circular food system whereby green technologies make use of food waste and reduce pressure on arable farmland.

Precision fermentation systems make use of bioreactors to produce ingredient proteins. These production systems are closed yet require the use of glucose to feed the cell cultures. Glucose most frequently comes from grain crops such as corn or sugar beet and thus takes up arable land. Recently, autotrophic microbes who do not require conventional sugar carbon sources have been developed as bioreactors for food proteins. For these microbes, the gases $CO_2$ or $CH_4$ can be used as a feed source, meaning that waste from industrial plants can become the feedstock.[13] A negative net use of greenhouse gas emissions can be achieved, and the environmental sustainability factor intensified. The tremendous advantage to this form of fermentation is that it is completely independent of outdoor agriculture in terms of food and biofuel and is independent of fossil fuel needs; thus, it is not subject to the same economic fluctuations that govern our food and energy system.[40] This method of alternative protein production is still in its infancy but is currently moving forward in leaps and bounds.[41–43]

## 4. Impact of the alternative protein on land use, food sovereignty, and social justice

The impact of the alternative protein movement on global agriculture remains unclear.[44] While intensive farming in general, but particularly of livestock, which is considered by many to be leading to an environmental catastrophe, how transitioning to animal protein replacements would impact the lifestyle of farmers and those involved in the animal protein industry, as well as those who produce the feed for it, is uncertain. Disruptive technologies such as precision fermentation and cultivated meat appear to be aligned to resolve both ethical and environmental issues of the food system as we currently know them.[1] It is predicted that alternative protein and cultivated meat will lead to rewilding, ecosystem restoration and reforestation to replace land previously used for farming. How will our current world market change? What changes in livelihoods and infrastructure in rural settings will be necessary to make this a more inclusive reality for all?

Acceptance of alternative protein will most likely vary from country to country, from culture to culture. Emerging markets in Asia provide much optimism, and so far, there has been higher consumer acceptance for alternative protein in China and India than in the United States.[45] While much of India embraces a meatless culture, the acceptance of new food technologies described here will most likely be subjected to a dichotomy of mixed reviews. The political landscape is also important, and issues such as the impact of greenhouse gases on the environment as well as the welfare of animals will change conventional livestock production in some countries but perhaps not in others. The role of livestock in rural agriculture in India or Latin America, for example, differs significantly from that found in the United States or European Union. While livestock owners in sub-Saharan Africa may be nomadic and care for small herds or flocks, livestock owners in the United States may manage tens of thousands of animals under industrialized settings. Each group will manage stresses such as infectious disease of their livestock or food safety of their animal products very differently.[44]

Industrialized countries have demonstrated the greatest readiness so far to accept alternative proteins. A long history of colonialism has illustrated that developed countries can strongly influence food production beliefs and behavior onto their low- and middle-income trading partners.[46] Thus, another issue that may arise is that of neocolonialism; will the Global South be expected to follow the trends and expectations of industrialized countries or be permitted to maintain their own cultural values associated with livestock production[1]? How will these differences affect the global free market? Will these differences in movements create a greater imbalance between the Global North and South[47]? Or could alternative proteins become part of the solution to food security and offer a new source of economic development for low- and middle-income countries? Little research has been carried out to address these challenges.[48]

## 5. Consumer behavior toward the alternative protein movement

Flexitarianism has become increasingly popular over the last few years in developed countries. Transitioning to a low-meat diet could transform society, not only in terms of mitigating climate change but also in reducing chronic disease due to dietary changes. There are several factors that appear to influence the public's perspective on alternative protein as a replacement for meat; besides environmental and animal welfare considerations, these also include gender, age, and current health and educational status. Other influences include trust in science/neophobia, cost, convenience, and coverage in the media.[49] The most well-established substitutes to meat are the plant-based alternatives, as they are already well known to many consumers.[50]

There is a need to understand in more detail the drivers behind consumer acceptance of alternative protein.[51] While some research has so far focused on choices available for non-meat-sourced products, familiarity with alternative protein and neophobia toward new technologies used to produce them, the research is not extensive enough to provide firm conclusions as to which factors influence consumer acceptance and purchasing behavior.[52] Consumer acceptance is lowest for insect-based protein, moderate for cultured meat, and greatest for plant-based alternatives. Edible insects remain a challenge to Western norms and culture, whereas cultivated meat is largely unknown and not readily available on the market. Studies have demonstrated, however, that consumers who routinely eat large amounts of meat demonstrate higher levels of acceptance to cultivated meat[44] as well as products which are similar in appearance to meat.[53] This same cohort of consumers is less likely to display interest in plant-based proteins. On the other hand, plant-based products were more likely to be favored by consumers who do not consume high levels of meat in the first place.

A general picture of the largest group of consumers of alternative protein emerges as young, left leaning urbanites who are highly educated.[54] Drivers of acceptance for these consumers include convenience, health and environmental benefits, appearance and taste, and familiarity.[7] Drivers of rejection include disgust, neophobia, and perceived social norms.

## 6. Conclusion and future prospects for alternative protein development

This review discusses the three main branches of alternative protein production and explores their potential impact on environmental sustainability as well as on human health and nutrition. How alternative protein disruptive technologies may influence consumer behavior, trade, and disparities in the Global South are also discussed. The increase in per capita meat consumption has been most marked in countries that have undergone a strong economic transition upward, with a larger middle class who desires affluence, including animal sourced products. Consumer behavior and willingness-to-pay will be instrumental for the further development of alternative protein products. The future development of the three alternatives to meat proteins in the economic groups studied should focus on the perceived healthiness, safety and nutritional characteristics, and/or higher sustainability, taste, and lower price of the alternative dietary protein sources compared with meat (Fig. 9.3).

| Plant-based | Cultivated meat | Fermentation |
|---|---|---|
| Impossible Food | Memphis Meat | Perfect Day |
| Beyond Meat | Eat Just | Clara Foods |
| Perfect Day | Aleph Farms | Quorn |

**FIGURE 9.3 Examples of companies that produce alternative protein products.** Names of a few companies involved in the production of foods that are plant-based meat that is cultivated and products developed by fermentation.

While alternative proteins have captured global interest, much work remains to be accomplished at several points along the supply chain. For plant-based meat production, improved crop breeding to increase varieties with higher levels of high-quality protein are required, as are the removal of off-flavors and improved sensory characteristics in crops produced for plant-based meat production.[6] Better color indicators for plant-based meat are also greatly needed, to mimic the color change from red to brown during cooking. Such changes are feasible and will lead to higher consumer acceptance. Another large requirement will be reducing the financial investment necessary to produce plant-based meat. Decreasing costs would render plant-based meat more amenable for production in developing countries.

For the cultivated meat industry, facility design and operation will be critical. The global meat market currently is over $800 billion to produce volumes that will capture a portion of that will require intense scaling up of current infrastructure.[55] Producing cultivated meat products with taste and texture that closely resemble conventional beefsteaks, chicken breasts, or fish filets is a challenge. As these are newly emerging technologies, winning over consumers may require the availability of resources with educational information about cultivated meat.[6]

Fermentation of animal proteins using microbial bioreactors will also require advancement. The development of new microbial strains that can perform tasks with greater precision and with better taste will be needed, as will be the exploration of novel feedstocks that could be optimized to enable production pathways not based on fossil fuels or reliant upon crop production. Efforts are currently underway to use bioreactors to produce green industrial products that are not only independent of petroleum, but in fact make use of greenhouse gases such as $CO_2$ and $CH_4$ as feedstock, thus making them carbon negative in production.[13]

While precision fermentation and cultivated meat both require bioreactors for cell growth, animal cells proliferate much more slowly than microbial cells and can also produce growth-inhibiting catabolites such as ammonia.[16] Animal cells are also more likely to be damaged, as they lack a cell wall. Different types of animal cell culture will be required to make complex forms of meat, with tailored media requirements and bioreactor design being essential for this task.[56] Currently, the total capital investment for cultured meat using a perfusion bioreactor leads to an estimated total cost per kg of $51, while a bioreactor with a fedbatch design encompasses a total cost per kg of $37. To reach the market, the predicted target price of cultured meat that consumers have demonstrated a willingness to pay is $25 per kg. After further packaging and distribution, cultured meat may reach a minimum of $50 per kg in the supermarket, making the way forward a significant challenge.[57]

## References


1. Sexton AE, Garnett T, Lorimer J. Framing the future of food: the contested promises of alternative proteins. *Environment and Planning E: Nature and Space* 2019;**2**(1):47–72. https://doi.org/10.1177/2514848619827009.
2. Aimutis WR. Plant-based proteins: the Good, bad, and ugly. *Annual Review of Food Science and Technology* 2022;**13**:1–17. https://doi.org/10.1146/annurev-food-092221-041723.
3. Wolk A. Potential health hazards of eating red meat. *Journal of Internal Medicine* 2017;**281**(2):106–22. https://doi.org/10.1111/joim.12543.
4. Fegan N, Jenson I. The role of meat in foodborne disease: is there a coming revolution in risk assessment and management? *Meat Science* 2018;**144**:22–9. https://doi.org/10.1016/j.meatsci.2018.04.018.
5. Lee HJ, Kim DW, Kim C, Ryu HD, Chung EG, Kim K. Concentrations and risk assessments of antibiotics in an urban–rural complex watershed with intensive livestock farming. *International Journal of Environmental Research and Public Health* 2021;**18**(20). https://doi.org/10.3390/ijerph182010797.
6. Specht L. *Good Food Intitute.Org*. 2022.
7. Eckl MR, Biesbroek S, Van't Veer P, Geleijnse JM. Replacement of meat with non-meat protein sources: a review of the drivers and inhibitors in developed countries. *Nutrients* 2021;**13**(10). https://doi.org/10.3390/nu13103602.
8. Onwezen MC, Bouwman EP, Reinders MJ, Dagevos H. A systematic review on consumer acceptance of alternative proteins: pulses, algae, insects, plant-based meat alternatives, and cultured meat. *Appetite* 2021;**159**:105058. https://doi.org/10.1016/j.appet.2020.105058.
9. Allied Market Research. April 2022. www.alliedmarketresearch.com/press-release/protein-supplement-market.html.
10. Sexton AE. Eating for the post-Anthropocene: alternative proteins and the biopolitics of edibility. *Transactions of the Institute of British Geographers* 2018;**43**(4):586–600.
11. Ciani M, Lippolis A, Fava F, Rodolfi L, Niccolai A, Tredici MR. Microbes: food for the future. *Foods* 2021;**10**(5). https://doi.org/10.3390/foods10050971.
12. Cherta-Murillo A, Frost GS. The association of mycoprotein-based food consumption with diet quality, energy intake and non-communicable diseases' risk in the UK adult population using the National Diet and Nutrition Survey (NDNS) years 2008/2009–2016/2017: a cross-

sectional study. *British Journal of Nutrition* 2021:1–10. https://doi.org/10.1017/s000711452100218x.

13. Järviö N, Maljanen NL, Kobayashi Y, Ryynänen T, Tuomisto HL. An attributional life cycle assessment of microbial protein production: a case study on using hydrogen-oxidizing bacteria. *Science of the Total Environment* 2021:776. https://doi.org/10.1016/j.scitotenv.2021.145764.
14. Spectre M. *Test tube burgers*. The New Yorker; May 2011. www.newyorker.com/magazine/2011/05/23/test-tube-burgers.
15. O'Riordan K, Fotopoulou A, Stephens N. The first bite: imaginaries, promotional publics and the laboratory grown burger. *Public Understanding of Science* 2017;**26**(2):148–63. https://doi.org/10.1177/0963662516639001.
16. O'Neill EN, Cosenza ZA, Baar K, Block DE. Considerations for the development of cost-effective cell culture media for cultivated meat production. *Comprehensive Reviews in Food Science and Food Safety* 2021;**20**(1):686–709. https://doi.org/10.1111/1541-4337.12678.
17. Hanga MP, Ali J, Moutsatsou P, et al. Bioprocess development for scalable production of cultivated meat. *Biotechnology and Bioengineering* 2020;**117**(10):3029–39. https://doi.org/10.1002/bit.27469.
18. Yao T, Asayama Y. Animal-cell culture media: history, characteristics, and current issues. *Reproductive Medicine and Biology* 2017;**16**(2):99–117. https://doi.org/10.1002/rmb2.12024.
19. Kawai M, Okiyama A, Ueda Y. Taste enhancements between various amino acids and IMP. *Chemical Senses* 2002;**27**(8):739–45. https://doi.org/10.1093/chemse/27.8.739.
20. Bentzinger CF, Von Maltzahn J, Rudnicki MA. Extrinsic regulation of satellite cell specification. *Stem Cell Research and Therapy* 2010;**1**(3). https://doi.org/10.1186/scrt27.
21. Girón-Calle J, Vioque J, Pedroche J, et al. Chickpea protein hydrolysate as a substitute for serum in cell culture. *Cytotechnology* 2008;**57**(3):263–72. https://doi.org/10.1007/s10616-008-9170-z.
22. Lee KY, Loh HX, Wan ACA. Systems for muscle cell differentiation: from bioengineering to future food. *Micromachines* 2022;**13**(1). https://doi.org/10.3390/mi13010071.
23. Bhat ZF, Kumar S, Bhat HF. In vitro meat: a future animal-free harvest. *Critical Reviews in Food Science and Nutrition* 2017;**57**(4):782–9. https://doi.org/10.1080/10408398.2014.924899.
24. Li Z, Zhou Y, Li T, Zhang J, Tian H. Stimuli-responsive hydrogels: fabrication and biomedical applications. *View* 2021;**3**. 20200112.
25. Post MJ, Rahimi N, Caolo V. Update on vascularization in tissue engineering. *Regenerative Medicine* 2013;**8**(6):759–70. https://doi.org/10.2217/rme.13.74.
26. Kumar P, Sharma N, Sharma S, et al. In-vitro meat: a promising solution for sustainability of meat sector. *Journal of Animal Science and Technology* 2021;**63**(4):693–724. https://doi.org/10.5187/jast.2021.e85.
27. van der Valk J, Bieback K, Buta C, et al. Fetal bovine serum (FBS): past - present - future. *ALTEX* 2018;**35**(1):99–118. https://doi.org/10.14573/altex.1705101.
28. Bueno-Díaz C, Biserni C, Martín-Pedraza L, et al. Seed storage proteins, 2S albumin and 11S globulin, associated to severe allergic reactions after flaxseed intake. *Journal of Investigational Allergology and Clinical Immunology* 2021;**32**(5). https://doi.org/10.18176/jiaci.0713.
29. Seah JSH, Singh S, Tan LP, Choudhury D. Scaffolds for the manufacture of cultured meat. *Critical Reviews in Biotechnology* 2022;**42**(2):311–23. https://doi.org/10.1080/07388551.2021.1931803.

30. Renuka N, Guldhe A, Prasanna R, Singh P, Bux F. Microalgae as multi-functional options in modern agriculture: current trends, prospects and challenges. *Biotechnology Advances* 2018;**36**(4):1255–73. https://doi.org/10.1016/j.biotechadv.2018.04.004.
31. Ścieszka S, Klewicka E. Algae in food: a general review. *Critical Reviews in Food Science and Nutrition* 2019;**59**(21):3538–47. https://doi.org/10.1080/10408398.2018.1496319.
32. Grosshagauer S, Kraemer K, Somoza V. The true value of Spirulina. *Journal of Agricultural and Food Chemistry* 2020;**68**(14):4109–15. https://doi.org/10.1021/acs.jafc.9b08251.
33. Poma G, Cuykx M, Amato E, Calaprice C, Focant JF, Covaci A. Evaluation of hazardous chemicals in edible insects and insect-based food intended for human consumption. *Food and Chemical Toxicology* 2017;**100**:70–9. https://doi.org/10.1016/j.fct.2016.12.006.
34. Barennes H, Phimmasane M, Rajaonarivo C, Schooling CM. Insect consumption to address undernutrition, a national survey on the prevalence of insect consumption among adults and vendors in Laos. *PLOS ONE* 2015;**10**(8):e0136458. https://doi.org/10.1371/journal.pone.0136458.
35. Myers G, Pettigrew S. A qualitative exploration of the factors underlying seniors' receptiveness to entomophagy. *Food Research International* 2018;**103**:163–9. https://doi.org/10.1016/j.foodres.2017.10.032.
36. Hall F, Johnson PE, Liceaga A. Effect of enzymatic hydrolysis on bioactive properties and allergenicity of cricket (Gryllodes sigillatus) protein. *Food Chemistry* 2018;**262**:39–47. https://doi.org/10.1016/j.foodchem.2018.04.058.
37. de Castro RJS, Ohara A, Aguilar JG, Domingues MAF. Nutritional, functional and biological properties of insect proteins: processes for obtaining, consumption and future challenges. *Trends in Food Science and Technology* 2018;**76**:82–9. https://doi.org/10.1016/j.tifs.2018.04.006.
38. Lusk JL, Norwood FB. Some economic benefits and costs of vegetarianism. *Agricultural and Resource Economics Review* 2009;**38**(2):109–24. https://doi.org/10.1017/s1068280500003142.
39. Rubio NR, Xiang N, Kaplan DL. Plant-based and cell-based approaches to meat production. *Nature Communications* 2020;**11**(1). https://doi.org/10.1038/s41467-020-20061-y.
40. Verstraete W, Yanuka-Golub K, Driesen N, De Vrieze J. Engineering microbial technologies for environmental sustainability: choices to make. *Microbial Biotechnology* 2022;**15**(1): 215–27. https://doi.org/10.1111/1751-7915.13986.
41. Parodi A, Leip A, De Boer IJM, et al. The potential of future foods for sustainable and healthy diets. *Nature Sustainability* 2018;**1**(12):782–9. https://doi.org/10.1038/s41893-018-0189-7.
42. Pikaar I, Matassa S, Bodirsky BL, et al. Decoupling livestock from land use through industrial feed production pathways. *Environmental Science and Technology* 2018;**52**(13):7351–9. https://doi.org/10.1021/acs.est.8b00216.
43. Tuomisto HL. The eco-friendly burger: could cultured meat improve the environmental sustainability of meat products? *EMBO Reports* 2019;**20**(1). https://doi.org/10.15252/embr.201847395.
44. Stephens N, Sexton AE, Driessen C. Making sense of making meat: key moments in the first 20 Years of tissue engineering muscle to make food. *Frontiers in Sustainable Food Systems* 2019;**3**. https://doi.org/10.3389/fsufs.2019.00045.
45. Bekker GA, Tobi H, Fischer ARH. Meet meat: an explorative study on meat and cultured meat as seen by Chinese, Ethiopians and Dutch. *Appetite* 2017;**114**:82–92. https://doi.org/10.1016/j.appet.2017.03.009.
46. Paarlberg R. *Starved for Science: How Biotechnology is Being Kept out of Africa*. Harvard University Press; 2009.

47. Jarosz L. Defining world hunger: scale and neoliberal ideology in international food security policy discourse. *Food, Culture and Society* 2011;**14**(1):117–39. https://doi.org/10.2752/175174411X12810842291308.
48. Tilman D, Clark M. Global diets link environmental sustainability and human health. *Nature* 2014;**515**(7528):518–22. https://doi.org/10.1038/nature13959.
49. He J, Evans NM, Liu H, Shao S. A review of research on plant-based meat alternatives: driving forces, history, manufacturing, and consumer attitudes. *Comprehensive Review in Food Science and Food Safety* 2020;**19**(5):2639–56. https://doi.org/10.1111/1541-4337.12610.
50. Schösler H, de Boer J, Boersema JJ. Can we cut out the meat of the dish? Constructing consumer-oriented pathways towards meat substitution. *Appetite* 2012;**58**(1):39–47. https://doi.org/10.1016/j.appet.2011.09.009.
51. Slade P. If you build it, will they eat it? Consumer preferences for plant-based and cultured meat burgers. *Appetite* 2018;**125**:428–37. https://doi.org/10.1016/j.appet.2018.02.030.
52. Verbeke W, Sans P, Van Loo EJ. Challenges and prospects for consumer acceptance of cultured meat. *Journal of Integrative Agriculture* 2015;**14**(2):285–94. https://doi.org/10.1016/S2095-3119(14)60884-4.
53. Hoek AC, Luning PA, Weijzen P, Engels W, Kok FJ, de Graaf C. Replacement of meat by meat substitutes. A survey on person- and product-related factors in consumer acceptance. *Appetite* 2011;**56**(3):662–73. https://doi.org/10.1016/j.appet.2011.02.001.
54. de Boer J, Schösler H, Aiking H. Meatless days" or "less but better"? Exploring strategies to adapt Western meat consumption to health and sustainability challenges. *Appetite* 2014;**76**:120–8. https://doi.org/10.1016/j.appet.2014.02.002.
55. Statista.com. 2022. www.statista.com/topics/4880/global-meat-industry/Obtained Alrim3.
56. Ben-Shaul S, Landau S, Merdler U, Levenberg S. Mature vessel networks in engineered tissue promote graft-host anastomosis and prevent graft thrombosis. *The Proceeding of the National Academy of Science United States of America* 2019;**116**(8):2955–60. https://doi.org/10.1073/pnas.1814238116.
57. Humbird D. Scale-up economics for cultured meat. *Biotechnology and Bioengineering* 2021;**118**(8):3239–50. https://doi.org/10.1002/bit.27848.

Chapter 10

# Recent developments in agricultural biotechnology

## 1. Introduction

Since ancient times, humans have employed conventional methods to modify animals and crops to suit their dietary and medicinal needs, these methods being mutation breeding, selective breeding, and cross-breeding. The reasons for genetically engineering plants are for the achievement of augmented crop yields, minimized crop loss, better storage life, aesthetically favorable flavor and appearance, enhanced nutrition, or a combination of these characteristics. Besides, there is growing global demand for medicinal, food, and specialty products in addition to the perils posed by global warming and other environmental changes. Agricultural biotechnology offers a propitious strategy to meet these challenges. The recent emergence of GM and GE crops and animal foods for consumption and pharmaceutical use is enhancing food security and human health. Nevertheless, sociocultural and ethical concerns have to be factored in, to establish widespread public acceptance and uptake. To be viable, we need to formulate solutions that are socially and ethically responsible, suitable to denizens of various socio-cultural backgrounds and to convey information about genetically engineered crops in a forthright and convincing manner. The current chapter addresses recent developments in phytoremediation, gene drive management, biofortified crops, genome-edited crops, plant-based biopharmaceuticals, engineering of microbes for soil and planet health as well as strategies for addressing climate change: novel ways to generate food proteins using biotechnology to reduce greenhouse gas emissions. For the Chapter 8 on "engineering of crops resistant to biotic stresses," recent advancements have been included within the chapter. Recent progress in other topics of agricultural biotechnology will be presented in volume 2 of this book (please refer to the Introduction section of this book for the topics to be addressed in volume 2). In all, this book reveals new insights into the field of agricultural biotechnology that is rapidly advancing with many novel and exciting developments, particularly in this era of genome editing and tailor-made crops.

Agricultural Biotechnology. https://doi.org/10.1016/B978-0-323-88476-1.00003-5

## 2. Phytoremediation

The plant, *Lolium perenne* L. has elevated potential for the phytoremediation of Hg- and Cd-contaminated soils in tropical regions. Mycorrhizae inoculation modulates the phytoremediation mechanisms of *L. perenne* and influences the composition of microbes in the rhizosphere of the plant. These changes are associated with augmented tolerance of the plant to Hg and Cd, and therefore, phytoremediation enabled by microorganisms is an attractive option to improve soil recovery from these heavy metals.[1]

Liu et al., 2023[2] demonstrated that polyaspartic acid (PASP) application significantly augmented the *Pennisetum sinese* shoot biomass as well as the Cd extraction capacity and enabled the maintenance of soil health by improving the structure and composition of the bacterial communities in the rhizosphere. The resulting enhanced plenitude of microbes such as *Stenotrophomonas, Enterobacter, Pseudomonas,* and *Acetobacter* promote the detoxification of heavy metals in the soil in addition to augmenting plant growth as well as phytoremediation. PASP, being eco-friendly and inexpensive, is therefore an inimitable candidate for the improvement of phytoextraction efficiency and for promoting microbial communities in the soil to enable phytoremediation.

Kösesakal and Seyhan, 2023,[3] conducted a pioneering study of the potential of *Azolla filiculoides*, a freshwater fern, to phytoremediate phenanthrene in contaminated water. *A. filiculoides* was shown to efficiently phytoremediate low concentrations of phenanthrene within a brief time period. About 60% efficiency for biodegradation of this pollutant was observed, making this plant a favorable candidate for phytoremediation of phenanthrene in aquatic environments. The phytoremediation capabilities of the plants *Pongamia pinnata* and *Jatropha curcas* were investigated[4] wherein nickel was shown to be removed with 93%–90% and 82%–86% efficiency, respectively, from contaminated soils in municipal landfills and paper mills within 4 months. The efficiency of zinc removal was significantly lower compared to that of nickel as a majority of the zinc remained in the roots and soil. Additionally, augmented phytostabilization efficacy of the roots and the rhizosphere region was reported as against phytoaccumulation.

*Juncus effuses* and *Typha angustifolia* plants demonstrated high tolerance to micropollutants such as organic contaminants (propranolol) derived from pharmaceuticals and personal hygiene products.[5] Besides their efficiency in removing micropollutants, these plants could withstand low temperatures while being recalcitrant to micropollutant toxicity. Therefore, these plants can be used for phytoremediating constructed wetlands and wastewater containing these contaminants. Singh et al., 2023,[6] showed that the water hyacinth, *Eichhornia crassipes* (Mart.) Solms, can be used to remediate phosphorous and nitrogen loads of glass industry effluents and decrease environmental hazards posed by their unsafe disposal.

The contaminant, ciprofloxacin, is a global health issue endangering aquatic ecosystems. The efficiency of aquatic macrophytes including the submerged *Egeria densa* Planch and the floating *Salvinia molesta* D.S. Mitchell to phytoremediate ciprofloxacin was explored by Kitamura et al., 2023.[7] These plants showed high tolerance to ciprofloxacin and over 58% removal efficiency. Additionally, *E. densa* was shown to metabolize the chemical, while *S. molesta* bioaccumulated it within its tissues.

## 3. Gene drive management

Gene drives generated by genetic engineering technologies (geGD) are potentially robust tools for the suppression or even eradication of insect pest populations. Nevertheless, prior to the release of geGD insects into the environment, they must go through environmental risk evaluation to ensure that there is no unintended harm to nontarget species in the environment. The major concern is that these nontarget entities could acquire the active GD elements that could lead to decreased abundance or even complete elimination of those species resulting in detrimental effects on the ecosystem. The major channel for gene flow is via hybridization between the insect strain carrying the drive element and its closely related insect species that likely occur in the same area where the geGD strain is released followed by its potential dispersal.

Wolf et al., 2023[8] performed a case study using *Drosophila suzukii* wherein they demonstrate a broadly applicable method of how a combo of experiments involving interspecific hybridization, along with behavioral observations as well as molecular genetic investigations can be employed to evaluate the likelihood for interspecific hybridization. In their hybridization experiments, crosses between *D. suzukii* and the taxonomically closely occurring Drosophila species in Europe namely, *D. melanogaster, D. simulans,* and *D. ananas-sae*, did not result in any hybrid offspring. Nor did any cross between *D. suzukii* and *D. biarmipes*, the latter being a member of the *D. suzukii* subgroup prevailing in Southeast Asia and alien to Europe. In this study, viable hybrid offspring were produced from interspecific crosses between *D. melanogaster* and *D. simulans*. Upon crossing of the female *D. melanogaster* with the male *D. simulans*, only hybrid female offspring were generated. Notwithstanding, when the females of *D. simulans* were mated with male *D. melanogaster*, hybrid off-spring belonging to both sexes were produced, but with a male bias. These findings are in accordance with earlier reports confirming that the extent of mating success between any two species is variable between strains.

Meiborg et al., 2023[9] constructed a model matching the life history of Asian hornet, *Vespa velutina* nigrithorax, with the effect of various gene drive schemes on an established invasive species. In order to examine the wider pertinency and sensitivity of this model, they also incorporated *Polistes dominula*, the European paper wasp in this study. From this, they observed that the only gene drive that holds promise in population control is the one targeting the fertility of females accounted for by the haplodiploidy of social hymenopterans.

This investigation shows that despite the ability of a gene drive to suppress a population of social wasps, it is able to do so only under stringent conditions specific to the gene drive. This is because a majority of surviving offspring produced by the social wasp colonies make it likely that, even with very minimal emergence of resistance alleles, these alleles can rapidly disseminate and rescue the respective population. Also, due to the life history of social wasps, infertile individuals of this species do not compete with fertile equivalents, thus allowing fertile members to maintain a substantial population size despite conditions under which these drive alleles are widely disseminated. Notwithstanding, continued innovations in gene drive technology could make it a favorable method to control invasive social insects in the imminent future.

Invasive rodents incur widespread damage to the environment and result in loss of biodiversity, especially on islands. Unlike that of insects, strategies for genetic biocontrol including gene drives aimed at population suppression with biased genetic inheritance have not yet been established in mice. Gierus et al., 2023,[10] established a gene drive strategy ($t_{CRISPR}$) that leverages the *t* haplotype super-Mendelian transmission to spread inactivating gene mutations in a haplosufficient female fertility gene (*Prl*). By means of in silico modeling of spatially explicit individuals, it was demonstrated that $t_{CRISPR}$ could eliminate island populations within a range of practicable field-based criteria. This study also engineered transgenic $t_{CRISPR}$ mice that importantly exhibited the biased transmission of the altered *t* haplotype as well as *Prl* mutations at ranges their modeling predicted would be enough for elimination of this species. Therefore, this study demonstrates the feasibility of using a gene drive system for controlling the populations of invasive alien rodents.

Homing gene drives based on CRISPR could enable the curtailment of vector-borne diseases by control of invasive species and crop pests on account of an inheritance rate surpassing Mendelian laws. Nevertheless, this technology has drawbacks such as the emergence of resistance alleles generated during the error-prone repair of the drive-elicited DNA break which causes mutations that disrupt the recognition sequence of the gRNA and precludes the further propagation of the gene drive. Bishop et al., 2022,[11] performed a study to circumvent this by encoding more gRNAs targeting the most commonly occurring resistance alleles into this gene drive, thus allowing another attempt at gene drive conversion. Using the recycling of resistance alleles, they demonstrated a double-tap genetic strategy to augment the efficacy of the gene drive. This double-tap gene drive effectively disseminates in caged populations surpassing the control drive and therefore can be readily used in any gene drive based on CRISPR to augment the drive performance. Additionally, similar strategies could be adopted for other systems that suffer from diminished HDR frequencies such as mouse germlines or mammalian cell transformations.

## 4. Biofortified crops

In soybean roots, a combination of low Fe and high Zn was found to highly activate the expression of *FIT1* (*Fe-deficiency-induced transcription factor 1*) and *FRO2* (*FERRIC REDUCTASE OXIDASE 2*) genes.[12] Interaction between Fe and Zn controls the antioxidant enzyme activities and uptake, accretion, and homeostasis of additional essential micronutrients such as manganese and copper in these plants. Zn uptake rates and Zn content were greatly diminished with increased supply of Fe. Nevertheless, accumulation of Fe was not greatly influenced by increasing Zn supplies.

The ZRT, IRT-like protein (ZIP) family is associated with the uptake and transport of several important metal ions in capsicum plants. Ahmad et al., 2023[13] report the identification of the *Capsicum annuum, C. baccatum,* and *C. Chinese* ZIP transporter genes and analysis of the ZIP gene expression under Fe and Zn stress. Under these conditions, root morphology was altered along with differential accretion of many metabolites such as amino acids, fatty acids, carboxylic acids, and sugars in the roots and leaves of these plants. Additionally, important metabolites such as 2-ketoglutaric acids, L-aspartic acid, aucubin, β-L-fucopyranose, chlorogenic acid, and quininic acid were highly expressed in leaves and roots under conditions of deprived Zn/Fe. Also, the expression of genes including CaZIP4-5, CaZIP14, and CaZIP7-9 was induced in leaves and roots. When Fe was absent, CaZIP4-5 and CaZIP8 were upregulated along with enhanced uptake of minerals such as P, S, Mg, Zn, and Cu in the roots, which suggested the potential role of these ZIP genes in influencing metal ion uptake in these plants. Therefore, these genes can enable further investigations of mineral uptake as well as their biofortification to augment the nutritional values of these commercially important plants.

*Allium* plants are deemed as a "natural dietary supplement" as they are a rich source of several compounds having beneficial and healing effects. Biofortification of the edible parts of these plants with microelements is considered as an effectual agrotechnical measure. The results of the foliar application of ($Na_2SeO_4$) Se fertilizer in different doses on the levels of micro- and macro-elements as well as secondary metabolites such as flavonoids, phenolics, and hydroxycinnamic acid were investigated in the edible parts of *A. schoenoprasum* and *A. odorum.*[14] The biofortification of Se was observed along with the accumulation of P, K, and S in the *Allium* leaves. Plants treated with Se contained higher secondary metabolite contents without toxic metal accumulation rendering these plants safe for human consumption.

Harrington et al., 2023,[15] demonstrated a transgenic strategy to combine the endosperm-specific expression of the *vacuolar iron transporter* gene *TaVIT2-D* in wheat with the constitutively expressed *Oryza sativa nicotianamine synthase* gene *OsNAS2,* which led to increased zinc concentration and relocation of iron in the grain and caused a 3-fold augmentation of iron in roller-milled, highly pure white flour. Intriguingly, *OsNAS2* partly restored the translocation of iron to the aleurone, that is depleted of iron in grains that overexpress the *TaVIT* alone.

Dalla Vecchia et al., 2023,[16] showed that the plants *Stanleya pinnata* and *Brassica juncea* possess transport mechanisms having differential affinities for the metals, selenate, and chromate. Selenate administration diminished the accumulation of chromate in *B. juncea,* while in *S. pinnata,* such an effect was found only in the roots. *B. juncea* was capable of storing more selenium and chromium compared to *S. pinnata* because of its higher biomass but stored lower amounts of selenium upon treatment with both elements. Both metals reduced the accumulation of sulfur in *S. pinnata* and *B. juncea*, the latter being more sensitive to the toxicity of these metals. Therefore, both species can used for the phytoextraction of selenium and chromium since these elements are less available in the soil compared to hydroponics. In soils rich in both elements, or with lower selenium content, *S. pinnata* is favored for the phytoextraction of selenium, and the biomass can be employed for biofortification of crops because of its minor chromium content.

Tsakirpaloglou et al., 2023,[17] developed biofortified rice having high densities of Zn and Fe in polished grains. These plants specifically expressed the rice *nicotianamine synthase 2* (*OsNAS2*) gene and dicot ferritins. This rice variety therefore has potential to decrease the prevalence of insufficient intake of Zn in women of childbearing age in Philippines and Bangladesh by 50% and 30%, respectively, when compared to the present prevalence rates.

Rice varieties biofortified for Zn could be valuable for Zn intake in populations that are deficient, but processing of this rice into parboiled rice could compromise this potential advantage. Parboiling conditions employed to produce milled rice that is targeted for consumption in own households resulted in higher Fe and Zn compared to that used for milled rice meant for markets.[18] More Zn content moved from the inner endosperm into the outer layers at elevated soaking temperatures that resulted in lower true retention of Zn for the milled parboiled rice that was soaked in hotter water. On the other hand, parboiled rice that was soaked at temperatures commonly used in households could enable the provision of more Zn into diets when compared to that soaked in water at higher temperatures used in large rice mills.

## 5. Genome-edited crops

Transformation of genes into grass species is circumscribed by the use of only immature embryos in these Poaceae crops. Other explants for transformation such as the leaf tissues have met with limited success, and this hinders the broad applicability of genome editing for enhancement of these crops. The morphogenic genes Babyboom (*Bbm*) and Wuschel2 (*Wus2*) were introduced into these crops via leaf transformation and Cas9-based mutagenesis; however, the complexity of genome editing applications that require screening of vast numbers of regenerated plants makes this elusive. Wang et al., 2023,[19] show that augmented *Wus2/Bbm* expression driven by strong promoters greatly enhances leaf transformation in sorghum and maize, enabling the retrieval of plants having Cas9-mediated gene dropouts as well as targeted gene insertion. Furthermore, using the *Wus2/Bbm* construct optimized for maize, it was demonstrated that embryonic callus and regenerated plantlets were generated successfully in eight species of Poaceae. This indicates that this technology could be used as a universal family-wide strategy for genome editing and transformation across the Poaceae.

Following the CRISPR-associated Cas9 and Cas12a systems, the Cas12b is the third more efficacious CRIPSR tool available for targeted genome editing of plants. Gurel et al., 2023,[20] generated sgRNA complexes having various RNA scaffolds to augment editing frequency in rice plants, wherein the *GRAIN SIZE 3 (OsGS3)* and *EPIDERMAL PATTERNING FACTOR LIKE 9 (OsEPFL9)* genes were respectively targeted with ATTC and GTTG protospacer adjacent motifs. Among the scaffolds used, two of the *Alicyclobacillus acidoterrestris* scaffolds (Aac and Aa1.2) greatly enhanced the targeted mutagenesis frequency. Subsequently, the stably transformed $T_0$ rice plants were subjected to whole genome sequencing (WGS) to examine any off-target mutations. Results of WGS showed that background mutations were generated in both noncoding and coding regions of the edited genome with no indication of any off-target effects. Also, Mendelian segregation of deletion and insertion mutations was observed in the $T_1$ generation indicating germline transmissibility, thereby proving that both Aac and Aa1.2 scaffolds enabled precise, heritable genome editing in rice.

Fruit color is a valuable horticultural attribute which greatly influences consumer preferences. Yang e al., 2022,[21] put forth a rapid breeding program to develop tomato lines having different colored fruits other than the red-colored ones using CRISPR/Cas9-based multiplex gene editing targeting three fruit color-associated genes (*PSY1*, *MYB12*, and *SGR1*). The "Ailsa Craig" red-fruited cultivar has been genetically engineered to obtain tomato genotypes having various fruit colors inclusive of pink, pink-brown, light-green, yellow-green, yellow, and light-yellow. When compared to conventional crossbreeding, this technology requires much less time and yields transgene-free plants having different fruit colors within a span of less than a year. This has been achieved without any impacts on other agronomically important traits such as fruit quality and yield and, therefore, holds great promise for tomato breeding and opens channels to likewise improve other horticultural crops having multigene-controlled traits.

Mutations causing loss-of-function in the susceptibility (S) gene, *Mildew resistance locus O* (*MLO*), bestow broad-spectrum, durable resistance to powdery mildew in several plant species. Nevertheless, *mlo*-associated resistance impacts plant growth and yield, thus circumscribing its widespread application in agricultural practices. Li et al., 2022,[22] describe a mutant, Tamlo-R32, of wheat having a targeted 304-kilobase pair deletion within the *MLO-B1* locus that preserves crop growth and yield while at the same time conferring resistance to powdery mildew. Specifically, this deletion alters the local chromatin landscape, resulting in ectopic induction of *the Tonoplast monosaccharide transporter 3* (*TaTMT3B*), which ameliorates yield and growth penalties associated with the disruption of *MLO* gene. Thus, precision genome editing accelerates the introduction of the *mlo* resistance allele into elite varieties of wheat. This work provides proof-of-concept for stacking genetic alterations to rescue defects in growth caused by recessive alleles, which is important in the development of crop varieties that are high-yielding in addition to having durable and strong disease resistance.

In the past two decades, use of transgenic technology and genome editing has greatly enabled the improvement of cotton crops. Nevertheless, genotype dependence is a major bottleneck in engineering gene-edited and transgenic cotton plants through *Agrobacterium*-mediated transformation or through particle bombardment. Ge et al., 2022,[23] generated a transformation system mediated by the shoot apical meristem cells (SAM) that facilitated the transformation of resistant cotton genotypes including the widely cultivated upland cotton (*Gossypium hirsutum*), Asiatic cotton (*Gossypium arboreum*), and the Sea island cotton (*Gossypium barbadense*). Through this SAM transformation system, they incorporated two foreign genes, RUBY and GFP into the genome of SAM cells belonging to some of the recalcitrant genotypes of cotton. Within a span of 2–3 months, adventitious roots transgenic for both genes were developed from the axillary meristem region which could be retrieved and grown into whole plants. Fluorescent signals produced by the GFP and accumulation of betalain (RUBY-produced bright red molecule) were observed in several tissues of the GFP- and RUBY-containing pants as well as their progenies, showing stable integration of the transgenes followed by transmission to the next generation. Additionally, by means of the SAM transformation system, genome-edited cotton plants were successfully generated containing targeted, inheritable mutations in the *GhRCD1* and *GhPGF* genes mediated by the CRISPR/Cas9 system. In all, this study proved that the SAM transformation system is capable of bypassing the embryogenesis process at the time of tissue culture in a traditional transformation procedure and greatly expedites the generation of gene-edited and transgenic plants for the genetic enhancement of recalcitrant cotton varieties.

CRISPR nucleases such as Cas9 and Cas12a have revolutionized crop breeding, the main advantage of this being their capability to perform multiplexed genome editing. Notwithstanding, it suffers from the drawback of inducing off-target effects during which concurrent multiple DNA double-strand breaks (DSBs) are generated in the crop genome. Zhang et al., 2022,[24] used whole genome sequencing to successfully demonstrate the specificity of the MbCas12a-based genome editing and the promiscuity of the protospacer adjacent motif of LbCas12a in rice (*Oryza sativa*). From this study, they could identify that rice plants tolerating several simultaneous DSBs contained larger chromosomal rearrangements when compared to plants which withstood lower order DSBs. This shed valuable insight into the regulation of genetically engineered crops derived from multiplexed genome editing mediated by the CRISPR/Cas system.

## 6. Plant-based biopharmaceuticals

Plant-derived natural products offer a good source of remedies against several pathogenic microbes. Yirgu et al., 2023,[25] report the anti-inflammatory and antimicrobial potential of aqueous extracts of *Viscum tuberculatum* against microbes such as *Staphylococcus aureus, Pseudomonas aeruginosa,* and *Escherichia coli,* as demonstrated in mouse models. Saleem et al., 2023[26] report that *Alternanthera pungens*-based copper nanoparticles (CuNPs) conjugated to the antibiotic gentamycin augment the antibacterial capability of the drug in bioassays conducted against *Salmonella typhimurium* and *Klebsiella pneumoniae.*

Highly pathogenic avian influenza viruses (HPAIV) are responsible for many severe influenza outbreaks worldwide. For protection of poultry farms and to preclude the potential spread of new influenza-generated pandemics, there is a compelling need for effective and low-cost vaccines. Phan et al., 2023,[27] generated large, stable hemagglutinin H5 oligomers in plants in combination with S-Tag-S-Protein interaction. Crude extracts elicited robust humoral immune responses and potent neutralizing antibody reactions and additionally conferred resistance in chickens following challenge with the wild-type strain, HPAIV H5. Crude extracts of plants expressing H5 oligomers generated better immune responses when compared to those expressing trimers. Neutralizing antibodies elicited by two-dose and single-dose immunization with the adjuvanted H5 oligomer crude extract afforded protection against two lethal H5N1 virus strains in vaccinated chickens with efficacies of 92% and 100%, respectively. Therefore, this H5 oligomer crude extract vaccine is comparable in performance to the currently available commercial vaccines. The low-cost and increased stability following long-term storage makes this plant-based vaccine very promising for further use of this technology in the development of veterinary vaccines.

Anupong et al., 2023,[28] showed the potential for antioxidant, antifungal, and antibacterial activities of *Coleus aromaticus*-based Titanium dioxide nanoparticles ($TiO_2$ NPs). These $TiO_2$ NPs exhibited concentration-dependent activities against pathogenic bacteria such as *Proteus mirabilis, Escherichia coli, Salmonella typhimurium, Staphylococcus aureus, Vibrio cholerae,* and *Pseudomonas aeruginosa* as well as antifungal activities against *Aspergillus flavus, A. nidulans, A. fumigatus, A. terreus,* and *A. niger.* At 100 $\mu g\ mL^{-1}$ dosage, these TiO2 NPs could inhibit the enzyme activities of α-glucosidase (37%) and α-amylase (17%). The typical 2,2-diphenyl-1-picrylhydrazyl (DPPH) free radical scavenging assay showed that these $TiO_2$ NPs could efficiently scavenge 89% of DPPH free radicals at 100 $\mu g\ mL^{-1}$ concentration, thus demonstrating its antioxidant activity. Therefore, these plant-based $TiO_2$ NPs could be considered for further investigations to explore their biomedical properties.

Mosquito repellents composed of essential oils from *Citrus reticulata* (leaves), *C. reticulata* (fruit peels), *Dysphania ambrosioides, Mentha longifolia,* and *Perovskia atriplicifolia* were shown to exhibit repellent activities against the mosquito, *Aedes aegypti* in bioassays.[29] The *Perovskia atriplicifolia* EO demonstrated the longest repellent effect, followed by that of *C. reticulata* (fruit peels), *M. longifolia,* and *C. reticulata* (leaves). Importantly, the *P. atriplicifolia* EO afforded complete protection for a time span of 60 min at 330 μg/cm$^2$ which was comparable to that of the commercial DEET mosquito repellent. These EOs upon GC-MS analyses showed that they contained camphor (19.7%), sabinene (24.9%), limonene (92.7%), *trans*-ascaridole (38.8%), and carvone (82.6%), as the major constituents in *P. atriplicifolia*, *C. reticulata* (leaves), *C. reticulata* (peels), *D. ambrosioides*, and *M. longifolia,* respectively. The findings of this study could aid the development of plant-based mosquito repellents on a commercial scale for protection of humans from mosquitoes transmitting deadly diseases such as dengue, malaria, and other diseases.

Malaria is estimated to kill about 409,000 of the human population in a year. Vaccines capable of blocking malarial transmission to reduce disease prevalence in the community could have the potential to develop a multifaceted strategy toward eliminating the malarial parasite responsible for causing this disease. The Pfs25 antigen, a lead molecule in blocking malarial transmission, has been generated successfully in plants both as a VLP and a subunit vaccine. Tottey et al., 2023,[30] developed an improvised version of the VLP-based antigen display molecule by removal of known protease sites from the initial A85 variant. This B29 re-engineered molecule exhibits 3× higher number of the Pfs25 antigen molecules per VLP when compared to that of the initial Pfs25 VLP. Additionally, an enhanced purification procedure was also developed that showed higher VLP yield and purity. Studies carried out in a mouse model demonstrated improved activity against parasite transmission at lower doses and longer time periods compared to that of the original molecule.

Medicago, a Canadian company, produced a SARS-CoV-2 plant-based VLP vaccine which has successfully gone through clinical trials and has been approved for use in Canada. Hager et al., 2022 (Medicago),[31] reported the production of coronavirus-like particles (CoVLPs) in plants, wherein the CoVLPs displayed the prefusion spike glycoprotein of SARS-CoV-2, and when this was combined with the AS03 adjuvant, it generated a candidate SARS-CoV-2 vaccine. This CoVLP + AS03 vaccine precluded COVID-19 infection caused by a gamut of variants, with effectiveness ranging from as much as 69.5% against symptomatic disease to 78.8% against moderate-to-severe infection.

To confront the threat of bioterrorism associated with the probable use of *Bacillus anthracis*, an effectual and safe vaccine is necessary that can be used immediately to enable mass vaccination. In this regard, the protective antigen (PA), a major component of the *B. anthracis* virulence factors edema toxin as well as lethal toxin, has been well-studied. Earlier, the full-length PA83 protein was expressed transiently in a plant-based expression system. Administration of PA83 antigen along with the alhydrogel adjuvant induced robust neutralizing immune reactions in rabbits and mice. The rabbits were protected 100% from challenge with the lethal, aerosolized *B. anthracis*. Paolino et al., 2022[32] conducted a phase 1 study evaluating the immunogenicity and safety of this vaccine in healthy adults. Results showed that it was immunogenic and safe with no associated serious adverse outcomes. This is the first report of a successful plant-based subunit anthrax vaccine.

## 7. Engineered microbes for improved soil and planet health

Biomass conversion by microbes into a variety of bioproducts including fuels or chemical building blocks such as acetic acid via fermentation and anaerobic digestion will remain important for promoting planetary health in the near future.[33] The use of agricultural residues will be the most available resource in the world. Efforts must be made to develop microbes which can catalyze these enzymatic reactions with increased productivity and at reduced cost. Other issues to be mitigated will include the high cost of transporting biomass to sites of conversion and the sustainability of producing these feedstocks. The use of plants which photosynthesize and capture $CO_2$ to produce biodegradable plastics and biopolymers is a good start.[34] Microalgae that have been developed to produce bioplastics with an appropriate life cycle assessment is just one example.[35–39]

Furthermore, the optimization of fertilizers use can lower agricultural greenhouse gas emissions substantially, by, for example, using engineering microbes to help crops fix nitrogen.[40] Other strategies include using microbes to utilize $CO_2$, $CH_4$, and other energy feedstocks to produce essential amino acids and thus food proteins.[41,42] The use of microbes to provide these sorts of products in addition to others such as plastics, paints, and other valuable commodities at low carbon emissions will require raising public awareness and approval.[43,44] Hopefully, consumers will appreciate how microbes can utilize biomass and offer a truly sustainable production system that is carbon neutral in nature.

## 8. Strategies to address climate change: novel ways to generate food proteins using biotechnology to reduce greenhouse gas emissions

The alternative protein industry is growing exponentially.[45] For the plant-based food industry, emulsion gels are commonly used. For this, different plant-derived proteins can also be explored, such as quinoa. Quinoa protein possesses a well-balanced amino acid profile and illustrates good emulsifying and gelling capabilities.[46] Differences in composition and structure of plant proteins, in addition to the wide range of processing steps and conditions, can have positive or negative effects on protein digestibility and the presence of antinutritional factors, depending on the crop type. Processing, protein sustainability, bioavailability, and digestibility of plant-based foods are issues that must be kept in mind to improve food system.[47]

Insects also show enormous potential for the food industry and will increase exponentially over the next few years.[48] The main insects that are being explored are silkworms, bees, beetles, mealworms, crickets, and cicadas. The global market for edible insects will reach US 8 billion by 2030.[49] Other sources of alternative proteins, such as fungal and algal proteins, have their own unique advantages.[50] Finally, the use of microbes and synthetic biology will enable food to be developed rapidly and in a way that is more environmentally friendly.[51,52]

## 9. Conclusion

Molecular breeders have been anticipating a watershed moment toward sustainable agriculture, and the advent of CRISPR technologies has duly delivered. Several decades-long investments in discerning plant genetics, mapping of desirable traits and hybrid breeding have fruitfully yielded enormous gains in agricultural productivity and resistance to pathogens and pests. We are presently in a transitioning mode toward greater focus on the quality of crops, sustainability, and plant-based biologicals with plant resilience at the limelight. The peasant spirit of obtaining a lot with comparably little is the ideology of agriculture-focused organizations committed to ensure healthier, safer, and more viable food supply. Foresters and farmers have increasingly embarked on next-generation plant breeding schemes and augmented carbon capture technologies. This pursuit has already attracted high-level investments and tremendous scientific talent. Strategic stakeholders and adventurous investors have started to seek such endeavors.

The crop breeding scenario has transformed from "Green revolution to Gene revolution" and the seeds for genome-edited crops have already been sown. Genome-edited rice is emerging from the laboratory level to the field. Promisingly, the volume of literature on genome editing in crops is continually increasing along with the developments in agriculturally important traits at the greenhouse level. The interest of industries in this area worldwide is estimated to reach significant commercial milestones this year, with the imminent launch of edited vegetables, fruits, and other crops across the world markets. The landscape of agriculture and farming is evolving tremendously, and advances are on-going regarding regulatory approvals and public awareness of genome-edited foods.

Despite some of the negative antiscience apprehension, awareness of the CRISPR technology with the necessity to more sustainably and responsibly regulate our agricultural practices affords inimitable opportunities to focus translating editing at the laboratory level to the farm. We foresee significant advancements in the editing of crops, livestock, and forestry areas during 2023. Given the constantly increasing food demands of an ever-growing population, usage of lands for cultivation and trends in globalization, the evolution of the bioeconomy toward providing worldwide food supply chain has become of prime importance. How well and fast these genome-edited crops will advance remains to be seen.

## References


1. Saldarriaga JF, López JE, Díaz-García L, Montoya-Ruiz C. Changes in Lolium perenne L. rhizosphere microbiome during phytoremediation of Cd- and Hg-contaminated soils. *Environ Sci Pollut Res Int* February 13, 2023. https://doi.org/10.1007/s11356-023-25501-y.
2. Liu Y, Zhou J, Sun D, Chen H, Qin J, Chen G, Qiu R. Polyaspartic acid assisted-phytoremediation of cadmium-contaminated farmland: phytoextraction efficiency, soil quality, and rhizosphere microbial community. *Sci Total Environ* March 1, 2023;**862**:160736. https://doi.org/10.1016/j.scitotenv.2022.160736.
3. Kösesakal T, Seyhan M. Phenanthrene stress response and phytoremediation potential of free-floating fern *Azolla filiculoides* Lam. *Int J Phytoremediation* 2023;**25**(2):207–20. https://doi.org/10.1080/15226514.2022.2069224.
4. Borah P, Rene ER, Rangan L, Mitra S. Phytoremediation of nickel and zinc using Jatropha curcas and Pongamia pinnata from the soils contaminated by municipal solid wastes and paper mill wastes. *Environ Res* February 15, 2023;**219**:115055. https://doi.org/10.1016/j.envres.2022.115055.
5. Lei Y, Carlucci L, Rijnaarts H, Langenhoff A. Phytoremediation of micropollutants by *Phragmites australis*, *Typha angustifolia*, and *Juncus effuses*. *Int J Phytoremediation* 2023;**25**(1):82–8. https://doi.org/10.1080/15226514.2022.2057422.
6. Singh J, Kumar P, Eid EM, Taher MA, El-Morsy MHE, Osman HEM, Al-Bakre DA, Kumar V. Phytoremediation of nitrogen and phosphorus pollutants from glass industry effluent by using water hyacinth (*Eichhornia crassipes* (Mart.) Solms): application of RSM and ANN techniques for experimental optimization. *Environ Sci Pollut Res Int* February 2023;**30**(8):20590–600. https://doi.org/10.1007/s11356-022-23601-9.
7. Kitamura RSA, Brito JCM, Silva de Assis HC, Gomes MP. Physiological responses and phytoremediation capacity of floating and submerged aquatic macrophytes exposed to ciprofloxacin. *Environ Sci Pollut Res Int* January 2023;**30**(1):622–39. https://doi.org/10.1007/s11356-022-22253-z.
8. Wolf S, Collatz J, Enkerli J, Widmer F, Romeis J. Assessing potential hybridization between a hypothetical gene drive-modified Drosophila suzukii and nontarget Drosophila species. *Risk Anal* January 24, 2023. https://doi.org/10.1111/risa.14096.
9. Meiborg AB, Faber NR, Taylor BA, Harpur BA, Gorjanc G. The suppressive potential of a gene drive in populations of invasive social wasps is currently limited. *Sci Rep* January 30, 2023;**13**(1):1640. https://doi.org/10.1038/s41598-023-28867-8.
10. Gierus L, Birand A, Bunting MD, Godahewa GI, Piltz SG, Oh KP, Piaggio AJ, Threadgill DW, Godwin J, Edwards O, Cassey P, Ross JV, Prowse TAA, Thomas PQ. Leveraging a natural murine meiotic drive to suppress invasive populations. *Proc Natl Acad Sci U S A* November 15, 2022;**119**(46). e2213308119. https://doi.org/10.1073/pnas.2213308119.
11. Bishop AL, López Del Amo V, Okamoto EM, Bodai Z, Komor AC, Gantz VM. Double-tap gene drive uses iterative genome targeting to help overcome resistance alleles. *Nat Commun* May 9, 2022;**13**(1):2595. https://doi.org/10.1038/s41467-022-29868-3.
12. de Oliveira NT, Namorato FA, Rao S, de Souza Cardoso AA, de Rezende PM, Guilherme LRG, Liu J, Li L. Iron counteracts zinc-induced toxicity in soybeans. *Plant Physiol Biochem* January 2023;**194**:335–44. https://doi.org/10.1016/j.plaphy.2022.11.024.

13. Ahmad I, Rawoof A, Priyanka IK, Momo J, Anju T, Kumar A, Ramchiary N. Diversity and expression analysis of ZIP transporters and associated metabolites under zinc and iron stress in Capsicum. *Plant Physiol Biochem* February 4, 2023;**196**:415–30. https://doi.org/10.1016/j.plaphy.2023.01.060.
14. Vuković S, Moravčević D, Gvozdanović-Varga J, Dojčinović B, Vujošević A, Pećinar I, Kilibarda S, Kostić AŽ. Elemental profile, general phytochemical composition and bioaccumulation abilities of selected *Allium* species biofortified with selenium under open field conditions. *Plants* January 11, 2023;**12**(2):349. https://doi.org/10.3390/plants12020349.
15. Harrington SA, Connorton JM, Nyangoma NIM, McNelly R, Morgan YML, Aslam MF, Sharp PA, Johnson AAT, Uauy C, Balk J. A two-gene strategy increases iron and zinc concentrations in wheat flour, improving mineral bioaccessibility. *Plant Physiol* January 2, 2023;**191**(1):528–41. https://doi.org/10.1093/plphys/kiac499.
16. Dalla Vecchia F, Nardi S, Santoro V, Pilon-Smits E, Schiavon M. Brassica juncea and the Se-hyperaccumulator Stanleya pinnata exhibit a different pattern of chromium and selenium accumulation and distribution while activating distinct oxidative stress-response signatures. *Environ Pollut* March 1, 2023;**320**:121048. https://doi.org/10.1016/j.envpol.2023.
17. Tsakirpaloglou N, Bueno-Mota GM, Soriano JC, Arcillas E, Arines FM, Yu SM, Stangoulis J, Trijatmiko KR, Reinke R, Tohme J, Bouis H, Slamet-Loedin IH. Proof of concept and early development stage of market-oriented high iron and zinc rice expressing dicot ferritin and rice nicotianamine synthase genes. *Sci Rep* January 12, 2023;**13**(1):676. https://doi.org/10.1038/s41598-022-26854-z.
18. Taleon V, Hasan MZ, Jongstra R, Wegmüller R, Bashar MK. Effect of parboiling conditions on zinc and iron retention in biofortified and non-biofortified milled rice. *J Sci Food Agric* January 30, 2022;**102**(2):514–22. https://doi.org/10.1002/jsfa.11379.
19. Wang N, Ryan L, Sardesai N, Wu E, Lenderts B, Lowe K, Che P, Anand A, Worden A, van Dyk D, Barone P, Svitashev S, Jones T, Gordon-Kamm W. Leaf transformation for efficient random integration and targeted genome modification in maize and sorghum. *Native Plants* February 2023;**9**(2):255–70. https://doi.org/10.1038/s41477-022-01338-0.
20. Gurel F, Wu Y, Pan C, Cheng Y, Li G, Zhang T, Qi Y. On- and off-target analyses of CRISPR-Cas12b genome editing systems in rice. *CRISPR J* February 2023;**6**(1):62–74. https://doi.org/10.1089/crispr.2022.0072.
21. Yang T, Ali M, Lin L, Li P, He H, Zhu Q, Sun C, Wu N, Zhang X, Huang T, Li CB, Li C, Deng L. Recoloring tomato fruit by CRISPR/Cas9-mediated multiplex gene editing. *Hortic Res* September 19, 2022;**10**(1):uhac214. https://doi.org/10.1093/hr/uhac214.
22. Li S, Lin D, Zhang Y, Deng M, Chen Y, Lv B, Li B, Lei Y, Wang Y, Zhao L, Liang Y, Liu J, Chen K, Liu Z, Xiao J, Qiu JL, Gao C. Genome-edited powdery mildew resistance in wheat without growth penalties. *Nature* February 2022;**602**(7897):455–60. https://doi.org/10.1038/s41586-022-04395-9.
23. Ge X, Xu J, Yang Z, Yang X, Wang Y, Chen Y, Wang P, Li F. Efficient genotype-independent cotton genetic transformation and genome editing. *J Integr Plant Biol* December 7, 2022. https://doi.org/10.1111/jipb.13427.
24. Zhang Y, Wu Y, Li G, Qi A, Zhang Y, Zhang T, Qi Y. Genome-wide investigation of multiplexed CRISPR-Cas12a-mediated editing in rice. *Plant Genome* September 30, 2022:e20266. https://doi.org/10.1002/tpg2.20266.
25. Yirgu A, Mekonnen Y, Eyado A, Staropoli A, Vinale F. Biological activities of *Viscum tuberculatum* aqueous leaf extract. *Pharm Biol* December 2023;**61**(1):30–6. https://doi.org/10.1080/13880209.2022.2151021.
26. Saleem F, Safdar N, Fatima I, Yasmin A, Hussain W. Functionalization of ampicillin and gentamicin with biogenic copper nanoparticles (CuNPs) remodel antimicrobial and cytotoxic outcome against MDR clinical isolates. *Arch Microbiol* February 13, 2023;**205**(3):88. https://doi.org/10.1007/s00203-023-03425-y.

27. Phan HT, Tran HX, Ho TT, Pham VT, Trinh VT, Nguyen TT, Pham NB, Chu HH, Conrad U. Plant crude extracts containing oligomeric hemagglutinins protect chickens against highly Pathogenic Avian Influenza Virus after one dose of immunization. *Vet Res Commun* January 2023;**47**(1):191–205. https://doi.org/10.1007/s11259-022-09942-3.
28. Anupong W, On-Uma R, Jutamas K, Salmen SH, Alharbi SA, Joshi D, Jhanani GK. Antibacterial, antifungal, antidiabetic, and antioxidant activities potential of Coleus aromaticus synthesized titanium dioxide nanoparticles. *Environ Res* January 1, 2023;**216**(Pt 3):114714. https://doi.org/10.1016/j.envres.2022.114714.
29. Haris A, Azeem M, Abbas MG, Mumtaz M, Mozūratis R, Binyameen M. Prolonged repellent activity of plant essential oils against dengue vector, *Aedes aegypti*. *Molecules* January 31, 2023;**28**(3):1351. https://doi.org/10.3390/molecules28031351.
30. Tottey S, Shoji Y, Mark Jones R, Musiychuk K, Chichester JA, Miura K, Zhou L, Lee SM, Plieskatt J, Wu Y, Long CA, Streatfield SJ, Yusibov V. Engineering of a plant-produced virus-like particle to improve the display of the Plasmodium falciparum Pfs25 antigen and transmission-blocking activity of the vaccine candidate. *Vaccine* January 23, 2023;**41**(4):938–44. https://doi.org/10.1016/j.vaccine.2022.12.048.
31. Hager KJ, Pérez Marc G, Gobeil P, Diaz RS, Heizer G, Llapur C, Makarkov AI, Vasconcellos E, Pillet S, Riera F, Saxena P, Geller Wolff P, Bhutada K, Wallace G, Aazami H, Jones CE, Polack FP, Ferrara L, Atkins J, Boulay I, Dhaliwall J, Charland N, Couture MMJ, Jiang-Wright J, Landry N, Lapointe S, Lorin A, Mahmood A, Moulton LH, Pahmer E, Parent J, Séguin A, Tran L, Breuer T, Ceregido MA, Koutsoukos M, Roman F, Namba J, D'Aoust MA, Trepanier S, Kimura Y, Ward BJ, CoVLP Study Team. Efficacy and safety of a recombinant plant-based adjuvanted Covid-19 vaccine. *N Engl J Med* June 2, 2022;**386**(22):2084–96. https://doi.org/10.1056/NEJMoa2201300. Epub 2022 May 4. PMID: 35507508; PMCID: PMC9127773.
32. Paolino KM, Regules JA, Moon JE, Ruck RC, Bennett JW, Remich SA, Mills KT, Lin L, Washington CN, Fornillos GA, Lindsey CY, O'Brien KA, Shi M, Mark Jones R, Green BJ, Tottey S, Chichester JA, Streatfield SJ, Yusibov V. Safety and immunogenicity of a plant-derived recombinant protective antigen (rPA)-based vaccine against Bacillus anthracis: a phase 1 dose-escalation study in healthy adults. *Vaccine* March 15, 2022;**40**(12):1864–71. https://doi.org/10.1016/j.vaccine.2022.01.047. Epub 2022 Feb 10. PMID: 35153091.
33. Wang F, Harindintwali JD, Yuan Z, Wang M, Wang F, Li S, Yin Z, Huang L, Fu Y, Li L, Chang SX, Zhang L, Rinklebe J, Yuan Z, Zhu Q, Xiang L, Tsang DCW, Xu L, Jiang X, Liu J, Wei N, Kästner M, Zou Y, Ok YS, Shen J, Peng D, Zhang W, Barceló D, Zhou Y, Bai Z, Li B, Zhang B, Wei K, Cao H, Tan Z, Zhao LB, He X, Zheng J, Bolan N, Liu X, Huang C, Dietmann S, Luo M, Sun N, Gong J, Gong Y, Brahushi F, Zhang T, Xiao C, Li X, Chen W, Jiao N, Lehmann J, Zhu YG, Jin H, Schäffer A, Tiedje JM, Chen JM. Technologies and perspectives for achieving carbon neutrality. *Innovation* October 30, 2021;**2**(4):100180. https://doi.org/10.1016/j.xinn.2021.100180. PMID: 34877561; PMCID: PMC8633420.
34. Jiang T, Duan Q, Zhu J, et al. Starch-based biodegradable materials: challenges and opportunities. *Adv Ind Eng Polym Res* 2020;**3**:8–18.
35. Chong JWR, Khoo KS, Yew GY, et al. Advances in production of bioplastics by microalgae using food waste hydrolysate and wastewater: a review. *Bioresour Technol* 2021;**342**:125947.
36. Atiwesh G, Mikhael A, Parrish CC, et al. Environmental impact of bioplastic use: a review. *Heliyon* 2021;**7**:e07918.
37. Nandakumar A, Chuah JA, Sudesh K. Bioplastics: a boon or bane? *Renew Sustain Energy Rev* 2021;**147**:111237.
38. Lim C, Yusoff S, Ng CG, et al. Bioplastic made from seaweed polysaccharides with green production methods. *J Environ Chem Eng* 2021;**9**:105895.

39. Bishop G, Styles D, Lens PNL. Environmental performance comparison of bioplastics and petrochemical plastics: a review of life cycle assessment (LCA) methodological decisions. *Resour Conserv Recycl* 2021;**168**:105451.
40. Shang Z, Abdalla M, Xia L, et al. Can cropland management practices lower net greenhouse emissions without compromising yield? *Global Change Biol* 2021;**27**:4657–70.
41. Acosta N, Sakarika M, Kerckhof F-M, et al. Microbial protein production from methane via electrochemical biogas upgrading. *Chem Eng J* 2020;**391**:123625.
42. Jiang W, Hernández Villamor D, Peng H, et al. Metabolic engineering strategies to enable microbial utilization of C1 feedstocks. *Nat Chem Biol* 2021;**17**:845–55.
43. Matassa S, Boon N, Pikaar I, Verstraete W. Microbial protein: future sustainable food supply route with low environmental footprint. *Microb Biotechnol* 2016;**9**:568–75.
44. Li Y, Nie C, Liu YH, et al. Soil microbial community composition closely associates with specific enzyme activities and soil carbon chemistry in a long-term nitrogen fertilized grassland. *Sci Total Environ* 2019;**654**:264–74.
45. Otero DM, da Rocha Lemos Mendes G, da Silva Lucas AJ, Christ-Ribeiro A, Ribeiro CDF. Exploring alternative protein sources: evidence from patents and articles focusing on food markets. *Food Chem* November 15, 2022;**394**:133486. https://doi.org/10.1016/j.foodchem.2022.133486. Epub 2022 Jun 15. PMID: 35759839.
46. Lingiardi N, Galante M, de Sanctis M, Spelzini D. Are quinoa proteins a promising alternative to be applied in plant-based emulsion gel formulation? *Food Chem* November 15, 2022;**394**:133485. https://doi.org/10.1016/j.foodchem.2022.133485. Epub 2022 Jun 16. PMID: 35753255.
47. Rivera Del Rio A, Boom RM, Janssen AEM. Effect of fractionation and processing conditions on the digestibility of plant proteins as food ingredients. *Foods* March 18, 2022;**11**(6):870. https://doi.org/10.3390/foods11060870. PMID: 35327292; PMCID: PMC8955167.
48. Tavares PPLG, Dos Santos Lima M, Pessôa LC, de Andrade Bulos RB, de Oliveira TTB, da Silva Cruz LF, de Jesus Assis D, da Boa Morte ES, Di Mambro Ribeiro CV, de Souza CO. Innovation in alternative food sources: a review of a technological state-of-the-art of insects in food products. *Foods* November 24, 2022;**11**(23):3792. https://doi.org/10.3390/foods11233792. PMID: 36496600; PMCID: PMC9737383.
49. Liceaga AM, Aguilar-Toalá JE, Vallejo-Cordoba B, González-Córdova AF, Hernández-Mendoza A. Insects as an alternative protein source. *Annu Rev Food Sci Technol* March 25, 2022;**13**:19–34. https://doi.org/10.1146/annurev-food-052720-112443. Epub 2021 Oct 26. PMID: 34699254.
50. Zeng Y, Chen E, Zhang X, Li D, Wang Q, Sun Y. Nutritional value and physicochemical characteristics of alternative protein for meat and dairy—a review. *Foods* October 23, 2022;**11**(21):3326. https://doi.org/10.3390/foods11213326. PMID: 36359938; PMCID: PMC9654170.
51. Shi S, Wang Z, Shen L, Xiao H. Synthetic biology: a new frontier in food production. *Trends Biotechnol* July 2022;**40**(7):781–803. https://doi.org/10.1016/j.tibtech.2022.01.002. Epub 2022 Feb 1. PMID: 35120749.
52. Banks M, Johnson R, Giver L, Bryant G, Guo M. Industrial production of microbial protein products. *Curr Opin Biotechnol* June 2022;**75**:102707. https://doi.org/10.1016/j.copbio.2022.102707. Epub 2022 Mar 8. PMID: 35276510.

Chapter 11

# Conclusion

The 2030 UN sustainable development objectives scheme prioritizes energy and food security, wherein agricultural biotechnology plays an important role in accomplishing these objectives. The Food and Agricultural Organization (FAO) has declared that over 800 million of the global population face food insecurity and more than two billion suffer from malnutrition due to deficiency of micronutrients. There is intense pressure on food supply worldwide due to increasing growth of the world's population and the rising competition for important crops as food and as renewable sources of energy. This is exacerbated by an increase in environmental stresses including scarcity of water supply, unprecedented climatic changes, and competition for arable land.

Considering the crosstalk between plant biology and agriculture, multinational research activities have led to the development of efficient, productive new cultivars. Upon fulfillment of screening variables, agricultural biotechnology products are released into the field and at the market level. Crop products include those that are meant for direct consumption such as fruits, grains, bulbs, tubers, corns, flowers, leaves, fibers, timber, and cork. Desired traits include taste, aroma, color, nutritional value, as well as marketability and storage capability. Important botanical traits include growth and elongation of the plants, architecture of the roots and shoots, and genetic regulation of flowering. Also desirable are crops that can tolerate and adapt to detrimental abiotic/climatic changes such as extremes of temperatures, salinity, drought and pollution, as well as biotic stresses such as infections by fungi, viruses, bacteria, insects, and herbivores. Seed manufacturers are investing great efforts into generating crops with augmented tolerance to abiotic and biotic stresses determined by their polygenic nature and trait complexity. Molecular genetic networks enabling plant adaptation to detrimental environmental conditions have been elucidated. All of these biotechnological applications are subject to scrutiny in regard to food security, sociological, ethical, economic, and legal considerations toward public acceptance.

Plants generate over 500,000 secondary metabolites with inexpensive investments. Novel genetic discoveries, availability of better metabolomics information, plant genetic engineering, and industrial production systems facilitate the augmented production of food as well as conventional plant-made products such as cork and fiber in addition to new, inimitable nonplant compounds such as favorable petrochemical substitutes raised in transgenic plants.

**Agricultural Biotechnology. https://doi.org/10.1016/B978-0-323-88476-1.00011-4**

Also, transgenic plants have been used to produce biodegradable thermoplastics, fibrous proteins such as elastin and silk, salt- and temperature-resistant enzymes like trypsin, cellulases, and amylases for the detergent, food, paper, and other industries.

Plants are being used as bioreactors for expressing non-plant, foreign biopharmaceuticals such as vaccines, bioactive peptides, proteins, antibodies, and several therapeutic products like collagen and hyaluronic acid. Genetically engineered plants producing edible vaccines have led to the development of molecular pharming involving efficacious production of biopharmaceuticals by means of plant-derived industrial platforms. Various new technologies including stable nuclear transformation, plastid transformation, and transient expression platforms have been used. Through such technologies, recently, the first plant-based recombinant biopharmaceutical called Elelyso against Gaucher disease has been approved by the FDA in addition to other clinically-standard proteins.

Often called nutraceuticals, several nutritional components of enhanced nutritional value and food quality have also been expressed in transgenic plants. These include flavonoids, vitamins, amino acids, antioxidants, and plant-derived essential minerals with improved bioavailability. Ethanol is the major biofuel generated from sugar or starch feedstocks like sugarcane or corn. Use of food crops with augmented biofuel production can be detrimental to food market costs, and hence, it is preferred that forest trees with increased lignocellulosic biomass and grasses such as *Miscanthus giganteus*, *Panicum virgatum,* and sorghum growing on marginal soils are used. Additionally, energy obtained from plants has significantly lower $CO_2$ footprint compared to that of gas or petroleum and therefore plants are considered "$CO_2$ mitigators." Biodiesel is currently generated principally from canola and soybean. However, seeds obtained from non-food plants like the castor bean (*Ricinus communis*) are an attractive alternative. Novel plant biomaterials are increasingly being generated by reliable production without diminishing crop yields, with superior physical and chemical properties, in addition to the generation of bulk materials at lower costs through photosynthesis. Bioenergy crops play a central role in mitigating climate changes by utilizing biogenic carbon in addition to the adoption of biomass for power, heat, biorefining, and biofuels. Despite the recognition of the efficacy of negative emission technologies in combating climate change effects, advancements in the production of bioenergy crops have been rather slow. Moreover, by and large, bioenergy crops showing great promise have circumscribed germplasm resources and are undomesticated.

Modern genetic modification (GM) and genome editing (GE) technologies afford great potential for accelerated generation of novel bioenergy and food crops for farmers. The significant precision now enabled by genome editing systems represents a major change from GM technologies, which involve unpredictable introduction of individual genes into the crop genome. Also, GE technologies are far superior to conventional breeding practices, which depend on random, unbridled radiation-, and chemical-induced mutagenesis. Products of plant biotechnology have become a reality in our modern lives as food, medicinal, and industrial applications.

Genome editing technology has undergone enormous development in recent times starting with plant transformation followed by the advent of TALENs, ZFNs, and engineered meganucleases. Presently, the CRISPR/Cas9 technology has brought about significant transformation in crop improvement. This technology affords several important advantages when compared to earlier equivalents. Its RNA-based strategy affords augmented specificity and increased flexibility combined with ease of use. This highly efficient precision and specificity is largely determined by the gRNA protospacer sequence and therefore the complimentary sequence of the DNA such that it avoids the costly and complex protein engineering. Genome modifications engineered by the CRIPSR/Cas9 technology are evidently highly efficient, and there are no reports of the incidence of mosaicism, an observation associated with other editing tools such as TALENs and ZFNs. This, in addition to the augmented affinity of the Cas9 endonucleases toward heterochromatic regions in the genome provides a way for the introduction of DSBs at regions of the genome which are highly methylated. Therefore, CRISPR/Cas9 as an editing tool is greatly versatile while incurring low costs and being extremely efficient.

The CRISPR/Cas9 system has been successfully used to modify metabolic pathways, augment crop quality, and develop plant-derived drugs through gene silencing, gene mutation as well as transcriptional regulation. In a few studies, transgenes introduced in food crops have been deleted using CRISPR/Cas9 technology. The existence of transgenes in crops is contentious in regard to safety issues as the transgenic technology inserts foreign genes into plants along with the selection-enabling marker genes. This drawback of the transgene technology has been overcome using the CRISPR/Cas9 system. Thus, transgene-free lines can be generated by CRISPR/Cas9 editing with almost no probability of mutations arising at off-target sites. The use of the type II CRISPR/Cas9 technology has enormously impacted molecular biology and bioengineering.

The CRISPR/Cas9 technology has created alterations in genetic structures at targeted locations so as to introduce modifications in plant structure, resistance/tolerance to biotic/abiotic stresses, productivity, improvement of nutritional yield, generation of herbicide resistance, and discernment of the functions of individual or multiple genes. Through the CRISPR/Cas9 technology, homology-directed repair (HDR) has been used as a favorable repair pathway for generating precise knock-in of genes or enabling gene replacement. The CRISPR/Cas9 technology has been used to regulate promoters, enhancers as well as transcription/translation factors. This has enabled the enhancement or inhibition of the amounts of specific proteins in the host plants. Translation-associated and other plant host factors have been identified as proviral molecules, which enable viral RNA translation, thereby promoting viral infection of plants. Genes associated with replication, translation, and metabolism have been modified by CRISPR/Cas9 to contain viral infections in plants. Nucleotide insertions or deletions (indels) at specific target sites of genes such as the promoters or the coding regions have been introduced using the CRIPSR/Cas9 system thus disrupting gene function. By disabling disease susceptibility genes through the CRISPR/Cas9 machinery, virus-induced pathogenic effects have been mitigated. Genes determining improper crop resistance to viruses have been replaced with a functional resistance gene using the HDR mechanism. Promising potential gene-editing targets using the CRISPR/Cas9 include plant-generated inhibitors of cell wall degradation enzymes produced by fungal pathogens. Similarly, plant factors responsible for susceptibility to viruses/fungi have been disrupted using the CRISPR/Cas9 mechanism thereby engineering resistance of these plants to several pathogens.

The CRISPR/Cas9 system is a cutting-edge tool with strong potential and precision. This technology is still evolving with the introduction of variations, which facilitate genome modifications along with its ability to introduce single base changes through prime editing and base editing tools. The accuracy, simplicity, and specificity provided by the CRISPR/Cas9 are important highlights of the effectiveness of this technology in remodeling the genome and enabling genome-wide high-throughput screens in crop plants. With different modifications performed frequently, the reach of this tool and its functions are continuously expanding at an accelerated pace. CRISPR/Cas9 is more advantageous compared to artificial RNAi and microRNAs for engineering plant virus resistance as they cause disruption of important viral or host genes rather than just silence those genes at the RNA level.

There are many reports of the use of CRISPR/Cas9 system as a favorable tool to efficiently provide resistance to RNA and DNA viruses in plants, either through direct targeting and cleavage of the viral genome or by introducing modifications in the host genes. Additionally, because the CRISPR/Cas9 is originated from prokaryotes, the probability of plant viruses to evolve mechanisms to antagonize the modifications engineered by this technology is unlikely. Whereas the RNAi technology is driven by a plant-derived mechanism against which most of the plant viruses already harbor suppressors. Therefore, multitargeting guide RNAs or gRNAs within the CRIPSR/Cas9 genome editing tool enable efficient removal of mutant viruses and escape mutants.

At the time of the Green Revolution prevalent in the 1960 and 1970s, novel crop varieties were developed using conventional breeding, which were quickly adopted by the farming community. However, this is in contrast to today's more circumscribed use of a majority of biotechnologically enhanced crops. Even after over 2 decades into the period of commercialized GM crops, the practice and political economy of this advancement remains largely precautionary. Reports of new risks besieged this technology from its advent and exist even today in spite of the lack of scientific proof. Contentions about the risks and advantages of agricultural biotechnology as well as ethical dilemmas still continue. There exist serious challenges to agricultural productivity due to increasing global demands for bioenergy, food and specialty plant products in addition to the perils faced due to varying changes in climate and environment.

In this light, agricultural biotechnology affords a favorable avenue to mitigate such challenges. Nevertheless, sociocultural and ethical concerns would have to be first addressed in order to by and large enable public trust and acceptance. For agricultural biotechnology to develop and thrive in a commercially viable manner, modern biotechnological practices facilitated by genetic modifications or gene editing technologies would have to be combined with a careful balance of risks versus benefits analysis, ethical decision-making, educating the public about the science behind the respective technologies along with ethical reasoning.

Plant biotechnology can address important issues involved in ensuring food security, augment the nutritive value of food crops, improve forest sustainability and resilience, in addition to bettering the environment and protecting the biodiversity of threatened traditional and native species. Specifically for practicing the CRISPR/Cas technologies, we must be circumspect in allocating financial and manpower resources into ventures that are conceptually inviting but do not have the requisite maturity to translate into agricultural products that are beneficial to society. Currently, in developing nations, there exists a paucity of resources for agricultural research and development in addition to the compelling need for technological progress in agriculture toward achieving robust returns. The CRISPR/Cas9 technology is a versatile mechanism for targeted genome modification while minimizing off-target mutations and along with other modern technologies, enables gene replacement, allelic modifications, posttranslational modifications, and structural elucidation of the proteome. This rapidly evolving genetic engineering tool provides extraordinary control over the genetic constitution of plant genomes and is essential for discernment of the physiological, morphological, and metabolic traits of plants. It affords more control over the modification of plant structure and function. In contrast to laboratory investigations, the achievements of plant biotechnologies at the field level cannot be feasible for agricultural practices in the absence of regimental testing procedures as well as screening methods using definitive algorithms.

# Index

*Note*: Page numbers followed by "*f*" indicate figures and "*t*" indicate tables.

## C

## H



## I



## J



## K

## L



## M



## N



## O



## P

## Q



## R



## S

## T



## U



## V



## W

CPI Antony Rowe
Eastbourne, UK
June 12, 2024